SCIENTIFIC AND PRACTICAL STUDIES OF RAW MATERIAL ISSUES

PROCEEDINGS OF THE RUSSIAN - GERMAN RAW MATERIALS DIALOGUE: A COLLECTION OF YOUNG SCIENTISTS PAPERS AND DISCUSSION

United Nations
Educational, Scientific and
Cultural Organization

International Competence Centre
for Mining-Engineering Education
under the auspices of UNESCO

Scientific and Practical Studies of Raw Material Issues

Editor

Vladimir Litvinenko

Saint-Petersburg Mining University, Russia

CRC Press
Taylor & Francis Group
Boca Raton London New York

CRC Press is an imprint of the
Taylor & Francis Group, an **informa** business

A BALKEMA BOOK

CRC Press/Balkema is an imprint of the Taylor & Francis Group, an informa business

© 2020 Taylor & Francis Group, London, UK

Typeset by Integra Software Services Pvt. Ltd., Pondicherry, India

Library of Congress Cataloging-in-Publication Data

Applied for

Published by: CRC Press/Balkema
            Schipholweg 107C, 2316XC Leiden, The Netherlands

First issued in paperback 2023

ISBN: 978-1-03-257099-0 (pbk)
ISBN: 978-0-367-86153-7 (hbk)
ISBN: 978-1-003-01722-6 (ebk)

DOI: https://doi.org/10.1201/9781003017226

Publisher's Note
The publisher has gone to great lengths to ensure the quality of this reprint but points out that some imperfections in the original copies may be apparent.

Scientific and Practical Studies of Raw Material Issues – Litvinenko (Ed)
© 2020 Taylor & Francis Group, London, ISBN 978-0-367-86153-7

Table of contents

Preface   ix

Organizers   xi

Mining exploration, mining and processing

Chemical composition and genesis of serpentinite group minerals in nickeliferous weathering
crust of the Elov deposit (Urals)   3
E. Nikolaeva, I. Talovina, V. Nikiforova & G. Heide

Dependence between the parameters of storage of artificial soils from their specific properties   11
V.N. Kondakova & G.B. Pospekhov

Petroleum prospects of the Middle Paleozoic sequences in the southern part of the Viluyui
syneclise   18
G. Cherdancev

The problems of cryolithozone mining in Yakutia in conditions of global warming   25
A.A. Pomortseva & O.A. Pomortsev

Innovations in mechanical rock excavation at TU Bergakademie Freiberg   34
B. Grafe, T. Shepel & C. Drebenstedt

Technological scheme of development flooded fields of sands   46
D. Dzyurich & V. Ivanov

The assessment of the roof beam stability in mining workings   53
M. Vilner, T.T. Nguen & P. Korchak

Modern mine survey techniques in the process of mining operations in open pit mines
(quarries)   58
A. Blishchenko

Output prediction of Semi Mobile In Pit Crushing and Conveying Systems (SMIPPC)   63
R. Ritter & C. Drebenstedt

Arrangements for increase the efficiency of mining operations on the deep ore mines   71
A.V. Kholmskiy & D.V. Sidorov

Influence of technological factors on the formation of spontaneous combustion centers in
underground mining   75
D.D. Golubev

Evaluation problem of harmful effects in mining works during construction of subway
escalator tunnels with the help of soils freezing method   82
D. Mukminova & E. Volohov

Optimization of gas pipeline operation modes considering the condition of gas compressor
units   91
A.V. Kokorin & A.M. Schipachev

Studies of mixing high viscosity petroleum and pyrolysis resin to improve quality indicators 97
R. Sultanbekov & M. Nazarova

The use of secondary polymers to ensure environmental safety storage of mineral waste ore-dressing and processing enterprise in the Southern Urals 103
D. Babenko

Studying the possibility of improving the properties of environmentally friendly diesel fuels 108
A. Eremeeva, N. Kondrasheva & K. Nelkenbaum

Environmental technologies in the production of metallurgical silicon 114
M. Glazev & V. Bazhin

Challenges in processing copper ores containing sulfosalts 120
A. Kobylyanski, V. Zhukova, G. Petrov & A. Boduen

Mining services

Use of brown coal ash for the amelioration of dump soils 129
C. Drebenstedt

Engineering and ecological survey of oil-contaminated soils in industrial areas and efficient way to reduce the negative impact 135
M.V. Bykova & M.A. Pashkevich

Landscape geochemical consequences of ore mining 143
V.A. Alekseenko, N.V. Shvydkaya & A.V. Puzanov

Reduction of nitrous gases from blast fumes in underground mines using water based absorbency solutions 148
A. Hutwalker, T. Plett & O. Langefeld

Geoecological justification of reclamation of outer coal waste piles 157
A. Mukhina

Disinfection of waste waters of industrial enterprises by vibroacoustic method 165
G. Fedorov, Y. Agafonov & C. Drebenstedt

Mining 4.0 in developing countries 171
J.C. Bongaerts

Risk management and its contribution to sustainable development of mining enterprises 182
D. Ivanova

Holistik responsible mining approach 191
C. Drebenstedt

Modern condition and prospects for the development of forest infrastructure to improve the economy of nature management 200
A.A. Kitcenko, V.F. Kovyazin & A.Y. Romanchikov

Responsible mining 205
O. Langefeld & A. Binder

Legal issues of the use of digital technology in mining - data ownership and liability 213
M. Paschke

Optimization model for long-term cement quarry production scheduling 219
T. Vu & C. Drebenstedt

The application of MPC to improve the efficiency of an AC electric drive vector control 232
S. Erokhov

Features of the walking mechanism of a floating platform autonomous modular complex for the extraction and processing of peat raw materials 239
D. Fadeev & S. Ivanov

Simulation of combined power system with storage device 244
B. Garipov & D. Ustinov

Arc steel-making furnaces functionality enhancement 251
E. Martynova, V. Bazhin & A. Suslov

Modeling of operation modes of electrical supply systems with non-linear load 263
V.N. Kostin, A.V. Krivenko & V.A. Serikov

The analysis of modern regulation devices of power streams on the basis of FACTS - devices 272
E. Shafhatov & E. Zhdankin

Author index 278

Scientific and Practical Studies of Raw Material Issues – Litvinenko (Ed)
© 2020 Taylor & Francis Group, London, ISBN 978-0-367-86153-7

Preface

Dear friends:

Here is a collection of articles by young scholars from German and Russian mining universities, devoted to problematic issues of subsoil use. The objective need to bring out a separate publication with research results obtained by young scientists of the two countries has long been overdue.

The universities of Russia and Germany work hard to train highly qualified personnel, support scientific schools and improve the image of specialities related to the raw material base, explaining the indispensability of such professions.

Working with young researchers, supporting their individual professional development, and creating conditions for mobility and scientific collaboration are essential for Russian-German Raw Materials Forum founded in Dresden thirteen years ago.

In November 2019, the 12th Russian-German Raw Materials Forum will traditionally take place in St Petersburg; for the first time its program includes a separate day for youth discussions.

I would like to quote the words of Klaus Töpfer, the chairman of the organizing committee from the German side, which he said in 2018 at the Forum held in Potsdam: "We are moving by leaps and bounds to a world which 10 billion people will live in. When I was born, 2.6 billion people lived on planet Earth. We cannot but ask questions: what impact does this have or will have on resources, on the energy? The new generation will have to search for answers. It is essential that today 100 young scientists representing universities in Russia and Germany are present here because we focus not on the past, but in the future, we understand what is important for the future."

The Youth Agenda with master classes, a case-based contest of project presentations give researchers plenty of opportunities to manifest themselves and, possibly, find potential customers. The publication of a collection of scientific articles prepared by young participants is not only a mark of respect to the work they have done but also an objective assessment of its quality by international top experts. This is the way to success! Throughout the long-term cooperation among the leading scientific schools of Germany and Russia, a huge material of great value has been accumulated; it can be used to create modern technologies in the handling of minerals from exploration and production to deep processing.

Germany has the oldest mining schools in the world and a well-developed mining and engineering base. Maintaining traditional close ties with Russia enables Germany, on the one hand, to export its research and production potential, and, on the other hand, fighting deficit of its mineral resources, the country is interested in the reliable supply of raw materials.

The concept of sustainable development and escalating demands to reduce the environmental impact force radical changes in the strategy for improving the mineral and raw materials complex in Germany: abandon traditional technologies in favor of alternative sources of heat and energy, ensure a complete restoration of the environment at the end of field development, maintain a high professional staff potential in the industry.

In Russia, one of the main suppliers of mineral and energy resources to Germany, the modernization of the mineral and raw materials complex is rapidly advancing based on the latest national and international achievements. The Russian Federation is particularly interested in developing international scientific cooperation with foremost developed countries; Germany being one of the undisputed leaders of the world community.

The forum organizers will keep doing their work to support young scientists in Russia and Germany in the future. We wish everyone good luck, unalterable determination and daring decision-making.

Vladimir Litvinenko
Doctor of Engineering, Professor;
Rector of Saint Petersburg Mining University

Scientific and Practical Studies of Raw Material Issues – Litvinenko (Ed)
© 2020 Taylor & Francis Group, London, ISBN 978-0-367-86153-7

Organizers

International Competence Centre for Mining Engineering Education under the auspices of UNESCO

United Nations Educational, Scientific and Cultural Organization · International Competence Centre for Mining-Engineering Education under the auspices of UNESCO

Russian-German Raw Materials Forum

Rohstoff-Forum
Российско-Германский сырьевой форум

Saint Petersburg Mining University

Mining exploration, mining and processing

Scientific and Practical Studies of Raw Material Issues – Litvinenko (Ed)
© 2020 Taylor & Francis Group, London, ISBN 978-0-367-86153-7

Chemical composition and genesis of serpentinite group minerals in nickeliferous weathering crust of the Elov deposit (Urals)

E. Nikolaeva, I. Talovina & V. Nikiforova
Saint-Petersburg Mining University, Saint-Petersburg, Russian Federation

G. Heide
TU Bergakademie Freiberg, Freiberg, Germany

ABSTRACT: The extremely complex geological structure of the substrate of the Elov deposit (the Nothern Ural), the large variety of source rocks, the long development period of various weathering processes occurring in different geological eras and different climatic conditions caused a very complex structure of the deposit' weathering crust and a huge variety of weathering residues of the source rocks, the majority of which is nickel-containing with industrial concentrations of useful components in them.

The article is devoted to the analysis of chemical composition and genesis of antigorite on the Elov deposit. The features of the minerals of the serpentinite group and antigorite in particular are described, and the chemical composition of antigorite is given according to the research results. Experimental data confirming the formation of antigorite in the close vicinity to dikes is presented.

1 INTRODUCTION

During the last century, nickel industry of the Urals was based on numerous supergene oxide–silicate ore deposits, which supplied raw material to the Yuzhuralnikel and Ufaleinikel metallurgical plants, as well as Rezh and Buruktal mills. New metallurgical plants are under active construction at present on the basis of the lateritic nickel ores of such type (>70% of the world's nickel reserves) in Australia, New Caledonia, Cuba, Indonesia, Samoa-New Guinea, Brazil, Columbia, Venezuela, and other countries. In terms of mineral composition, and, partially, geological position, the Uralian ores are similar to the nickel-bearing lateritic ores in the Earth's modern tropical belt, but they are not their complete analogs. The majority of the Uralian deposits mainly contain low-grade ores with the Ni content of 0, 7–1, 0%. The wide compositional diversity of their ore and non-ore rock-forming minerals significantly affects the technology of nickel processing. The main nickel-carriers in the ores are Mg-silicates (serpentines, chlorites, and nontronites), with the minor contribution of iron and manganese hydroxides. The Ni content in these minerals is 1–3%, increasing in some chlorite and serpentine species to 7–12% or more.

Nickel-bearing ore mineralization of the Elov deposit (Figure 1) is confined to the weathering crust of ultramafic rocks of peridotite's formation which all nickel deposits of the Urals and the North Kazakhstan are related with.

The Elov deposit (Figure 2) contains a large number of minor intrusions in contrast with all other Urals' deposits and for the formation of the weathering crust with equivalent nickel content as in other deposits with similar composition it could be required twice the amount of nickel in the source serpentinites or double expansion of decayed and weathered serpentinites, from which fully leached nickel was redeposited in orebodies.

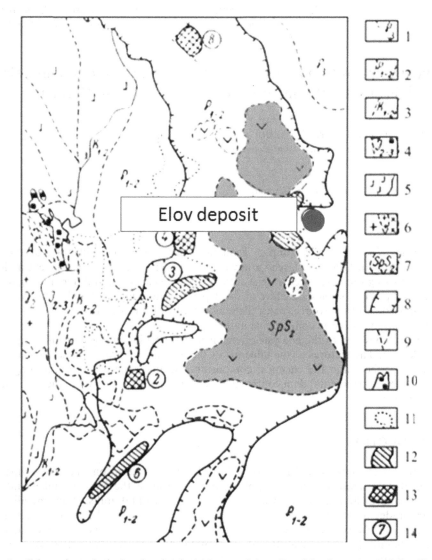

Figure 1. Schematic geological and technological map of deposits of the Serov group (after Vershinin et al., 1988).

1-4 Meso-Cenozoic deposits of the weathering crust; 5-6 Paleozoic formations; 7 day stones of serpentinites; 8 contours of serpentinites in accordance with magnetic survey and drilling; 9 zones of tectonic dislocations; 10 chalcopyrite-magnetite skarns; 11 nodular-conglomerate sedimentary iron ores in the deposits of the Myssov suite; deposits of supergene nickel ores of various technological types: 12 ores for alkaline hydrometallurgical process, 13 ores for blast smelting process; 14 numbers of deposits. The Serov group of deposits: 7 Elov, 6 Katasmin, 8 Ustey and 2,3,4 - others.

According to calculations of the ratio of minor intrusions height to the seam height of serpentinites in the central part of the Elov deposit minor intrusions estimated 40% of total rocks thickness, sometimes in some small areas volume of vein-shaped bodies exceeds volume of the serpentinites.

Serpentinites are the main rock-forming minerals in the Urals serpentinite zones for supergene deposits and according to the G. Brindley and Z. Maksimovich classification (Brindley & Maksimovich, 1978) relate to the two nickel series:

– Lizardite $Mg_6Si_4O_{10}(OH)_8$ - nepuite $Ni_3Si_2O_5(OH)_4$;
– Chrysotile $Mg_3Si_2O_5(OH)_4$ - pekoraite $Ni_3Si_2O_5(OH)_4$.

Figure 2. Geological record of the weathering crust of the Elov deposit (after Berkhin et al., 1970).

1 - nodular-conglomerate sedimentary iron ores (K_1), 2 - nontronited serpentinites, 3 - leached serpentinites, 4 - disintegrated serpentinites, 5 - unaltered serpentinites, 6 - chamosites, 7 - chlorite vein-rock, 8 - fractures.

Except for lizardite, chrysotile, pekoraite and nepuite in serpentinite zones of Urals deposits antigorite $Mg_3Si_2O_5(OH)_4$ occures permanently in small quantities, and in the iron-oxide zones manganese serpentine is noted, which is called kariopilite $Mn_3Si_2O_5(OH)_4$.

According to earlier studies it was considered that antigorite $Mg_3Si_2O_5(OH)_4$ is presented in the lizardite serpentinites of all the Urals deposits (besides the Elov deposit) in small quantities (less than 5%). Currently it is found that antigorite is one of the major rock-forming minerals for the disintegrated and leached serpentinites of serpentinite zones in the Elov deposit.

2 PREVIOUS STUDIES

Antigorite is a very prevalent (wide-spread) serpentine of antigorited harzburgites of Urals (shtubahites in accordance with A.Varlakov, 1986) including the Cheremshan deposit. In contrast with the other serpentine species in the Cheremshan deposit, antigorite is presented in the shape of large macroscopically well-diagnosed lamelliform squamous grains with a size up to 15 mm (Varlakov, 1986). Sometimes lamellas have a distinct plan-parallel orientation, this is evidenced by formation under pressure (so-called "stress-mineral") in the dynamometamorphic conditions.

Large form of antigorite lamellas allowed us to distinguish them from the serpentinites of the Cheremshan deposit and research both the lamellas and serpentinites are inclusive them by X-ray analysis, which consist of chrysotile $2Orc_1$ (65%), clinochlore (25%) and other minerals of serpentinites. As a result of analysis antigorite wasn't detected in the majority of lamellas and on the X-ray diagrams chrysotile (40%), lizardite (20%), talc (20%), clinochlore, forsterite, magnetite and kumingtonite in the amount of 1-3% were traced (Figure 3).

Antigorite $M_8A_1B_{1C}$ was found in lamella only. In other words, macroscopically diagnosed antigorite presents almost completed pseudomorphs of chrysotile, lizardite and talc by alteration of antigorite (Lazarenkov et al., 2000), namely the minerals which are forming the serpentinites. Consequently, antigorite mineralization clearly preceded a serpentinization of chrysotile and lizardite. Antigorite was the earliest formation among other serpentine species. In addition to these minerals sometimes antigorite was replaced by brucite, chlorite, nontronite, quartz and iron hydroxides. Diffraction diagrams of antigorite are distinguished with a high number of reflections.

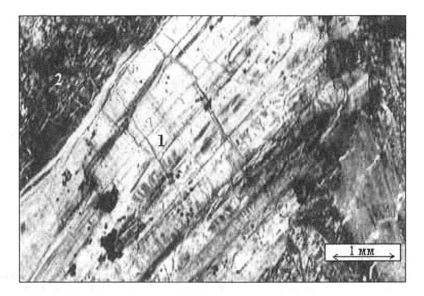

Figure 3.   Pseudomorphs of chrysotile 2Orc1 by alteration of antigorite (with the analyzer).

1 - lamelliform crystal of antigorite (in obedience to X-ray structural data chrysotile 2Orc1)
2 - chrysotile 2Orc1 of the serpentine mass.

3   RECENT DATA

In selected serpentine samples of the disintegrated and leached serpentinites of the Elov super-gene deposit serpentinite zones were mistakenly named as chrysotile, but several research tests have been conducted and this assumption has been disproved.

Antigorite of the Elov deposit is accumulations of radiating, "tubular" and needle-shaped aggregates, as well as nodular and opal-shaped masses (Figure 4). Calcite together with

Figure 4.   Nodular mass and needle-shaped crystals of antigorite in calcite.

antigorite is found to a greater extent, while talc, chlorite and quartz become apparent in a less degree.

Figure 4 shows that calcite accretes crystals of antigorite which allows to say that either a crystallization of the both minerals was at the same time or a crystallization of calcite took place after a crystallization of antigorite.

Using scanning electron microscopy in the mode of secondary and backscattered electrons a relief and a structure of antigorite were studied in details. Figure 5 represents that the central part of the "tube" aggregate is a dense non-crystalline mass, and there is a clear radiating structure on the peripheral areas. Within this framework it can be concluded that the hardening of antigorite was going very fast without the possibility of completed crystalline aggregates formation.

In Figure 6 we can see that well-defined center of crystallization is a "tubular" aggregate of antigorite axis, also the characteristic internal radiating structure of aggregate around the central axis can be observed.

Figure 7 illustrated an emission spectrum of elements of the Elov deposit antigorite; the following chemical composition was found: Mg - 22%, Fe - 3%, O - 56%, Si - 17%, C - 2% (which was similar to the composition of olivine for this deposit). As the spectrum which was obtained by scanning electron microscopy doesn't reflect the content of water and volatile components, the samples were subjected to XRD analysis for specification of the chemical composition. In the diffraction diagram (Figure 8) we can see the results of X-ray analysis - the X-ray peaks of the investigated sample coincided with the three possible versions of the database: 1 − olivine, 2 − antigorite, 3 − olivine-antigorite. Results variation could be testified by initially olivine of dunite-harzburgite composition rocks, which later was transformed into the antigorite by hydrothermal alteration.

Figure 5.   Cross section of "tubular" aggregate of antigorite.

Electron microscope JEOL JCM-5700 CarryScope,
TU Bergakademie Freiberg, an analyst V. Nikiforova.

Figure 6.   Radiating structure of "tubular" aggregate of antigorite.
Electron microscope JEOL JCM-5700 CarryScope,
TU Bergakademie Freiberg, an analyst V. Nikiforova.

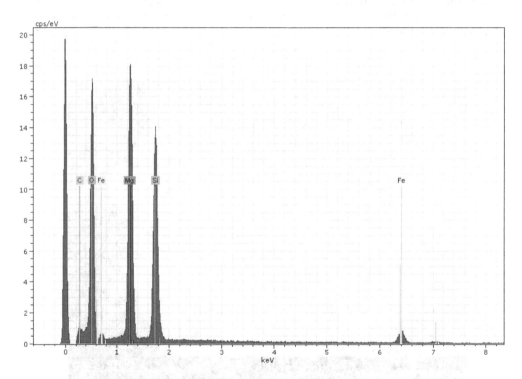

Figure 7.   X-ray spectrum of chemical elements of antigorite.
Electron microscope JEOL JCM-5700 CarryScope,
TU Bergakademie Freiberg, an analyst V. Nikiforova.

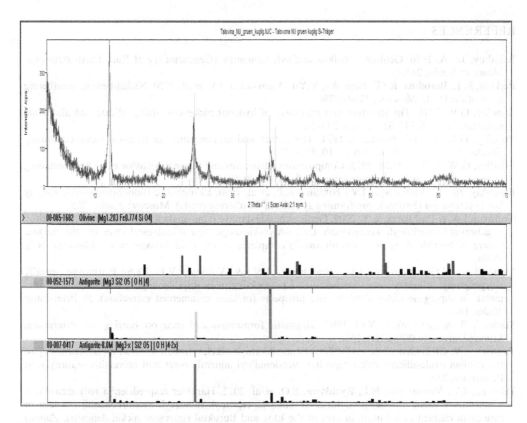

Figure 8. Diffraction diagram of antigorite of the Elov deposit.

Diffractometer URD-6 (Seifert-FPM),
X-ray laboratory of the Institute of Mineralogy, TU Bergakademie Freiberg, an analyst Dr. R. Kleeberg.

4 CONCLUSION

The ores of the Elov deposit were formed by transformation of the Serov dunite-harzburgite massif ultramafic rocks. Dunite-harzburgite bedrocks of the deposit are saturated with a dense network of parallel basalt dikes and as well as dikes of middle and acid composition, sometimes are making up to 40% or more of the substrate percentage. Joints of dikes are tectonically disturbed and the weathering crust was developed along them.

A large number of parallel basalt dikes in bedrocks of the Elov deposit establishes that the ultrabasites of Serov massif had strong tectonic stresses. Even in the period before the injection of the dikes they were exposed a contact-metasomatic transformation with emergence of serpentinites and also amphibole and chlorite metasomatic rocks. Cracks and breaks were renovated regularly filled by the injection of the dikes. The lower boundary of the weathering crust of the deposit is very uneven, with numerous wedge grooves, developed by tectonic series and were reached 85 m below the surface. Thus, the main reason for the heterogeneous substrate of the Elov deposit' formation was multiple premineral and postmineral processing of the Serov massif ultramafic rocks in a tectonically active zone conditions of the Serov-Mauksky fault.

Antigorite was found in veins and veinlets of the disintegrated and leached serpentinites in tectonically disturbed zones and failures in close vicinity to dikes. The presence of so many dikes proves a constant heating of the substrate, and this establishes that the serpentine was formed by cooling of the dikes complex.

REFERENCES

Balashov, Ju. A. 1976. Geohimija redkozemel'nyh elementov (Geochemistry of Rare Earth elements). Moscow: *Nauka*, 265.

Berkhin, S. I., Borodina, K.G., Bugelsky, Y.Yu., Vitovskaya, I.V. et al. 1970. Nickel-bearing weathering crust of the Urals. Moscow: *Nauka*, 286.

Brindley, G.W. 1978. The structure and chemistry of hydrous nickel-containing silicate and aluminate minerals. *Bull. BRGM*. Sec. 2, N 3, 233-245.

Brindley, G.W. & Maksimovic, Z. 1974. The nature and nomenclature of hydrous nickel-containing silicates. *Clay. Miner. Bull.*, G.B., v. 10, 271-277.

Brindley, G.W. & Wan, H.M. 1975. Compositions, structures and thermal behavior of nickel-containing minerals in the lizardite-nepouite series. *Amer. Miner.* v. 60, 863-871.

Bugelskiy, Ju. Ju., Vitovskaya, I.V., Nikitina, A.P. et al. 1990. Ekzogennye rudoobrazujushhie sistemy kor vyvetrivanija (Exogenic ore-forming systems of weathering crusts). Moscow: *Nauka*, 365.

Gottman, I.A. & Pushkarev, E.V. 2009. Geologicheskie dannye o magmaticheskoi prirode gornblenditov v gabbroul'tramafitovyh kompleksah Uralo-Alyaskinskogo tipa (Geological data on the magma nature of hornblendites in gabbroultramafic complexes of the Ural-Alaskan type). *Litosfera*. N 2, 78-86.

Lazarenkov, V.G., Talovina, I.V., Beloglazov, I.N. & Volodin, V.I. 2006. Platinovye metally v gipergennyh nikelevyh mestorozhdenijah i perspektivy ih promyshlennogo izvlechenija (Platinum metals in supergene nickel deposits and prospects for their commercial extraction). St Petersburg: *Nedra*, 188.

Marin, Y.B. & Lazarenkov, V.G. 1992. Magmatic formations and their ore-bearing. St. Petersburg, Mining University (SPbGGI), 166.

Talovina, I.V. 2012. Geohimija ural'skih oksidno-silikatnyh nikelevyh mestorozhdenij (Geochemistry of the Uralian oxide-silicate nickel deposits). Natsional'nyi mineral'no-syr'evoi universitet «Gornyi». St Petersburg, 270.

Talovina, I.V., Vorontsova, N.I., Ryzhkova, S.O. et al. 2012. Harakter raspredelenija redkozemel'nyh jelementov v rudah Elovskogo i Buruktal'skogo gipergennyh nikelevyh mestorozhdenij (Pattern of rare earth element distribution in ores of the Elov and Buruktal supergene nickel deposits). *Zapiski Gornogo instituta*. Vol.196, 31-35.

Volchenko, Y.A., Ivanov, K.S., Koroteev, V.A. et al. 2007. Strukturno-veshhestvennaja evoljuciya kompleksov Platinonosnogo pojasa Urala pri formirovanii hromit-platinovyh mestorozhdenii ural'skogo tipa (Structural-compositional evolution of the Ural Platinum Belt complexes during the formation of platinum-chromite deposits of the Ural type). Part 1. *Litosfera*. N 4, 73-101.

Vershinin, A.S. 1993. Geology, searches and investigation of supergene nickel deposits. Moscow: *Nedra*, 302.

Yudovich, Y.E. & Ketris, M.P. 2011. Geochemical indicators of lithogenesis (lithological geochemistry). Syktyvkar: *Geoprint*, 742.

Scientific and Practical Studies of Raw Material Issues – Litvinenko (Ed)
© *2020 Taylor & Francis Group, London, ISBN 978-0-367-86153-7*

Dependence between the parameters of storage of artificial soils from their specific properties

V.N. Kondakova & G.B. Pospekhov
Saint-Petersburg Mining University, Saint-Petersburg, Russian Federation

ABSTRACT: Nowadays the rates of the extraction of the mineral resources have an upward tendency. Hence, fast increase of the accumulation of the solid mining wastes, in this paper regarded as man-made soils, is an issue of growing importance. As it known, the mechanical properties of the man-made soils significantly differ from the natural soils and in many cases are difficult to study. In this paper the main geotechnical features of the man-made soils affected strength and stiffness and therefore the stability of rock slopes have been presented. Also, current issues related to studying of the mechanical properties of man-made soils, their storage and usage as recycled material have been described.

1 INTRODUCTION

Nowadays the rates of the extraction of the mineral resources have an upward tendency. Hence, fast increase of the accumulation of the solid mining wastes, in this paper regarded as man-made soils, is an issue of growing importance. According to the Russian government report about environmental data over 2016 published by Department of Environmental Resources, the area of the derelict territories affected by mining industry annually rises by 5%. Moreover, only 45% from the disturbed lands have been rehabilitated, while other part goes for the storage of the waste piles. About 10000 hectare of fertile land were taken for polygons (Table 1). Rock refuse not only occupies great areas of the potential fertile ground, but also affects the environment. Herein the points concerning the probable effects are highlighted. First of all, after the extraction and transportation chemically unstable minerals can be exposed to weathering, which leads to contamination of the air, water and the whole ecosystem by toxic elements and compounds. Moreover, in this case contaminated area is much more extensive than one which was developed. The landscapes are also transformed significantly caused by land movement, which leads to alteration of the microclimate. Due to extensive areas of barren mine sites and sometimes poor rehabilitation of them the problem of the storage of the mine wastes is relevant.

2 LITERATURE REVIEW

It is important to clarify which types of ground are related to the man-made soil. According to Russian national standard, artificial soils are the soils, which have been altered, transported or formed as a result of anthropogenic activities. So, industrial wastes like construction wastes, waste piles of overburden rocks, tailings, cakes, slugs and ash wastes may be regarded as man-made soils. Based on this, we assumed that wastes formed as a result if mining, ore-dressing and other actions related to mining industry, may be named as man-made soils. Tailings and tailing dumps, on which this paper is focused, are the one of the most specific soils and with the most unknown properties. This kind of soils is formed, for example, as a product of heap leaching process. This technology becomes more and more wide spread, especially in Russia, due to its effectiveness.

Table 1. Data on disturbed and reclaimed territories affected by mining industry.(governmental report "about the environmental state and protection...", 2018).

Parameter	Area, hectare
Disturbed territories on 01.01.2016 – total	993905.61
Disturbed during the reporting period	104968.42
Reclaimed territories – total	45493.67
for agricultural purpose	3005.96
forestation	35243.86
water basin	4457.89
Disturbed territories on 01.01.2017 – total	1053380.36

A detailed description of the heap leach process and the underlying mechanisms has been given elsewhere (Petersen and Dixon, 2007, Watling, 2006, Bartlett, 1998). Figure 1 shows in a schematic representation a typical heap leach processes.

A heap is a constructed pile of crushed, and in most cases agglomerated rock material built on an impermeable fitted with a solution collection system. Underneath the heap aeration pipes are placed. The heap is irrigated from the top surface, either by sprinklers, sprays or drip emitters. Heap heights are typically 6-10 m, but taller heaps are also common to reduce the footprint of the operation. Another option is to construct series of a piles on top of one another. In this case the height of a heap may be rather significant. The main aim of all operations is to provide appropriate filtration properties besides adequate construction characteristics. Optimally, heap leaching should be a low-cost technology for the extracting from low-grade ores, suitable mostly for remote mine sites. However, the complexity of leach process is often underestimated, resulting in underperforming heaps.

A typical generic copper heap leach curve versus time is shown in Figure 2 – this suggest that 60-70% extraction can be achieved quite rapidly, but beyond that the extraction curve begins to slow considerably. The multitude of processes taking place in a heap contribute to this slowing, but filtration through stagnant zones in the ore bed and within particle pores are perhaps the most critical. The latter can be countered by crushing technology and crushing to finer sizes, but this is likely to adversely affect the former with more fines, corresponding to overall lower heap permeability. Further, the length of time over which an ore is kept under leach appears to decrease heap permeability due to decrepitation of ore particles in the aggressive leach environment and precipitation of leached gangue minerals (Ahonen & Tuovinen, 1995). It proves the complicated behavior of the soils in the heap.

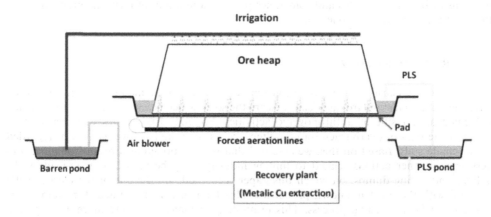

Figure 1. Scheme of a typical heap leach circuit (adopted from Petersen, Jochen, 2015).

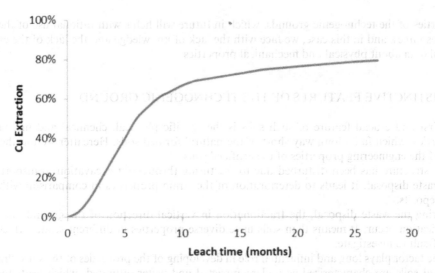

Figure 2.   A typical Cu sulphide heap extraction curve (based on data by Garcia et al., 1999).

Historically, treatment of low-grade ores by heap leaching was seen as an add-on process to recover additional values from a particular ore body without it determining the profitability of the operation as a whole. Hence, less care was applied in their design, and the understanding of the process was built more on operator experience than hard science. Here we faced with the main problem, related to the lack of knowledge about properties of the soils and their prediction for ensuring the proficient leaching. This field of study is the issue of growing importance for engineering geology. On practice in each particular case the properties of the soils, formed a heap, vary on each mine site. Hence, it is a complicated process to analyze them and make assumptions about the state of the soils in future.

Significant number of papers was devoted to the topic of the development of the technologies of the tailing dumps construction and environmental protection. The first who discussed the questions about environmental safety during waste storage was P.V. Beresnevich (1993). The properties of the technogenic ground, their impact on the atmosphere and surroundings. Also, researches made by Aksenov S.G. (2010), Sergeev S.V.(2006), Pashkevich M.A. (2010) should be mentioned. Those authors made a significant contribution to the investigation of the theory and applied knowledge about ecological aspects of the tailing dumps functioning.

The classification of the technogenic ground in accordance with their environmental threat was provided by M.A. Pashkevich (1999), in with the hazard is determined by physical, chemical and toxicological properties of the grounds. This classification may be used in findings of the environmental risks from the mine site. Trubetskoi K.N. (1989) also worked on classification. The author marked the importance of the secondary use of the areas of waste storage in order to evolving of the principles of the sustainable development. The key work in this field was provided by A.M. Galperin "Technogenic massives and the resources conservation". Here the basic principles of the environment protection and some fundamentals about the properties of the technogenic ground have been described.

All the authors characterized these massives by high environmental hazard. Also, they have noticed that there are several measurements aimed on the ensuring of the environmental safety which are common for the all types of grounds. Those are such actions as slope stability assessment, forecasting the compression of the massive and their base, shielding and reclamation. authors of the most publications argue that the most part of the negative effect is related to nonfulfillment of the regulatory documents, unsatisfying quality of the scheme of the works on the objects, low quality of the monitoring investigations.

Summing up, before mentioned aspects show the existence of the problem of the waste accumulation and storage, and the importance of the further complex investigations in this area. From the side of engineering geology, the task of prime importance is the study of the properties of the technogenic grounds, which in future will helps with rational use of the natural resources. and in this case, we face with the lack of knowledge and the lack of the experimental data about physical and mechanical properties.

3 DISTINCTIVE FEATURES OF THE TECHNOGENIC GROUND

The first and crucial feature of such soils is the specific physical, chemical and mechanical properties, which fall a long way short of the natural formed soils. Hereafter we put the key-note of the engineering properties of the artificial soils.

The structure has been disturbed due to fracturing throughout excavation, transportation and waste disposal. It leads to deterioration of the strain properties in comparison with natural deposits.

During the waste disposal, the fractionation in vertical direction of the ground and even stratification occur. It means than soils have diverse properties in different strata, which can be difficult to investigate.

Time factor plays long and influential role in developing of the properties of the soils. Primary artificial soils are characterized as undercompacted and water-saturated, which leads to high compressibility and extended period of the soil consolidation (10-20 years) (Figure 3a, 3b).

Water-saturated clay minerals are characterized by high pore pressure, which reduces slope stability in the waste piles and may be one of the factors to developing the failure. Mechanical properties of the soils significantly affected by heat and moisture transport in the massive.

Foregoing features of the artificial soils prove the fact of their dissimilarity with natural soils. Thus, standard methods used for latter do not work for former, for example, to assess the stability of the slopes, formed by artificial ground. Previous works indicated that strain and shear properties of some man-made soils vary vertically and the results of the investigation of their

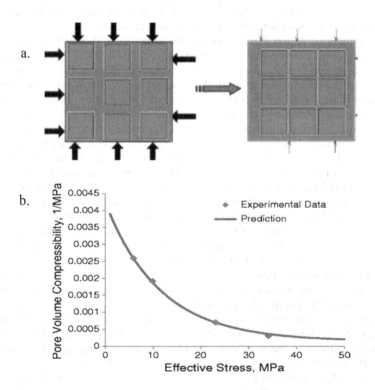

Figure 3. a. The principle of the ground compression under loading. b. The dependence between the stress and compressibility.

properties depend on applied method (statistical data manipulation of laboratory measurements versus results obtained with logging) (Pendin et al., 2018). According to this paper, the crucial point for obtaining the accurate results in the investigation of the waste piles, formed even by closely graded soil, is to mark certain strata, characterized by different cohesion and angle of the internal friction (Figure 4). However, nowadays, modelling and study of the spatial distribution of the mechanical properties of the man-made soils in the massive causes difficulties.

Mechanical properties of the rocks contained water are affected by temperature and water regime. Fundamental studies of the mechanical properties of ice have revealed that its strength and stiffness are a function of temperature [4]. In particular, element tests, conducted in the laboratory under controlled testing conditions, have shown that with decrease in temperature there is an increase in both stiffness and strength of ice. Thus, an investigation involving both laboratory experiments, using the technique of geotechnical centrifuge modelling, and simple geotechnical analysis has been conducted to investigate the effect of rise in ambient temperature on the stability of rock slopes containing ice filled discontinuities and affected by freeze-thaw process (Davies & Hamza, 2003). This work shows that as the temperature of ice increases the strength of an ice filled joint can be significantly less than that of an un-frozen joint. Such sophisticated correlations show the significance of thermometric surveys throughout geotechnical analysis of the soils. Nevertheless, thermal conditions still are not considered in a proper manner during geotechnical surveys. That is due to absence of such imperatives in the running regulatory documents.

The next problem related to the mining wastes is the lack of knowledge about properties of some types of soils, which is regarded as limiting factor for their applicability. One of the most comprehensive descriptions of the properties of artificial soils was made by M. Lychko (1982) and V.Trofimov (2005). However, artificial soils often characterized by composite spatial distribution of the properties, which may tend to change over time, and some consistent patterns and their behaviours can be difficult to determine. Moreover, some features appear only on a large scale, in the massive of rocks, which means impossibility to measure them in lab conditions.

Due to large areas have become derelict as a result of storage of the mine wastes and caused by this degradation of the environment, at the present time the question about directions of secondary usage of the wastes is under discussion. One of the most perspective methods is their application as a basement footing. Nevertheless, even from the one side there is a positive experience of their usage, on the other side with alteration of thermal and moisture

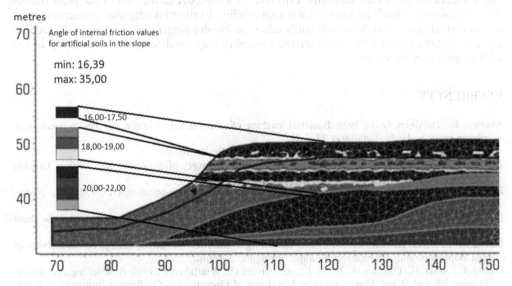

Figure 4. The example of differences in values of the angle of internal friction in the slope,formed by artificial soils (adopted from Pendin et al., 2018).

conditions significant settings of constructions on such footing have been observed. Also, technically there is an opportunity to extract some valuable elements from waste piles or tailing impoundments, for example. It means that not only the realization of the adequate waste management, but also probable financial benefits for mining company due to extraction itself and driving down cost of rehabilitation measurements. In spite of this, in many cases there are not appropriate protection measurements of the waste piles from the weathering, leaching and oxidation. It these processes occurred, with a course of time it becomes economically unprofitable and potential valuable elements turn into pollutants and the source for degradation of the environment. Therefore, the question about changes of the mechanical properties of the artificial soils should be studied properly, especially their behaviour under changed geotechnical and hydrogeological conditions of the area taken for storage. The attention is also should be paid on study of the possible existence of self-tightening processes. So, we argue that the scheme of protection measurements has to be developed in each particular case in a certain way with finding out the optimal solution.

In some cases, one of the methods to utilize the wastes may be the improvement of their strain and shear properties and further application. One of the most common use is installations on permafrost and collapsible loessial soils. Improved soils are characterized by identified and homogeneous properties, which makes them appropriate for using on constructing works. So, development of the methodology of improvement of the geotechnical properties of artificial soils also may be regarded as crucial task for the modern engineering geology.

4 CONCLUSION

It is obvious that the problem is complex and requires an integrated approach. All the contradictory factors connected with the development of marine oil and gas deposits and influence the effectiveness of the project need to be taken into account. Among the most important ones are Summing up, mechanical properties of the artificial soils are characterized by spatial and time variability, as well as comprehensive process of determination. Generally, on the state and properties of the man-made soils such parameters as mineralogical and grain content, depth of burial, thickness, temperature and water regimes, salinity of the subsurface water and age affect significantly. Artificial soils have a great potential and sometimes it is the valuable resource. In the present situation of the reduction of the natural resources, artificial soils can play a significant role in the economy. Low level of knowledge in the field of the properties of the artificial soils is the inhibiting to their applicability. Further investigation concerning mine wastes and reliable methods of their study allows to involve bigger part of them in the secondary use, to take rational decisions on the feasibility stage and to prevent occurring of the adverse geological processes.

REFERENCES

Ahonen, L., Tuovinen, O.H., 1995. Bacterial leaching of complex sulphide ore samples in bench-scale column reactors. Hydrometallurgy 37, 1-21.
Aksenov et al., 2010. Study guide about waste dumps, 90 p. [in Russian].
Bartlett, R.W., 1998. Solution mining, Leaching and fluid recovery of materials, 2nd Edition, Gordon and Breach Science Publishers, ISBN 90-5699-633-9.
Beresnevich V.P., 1993. Natural environment protection under waste dumps exploitation, Nedra, Moscow, 128 p. [in Russian].
Davies, M.C.R., Hamza, O., Lumsden, B.W. and Harris, C., 2000.Laboratory measurement of the shear strength of ice filled rock joints, Annals of Glaciology, Volume 31, 463-467.
Davies M. C. R., O. Hamza.. 2003. Physical modelling of the effect of climate change on the stability of rock slopes instability. Permafrost, Phillips, Springman&Arenson.
Garcia, C., Arias, H., Campos, J., Roco, J., San Martin, O., Whittaker, J., 1999. Bioleaching at Zaldivar. Biomine '99 and Water Management in Metallurgical Operations – Conference Proceedings. Perth, Australia; Glenside, Australia: Australian Mineral Foundation Inc., 72-81.

Government report "About environmental data and environment protection in Russian Federation over 2016". Department of Environmental Resources, 2017 [in Russian] URL: http://www.mnr.gov.ru/docs/o_sostoyanii_i_ob_okhrane_okruzhayushchey_sredy_rossiyskoy_federatsii/

Lychko U.M., 1982. Utilizing of the mine waste as bssement footing. VNIIS. [in Russian].

Russian National Standard "Soils. Classification", adopted 01.01.2013, – M. Standartinform, 2013. [in Russian].

Pashkevich M.A., 2010. Technogenic massives and their impact on the environment, SPPGI, 230 p. [in Russian].

Pashkevich M.A., 1999. Environmental risk assessment of the impact of the technogenic massives on the environment, MGGU, Moskow, 230 p. [in Russian].

Petersen, J. and Dixon, D.G., 2007. Modeling and Optimisation of Heap Bioleach Processes. In Biomining, D.E. Rawlings, D.B. Johnson (Eds.), Springer Verlag, Berlin, ISBN 978-3- 540-34909-9, pp 153-176.

Pendin V.V., I.K. Fomenko, D.N. Gorobtsov, M.E. Nikulina., 2018. Complex modelling of the slope stability of the waste piles. Mining Journal, vol.11, p. 92-96 [in Russian].

Petersen, Jochen, 2015. Heap leaching as a key technology for recovery of values from low-grade ores – a brief overview, Hydrometallurgy, doi: 10.1016/j.hydromet.2015.09.00

Sergeev S.V., Sinitsa I.V., 2006. The negative impact of the waste dumps on the environment and measurements for dust control. Journal of Tula State University, p. 159-163. [in Russian].

Trofimov V.T., 2005. Soils science. MSU, Moscow [in Russian].

Trubetskoi K.N. 1989. Classification of the technogenic deposits. Mining Journal # 12. P. 6-9. [in Russian].

Watling, H.R., 2006. The bioleaching of sulphide minerals with emphasis on copper sulphides-A review, Hydrometallurgy 84, 81-102.

17

Scientific and Practical Studies of Raw Material Issues – Litvinenko (Ed)
© 2020 Taylor & Francis Group, London, ISBN 978-0-367-86153-7

Petroleum prospects of the Middle Paleozoic sequences in the southern part of the Viluyui syneclise

G. Cherdancev

Saint-Petersburg Mining University, Saint-Petersburg, Russian Federation

ABSTRACT: Article is devoted to the forecast of petroleum potential of the Middle Paleozoic sequences in the southern part of the Viluyui syneclise. The updated scheme of structural and formation interpretation was produced. Furthermore, the scheme of petroleum prospects of the Middle Paleozoic sequences was renewed.

1 INTRODUCTION

The demands of modern industry cause active searches for new hydrocarbon deposits in the areas of Eastern Siberia, in particular – in the Viluyui Syneclise, which is located in the east of the Siberian Platform. One of the research directions is the study of its southern part, where the discovery of new hydrocarbons accumulations is expected in the Mesozoic and Paleozoic sections (Sitnikov et al, 2014). More than 10 gas and gas-condensate fields have already been discovered in the central part of the syneclise, mainly in the Mesozoic formations.

The Middle Paleozoic sequences have only a limited distribution within the Viluyui Syneclise – they are absent in a large part of the region, in other areas they are buried under significant thicknesses of the Upper Paleozoic and Mesozoic rocks. At the same time, closer to the edges of the syneclise, the Middle Paleozoic section occurs at the depths available for deep drilling. Moreover, the evidence of their oil and gas potential is observed in rock samples on the surface and in core samples from some wells in the east of the Siberian Platform (Fradkin et al, 2014).

Accumulation of petroleum in the Middle Paleozoic potential reservoirs was possible due to the hydrocarbons generated by the Early Paleozoic source rocks: Cambrian Kuonam Fm $Є_{1-2}$, Vendian–Lower Cambrian section: clayey part of the Nepa Fm (V_2np) (Late Proterozoic Kursov Fm and its analogues (V_2np)) and clayey-carbonate part of the Tirsk Fm (Late Proterozoic Buk Fm (V_2)) (Cherdancev & Golovin, 2018).

Different researchers have repeatedly pointed out the prospects of the possible petroleum accumulations in the Devonian and Carboniferous reservoirs of the Viluyui Syneclise since 50-60s of 20th century (Basharin et al, 2005; Fradkin et al, 2014; Safronov et al, 2003; Sivtsev et al, 2014). Recent composite of geological and geophysical research carried out on this territory by the Joint-Stock Company «ROSGEO» and All-Russia Petroleum Research Exploration Institute «VNIGRI», taking into account previous study of the territory, allowed to substantiate the prospects of sedimentary cover, including Middle Paleozoic ones.

2 STRATIGRAPHY AND RESERVOIR PROPERTIES OF THE MIDDLE PALEOZOIC SEQUENCES

According to the results of geophysical studies and well data, only Late Devonian and Early Carboniferous rocks were identified in the area of the geophysical surveys and adjacent territories. The Middle Paleozoic formations were formed in a variety of the sedimentation environments and are characterized by significant variation in their thicknesses.

The maximum thickness of the opened up Middle Paleozoic rocks reaches 2.5 km, the greatest thickness is identified in the Kempendyay (2350 m) and Ygyattin (1400 m) Depressions.

These characteristics allowed previous researchers to distinguish several structural-facial zones within the syneclise: 1) Kempendyay, corresponding to the same-named depression; 2) Ygyattin, corresponding to the same-named depression, and 3) Arbaysk-Sinsk - corresponding to the same-named uplift zone. These three zones together constitute the western part of the syneclise (Figure 1).

The stratigraphy and reservoir properties of the Middle Paleozoic sequences in these areas are the following (Table 1, Table 2):

1) Kempendyay Depression.

Kygyltuus Fm (D_3) is composed of argillites, siltstones, sandstones, marls and tuffs. The thickness varies up to 800 m (Fradkin, 1967). Sandstones make up 30 % of this formation in the Esselyahskaya-3010 well and can be considered as possible reservoirs. In the west (Kedepchikskaya-1 well) the sandstone layers are absent, probably because of its remoteness from the Suntar Uplift, which was the source area for sedimentation in the Devonian period. Thus, reservoirs can be expected only in the west of the depression.

Salt layers of this formation can serve as good caprocks. Their thickness is up to 90 m.

Namdyr Fm (D_3) is composed of sandstones, siltstones, dolomites, tuffs, gypsum. The maximum thickness is 300 m. Sandstones and siltstones of this formation are characterized by opened porosity of 7.7-34.3 % and permeability of up to 19.5 mD. These layers can be

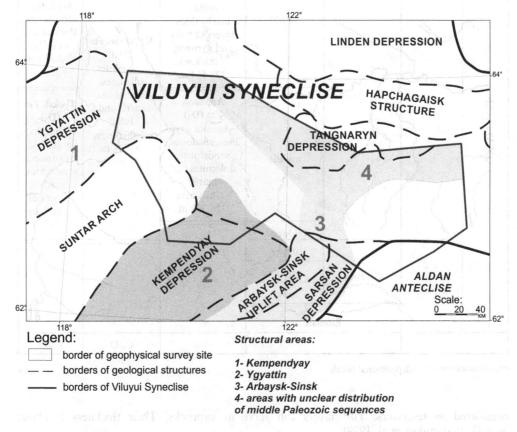

Legend:

☐ border of geophysical survey site
— — borders of geological structures
—— borders of Viluyui Syneclise

Structural areas:

1- Kempendyay
2- Ygyattin
3- Arbaysk-Sinsk
4- areas with unclear distribution of middle Paleozoic sequences

Figure 1. Scheme of the structural and formation interpretation of the Middle Paleozoic sequences of the southern part of the Viluyui Syneclise (Cherdancev & Golovin, 2018).

Table 1. Correlation scheme of the Devonian formations of the southern part of the Viluyui Syneclise (according to Fradkin et al, 2014).

System	Series	Stage			South of Viluyui syneclise		
					North-west of the Ygyattin Depression	North-west of the Kempen-dyay Depression	North-west of the Arbaysk-Sinsk Uplift area
					P_2	C_1-P	P_2
DEVONIAN	Upper	Famenian				**Namdyr Fm (D3)** sandstones, siltstones, dolomites; thickness 196-360 m	
		Frasnian	upper		**Viluchansk Fm (D$_3$)** dolomites, siltstones, argillites, marls, thick layers of tuffs and gypsum; thickness 323-554 m	**Kygyltuus Fm (D$_3$)** argillites, siltstones, sandstones, marls and tuffs; thickness >1948 m	**Altanottoh Fm (D$_3$)** gypsified siltstones, argillites, sandstones, thick layers of tuffs; thickness <339 m
			middle		**Appainsk Fm (D$_3$)** basalts, argillites, siltstones, sandstones, dolomites and marls; thickness 726-778 m		**Tisiksk Fm (D$_3$)** basalts, argillites, siltstones, sandstones, marls, dolomites; thickness <537 m
			lower				
	Middle	Givetian	upper				
			middle				
			lower				
		Eifelian					
					O-S	$Є_3$-O	$Є_3$-O

══════════ – depositional break

considered as reservoirs. Tuff layers can serve as caprocks. Their thickness is about 50 m (Kolodeznikov et al, 1979).

Kurunguryah Fm (C$_1$) is composed of limestones, dolomites, marls, clayey and aleuro-litic sedimentary rocks, anhydrites, sandstones, tuffs and tuffites. The thickness varies

Table 2. Correlation scheme of the Carboniferous formations of the southern part of the Viluyui Syneclise (according to Fradkin et al, 2014).

System	Series	Stage	South of Viluyui syneclise	
			East of the Kempendyay Depression	North of Kempendyay Depression
			P_1, J_1	P_2
CARBONIFEROUS	Middle	moscovian		
		bashkirian		
	Lower	serpukho-vian	**Kurunguryah Fm (C₁)** limestones, dolomites, marls, anhydrites, sandstones, tuffs and tuffites; thickness 100-624 m	**Kurunguryah Fm (C₁)** thickness 206-231 m
		visean		
		tournasian		
			D_3	D_3

═══════════════ – depositional break

from 380 m in the center of depression up to 624 m in the north-west (in Kedepchiks-kaya-1 well).

Layers of coarse siltstones and fine-grained sandstones are characterized by open porosity of 21.65 – 22.45 % (K-4 well) and permeability of 7.4-626.22 mD. These layers are also characterized by good reservoir properties in Kedepchikskaya-1 (open porosity 16.9 %, permeability 178 mD) and Eselyahskaya-3010 (open porosity 12-31 % according to the well logs petrophysics models) wells. Inflows of reservoir water were obtained during testing of these layers in 1 and 444-Kedepchikskaya wells. Flow rate was estimated at 26-224 m³ per day.

2) Ygyattin Depression.

The Middle Paleozoic formations in the Ygyattin Depression are characterized by a wide distribution of such sedimentary rocks as clayey siltstones, basalts, layers of salts, which may be considered as caprocks. Sandstone formations, as possible reservoirs, are less developed. In the most part of territory they are confined to the upper layers of the middle Paleozoic sequences. The layers of carbonate rocks are observed; however, they are unlikely to serve as oil and gas reservoirs due to their low thickness and lateral heterogeneity.

Appainsk Fm (D₃) is composed of basalts and variegated argillites, siltstones, sandstones, thick layers of dolomites and marls. Its thickness in the outcrops is up to 100 m. This formation is composed of impermeable rocks over most of the territory and is considered as a non-reservoir.

Viluchansk Fm (D₃) is composed of variegated carbonate and clayey sedimentary rocks, dolomites, siltstones, argillites, marls with thick layers of tuffs and gypsum. Thickness is up to 210 m.

Layer of feldspar–quartz sandstones distinguished in the basal layers of formation. It is traced from the marginal to the central parts of the depression over a distance of 250 km (Ust'-Meikskaya, Sygdahskaya, Ust'-Markhinskaya, Yuzhno-Sagytayskaya-290, 292 wells). The thickness of this sandstone layer varies from 30 m to 65-90 m. In Ust'-Markhinskaya, Yuzhno-Sagytayskaya-290 wells the thickness reaches 220 m and acquires a dual structure, separated by silty-clayey-basalt interlayer with the thickness of about 70 m. Porosity of sandstones across all wells is about 21-30 %. The sandstone layer was tested in Sygdahskaya, Ust'-Markhinskaya, Yuzhno-Sagytayskaya-290 (top layer) wells. The inflows of highly mineralized formation water were obtained; flow rate was about 307 m^3/day (Cherdancev & Golovin, 2018). This data prove the high reservoir characteristics of this level. Layers with improved reservoir properties also exist in the upper part of formation. According to the results of samples analysis from the Viluyui River, the open porosity is about 14.5-20.7 %, the permeability reaches hundreds of mD.

3) Arbaysk-Sinsk Uplift area.

Tisiksk Fm (D$_3$) is composed of basalts, argillites, siltstones, sandstones, marls, dolomites, rarely sandstones and tuffs. Thickness varies from 120-170 m in the west to 537 m in the east (Kumahskaya 2 well) (Cherdancev & Golovin, 2018). This formation is considered as non-reservoir due to low permeability.

Altanottoh Fm (D$_3$) consists of variegated gypsified siltstones, argillites sandstones and thick layers of tuffs. Some silty-sandstone layers in this formation have satisfactory reservoir properties; however, it is not possible to trace their spread in the investigated territory considering available information.

3 PETROLEUM POTENTIAL OF THE MIDDLE-PALEOZOIC SEQUENCES

The Middle Paleozoic formations include layers of permeable rocks (sands, sandstones, siltstones and carbonate rocks), and impermeable layers (layers of salts, clayey siltstones, basalt rocks and tuff layers). Direct signs of oil and gas potential are also obtained (bituminous calcite lodes, presence of higher methane homologues in dissolved gases of a number of springs of the Kempendyay Depression, etc.).

Formation of oil and gas accumulation zones of the Middle Paleozoic sections was possible due to the hydrocarbons coming from the Early Paleozoic and Late Proterozoic source rocks.

All this confirms the possibility of oil and gas deposits discovery in the rocks of the Middle Paleozoic structures. However, little attention was paid to the geological study of the Devonian and Carboniferous rocks during previous geological and petroleum exploration within the syneclise territory. Most of the deep wells have been drilled with extremely small core sampling. For example, only 6.9 m of the core was picked up from the 660 m thickness of the Middle Paleozoic formations during Ust'-Markhinskaya 1-P well drilling. Moreover, there was extremely low number of well testing of the prospective layers in these reservoirs. New seismic data in the southern and south-western parts of the Viluyui Syneclise together with reinterpretation of well logs allowed us to specify the depths and thickness of the Middle Paleozoic formations, mentioned above, and to identify promising areas (Figure 2).

In the research site unpromising areas include not only areas with the absence of the Middle Paleozoic sections, but also area bounded by the contour of 0.5 km (development of permanently frozen ground zone - permafrost area) on one side and the depth of 4 km of P1 seismic level (bottom of Permian sequences above Middle Paleozoic rocks) on the other side. In places where the bottom depth of the Permian section is more than 4 km, taking into account the forecasted thickness of the Middle Paleozoic section (about 2.5 km in the central part of the Kempendyay Depression, and 0.5-1.0 km on other territories), the probable oil and gas intervals in the Middle Devonian section occur at depths of more than 5 kilometers, therefore, well drilling is not profitable in this region at present day.

The permeable and impermeable layers distinguished in the Middle Paleozoic sequences are not equivalent. Thus, sandy layers of the Kygyltuus and Namdyr Fms are developed mostly locally on the west side of the Kempendyay Depression; perhaps, tuff layers are more widely spread and can serve as caprocks for these layers. The sandy layer of the Viluchansk structure, which is regionally spread in the Ygyattin Depression, is of greatest interest. The Appainsk and Tisiksk Fms (considered as the impermeable layers), composed of basalt and clay layers, together with the salt and clay layers of the Kygyltuus structure are also regionally developed.

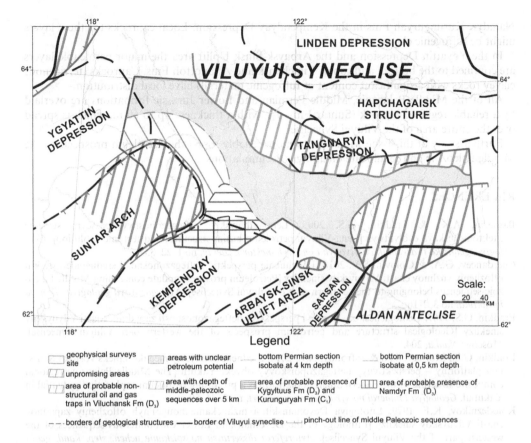

Figure 2. The scheme of the petroleum prospects of the middle Paleozoic sequences of the southern part of the Viluyui Syneclise (Cherdancev & Golovin, 2018).

The paleotectonic of the Middle-Paleozoic sedimentation should not be underestimated in oil and gas potential forecasting. The thicknesses analysis showed that mobile and relatively stagnant zones were present on the territory of syneclise at the end of the Middle Paleozoic period. The first ones included Kempendyay and Ygyattin Depressions with an amplitude of downwarping up to 3 km and more; while the Suntar Arch was relatively stable area.

Initially, the zones of maximum downwarping (Kempendyay and Ygyattin Depressions) with compensated sedimentation of the Middle Paleozoic section were the most favorable structures for hydrocarbon formation and accumulation, particulary their central axial parts. Marginal parts of these depressions, including those related to the current geophysics surveys site, where Middle Paleozoic layers have reduced thickness and some of levels wedge out, are considered as possible areas for hydrocarbon accumulation. Migration could occur from the deep parts of structures into anticlinal and non-anticlinal traps. These areas include western slope of the Kempendyay Depression adjacent to the Suntar Arch, its northern part, where sandstone layers of the Kygultuus Fm are developed, and also eastern slope of the Ygyattin Depression, adjacent to the Suntar Arch (Figure 2).

4 CONCLUSION

Taking all this into account the oil and gas prospects of the Middle Paleozoic rocks depend on the presence of strata with high reservoir properties. These layers are located in the Kygyltuus,

Namdyr, Kurunguryah Fms in the Kempendyay Depression. Local caprocks for these layers might be tuffogenic clayey rocks.

In the Ygyattin Depression and the Arbaysk-Sinsk Uplift area the major permeable layers are confined to the bottom and top of Viluchansk and Altanottoh Fms. Caprocks (terrigenous clayey rocks with a significant content of tuffogenic material) have local distribution.

All of the Middle Paleozoic, Middle Permian and Lower Jurassic formations are overlaid by a reliable regional caprock (Suntar Fm - J_1), with a thickness up to 60 m, which is spread over the entire area of the Viluyui Syneclise.

Further study of this issue will allow to more reliably assess the petroleum prospects of the Middle Paleozoic and to allocate oil and gas accumulations.

REFERENCES

Basharin, A.K. & Fradkin, G.S. 2005. Leno-Vilyuyskiy neftegazonosnyy basseyn: stroenie i tektonicheskaya evolyutsiya [Lena-Viluyui petroleum basin: structural and tectonic evolution]. *Geologiya, geofizika i razrabotka neftyanyh i gazovyh mestorozhdeniy*, no I, 22–33.

Cherdancev, G.A. & Golovin, S.V. 2018. Utochnenie perspektiv neftegazonosnosti srednepaleozoyskih otlozheniy yuzhnoy chasti Vilyuyskoy sineklizy [Petroleum prospects update concerning Middle Paleozoic sequences belonging to the southern part of Viluyui Syneclise]. *Neftegazovaya geologiya. Teoriya i praktika*, no 3, 1-16.

Fradkin, G.S. 1967. Geologicheskoe stroenie i perspektivy neftegazonosnosti zapadnoy chasti Vilyuyskoy sineklizy [Geological structure and petroleum prospects of the western part Viluyui Syneclise]. Moscow: *Nauka*, 204.

Fradkin, G.S., Moiseev, S.A. & Safronov, A.F. 2014. Srednepaleozoyskiy megakompleks vostoka Sibirskoy platformy – perspektivnyy neftegazopoiskovyy obyekt Yakutii [The Middle Paleozoic regional composite structure of the eastern Siberian platform is a promising petroleum exploration potential in Yakutia]. *Geologiya I mineral'no-syr'evye resursy Sibiri*, no 2, 44–59.

Kolodeznikov, K.E. 1979. Litologiya Devonianskih i nizhnekamennougol'nyh otlozheniy zapadnoy chasti Vilyuyskoy sineklizy [Lithology of the Devonian and Lower Carboniferous sequences of the western part of the Viluyui Syneclise]. *Avtoreferat dissertatsii na soiskanie uchen. step. kand. geol.-mineral. nauk* - Novosibirsk, 26.

Safronov, A.F., Berzin, A.G. & Fradkin, G.S. 2003. Tektonicheskaya priroda lokal'nyh podnyatiy Vilyuyskoy sineklizy [Tectonic character of local uplifts of the Viluyui Sineclise]. *Geologiya nefti i gaza*, no 4, 20–28.

Sitnikov, V.S., Prishchepa, O.M., Kushmar, I.A. & Pogodaev, A.V. 2014. Perspektivy neftenosnosti yuzhnoy chasti Vilyuyskoy sineklizy [Petroleum prospects in the southern part of the Viluyui syneclise]. *Razvedka i okhrana nedr*, no 7, 22–28.

Sivtsev, A.I. & Aleksandrov, A.R. 2014. Galokinez v tektonicheskom stroenii Kempendyayskoy vpadiny [Salt tectonic of the Kempendyai depression]. *Elektronnyy nauchnyy zhurnal «Neftegazovoe delo»*, no 5, 54-70.

Scientific and Practical Studies of Raw Material Issues – Litvinenko (Ed)
© 2020 Taylor & Francis Group, London, ISBN 978-0-367-86153-7

The problems of cryolithozone mining in Yakutia in conditions of global warming

A.A. Pomortseva
Saint-Petersburg Mining University, Saint-Petersburg, Russian Federation

O.A. Pomortsev
North-East Federal University, Yakutsk, Russian Federation

ABSTRACT: The article considers the problem of global warming threatening most of the area of the Northern hemisphere lying on permafrost. In North Asia, the most likely to be affected is the Republic of Sakha (Yakutia) located at the center of the cryolithozone. The objective of the study is identifying patterns in the development of the present-day warming surge in Yakutia and in hazardous changeability of climate modulated engineering-geological conditions in the locations where cryolithozone mining takes place, as well as climate prediction. Methods of mathematical modeling, rhythmic-morphological and comparative geographic analysis were used. The materials the study are based on the climatological, geographical, paleographical and engineering-geological investigations of natural systems of Yakutia (Russia). The study presents the authors concept of the hundred-year rhythm for Yakutian climate. Structural coherence in the development of climate phases has been identified in the model of the hundred-year and long-term multi-thousand-year rhythms.

1 INTRODUCTION

Located in north-east Eurasia, the Republic of Sakha (Yakutia) provides strategic natural resources for Russia, such as diamonds, gold, tin, coal, oil, gas, uranium and others. The unique feature of the area is the fact that it lies wholly in the permafrost zone, and it is here that the outperforming rate of global warming has been registered (Gavrilova, 2003; Pomortsev et al., 2015; Pomortsev et al., 2015). For today, there are two inescapable facts: first, the progress of global warming and second, the scale of its possible implications, which have not yet been explored or assessed. The presence in permafrost sections of natural ice with the trend towards phase transformation, huge mass, and resistance to melting is a serious threat that can come into reality at the worst possible time. A good example is the current Arctic ice melting, in which case after the decades of warming the Arctic Ocean lost 25% of its ice cover in 2007 alone. The process reached the critical point notwithstanding the strong resistance to the melting of the Arctic, but after that, it has advanced with giant strides. Due to the fact that all the mining infrastructure in Yakutia, namely mines and quarries, processing plants, wells, pipelines, access ways, power lines, A-strips, shift camps, etc. have all been developed in cryolithozone, the issue of warming deserves the fullest consideration. Nowadays, there is neither reliable informed prediction about the performance of deep thickness of permafrost at growing warming nor any reliable evidence of the warming trend. The leading specialists have absolutely various opinions. Some experts predict global warming with "global flood" (Balobaev et al., 2009; Diskey et al., 2011; Maksimov, 1972; Maksimov, 1995; Shnitnikov, 1957) for the next decades, the others say about the same scale of cooling with the onset of a new ice age (Balobaev et al., 2009, Bolshiyanov & Verkulich, 2018).

2 METHODOLOGY

The main purpose of the study is identifying patterns in the development of the present-day warming surge in Yakutia and of climatic modulation of hazardous changeability in engineering-geological conditions in the locations of mining operations and economic activity in cryolithozone, and climate prediction.

The objectives:

(1) To investigate the process of global warming affecting the climate of Yakutia and its connection with the runs of global rhythms of the landscape layers;
(2) To assess the impact of the current warming on the geological environment of cryolithozone in the locations of mining activity;
(3) To identify the tendencies of climate change and the response of cryogenic bedding for the next few decades and more distant future.

It is important to note, that problem solving within the scope of this study based on the search of patterns in the variability of climatic conditions in the cryolithozone of Yakutia. The solution is implemented with the help of visualization and analysis of the global warming wave according to the scale of the global rhythms of the landscape shell: 1850-year Shnitnikov rhythm and 40700-year Milankovich rhythm and century rhythm. This is a new solution to justify the issue of the model of the long-term climate and geological forecast in a dangerous Northern permafrost area.

The theoretical base for the study is presented by Evgeny V. Maksimov's concept of rhythms pertaining to nature (Maksimov, 1972; Maksimov, 1977; Pomortsev et al., 2015). The key investigation methods are rhythmic-stratigraphic and comparative-geographic, providing the opportunity for extensive comparison and structuring of the processes and phenomena under study on the time-space basis (Kalesnik, 1975; Maksimov, 1972). As the working tool, we have applied the method of mathematical simulation and graphical analysis. In the key locations, the authors employed the methods of geomorphological and engineering-geological research along with validation of the information obtained through route monitoring observations.

3 ANALYSIS

The current warming, in our opinion, was most accurately predicted by A. Shnitnikov and E. Maksimov (Maksimov, 1977; Shnitnikov, 1957). The former relied on the self-discovered long-term 1850-year rhythm, which changes the moisture content on the planed, while the latter also considered the rhythm of M. Milankovich (40,700 years) which underlies the warming and cooling cycles in the history of the Earth and is rigidly connected with the 1850-year rhythm (Shnitnikov, 1972).

According to Arseny V. Shnitnikov rhythm (Figure 1b), the global warming is to continue for about 700 years, developing under the conditions of the warm-and-dry phase that dominates in this rhythm (it takes 1200 years or around 60% of the length of the rhythm). The temperature rises to the level double the current increase is to happen within the next 200 years on the base of the hundred-year rhythm. The further warming increase will be curbed by the decrease of heat supply and the increase in humidity in the warm-and-humid Milankovich rhythm phase (see Figure 1a), which checks the 1850-year Shnitnikov rhythm. The increase in moisture content along the path of the two leading long-term rhythms will be moderate and stable. Considering the term of the warm-and-humid phase of Milankovich rhythm, the warm climate will dominate for about another 7,000 years (Pomortsev et al., 2015).

The long-term rhythms determine the conditions for the development of the hundred-year rhythm (80 to 90 years), as well as the short-term interdecadal ones (44, 22 and 11 years) (Diskey et al., 2011; Grichuk, 1961; Kononova, 2003; Maksimov, 1972; Maksimov, 1995; Pomortsev et al., 2015, Trofimtsev et al., 2017).

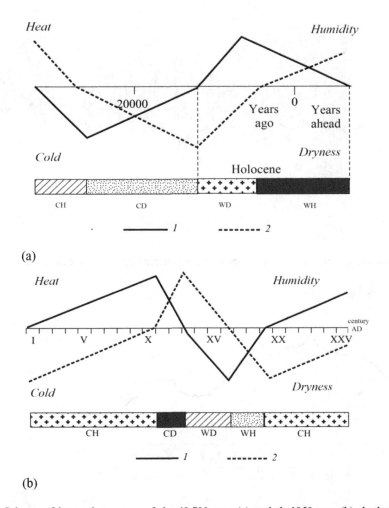

(a)

(b)

Figure 1.   Scheme of internal structure of the 40,700-year (a) and the1850-year (b) rhythms 1 = heat supply, 2 = moisture content. Climate phases: CH = cold-and-humid, CD = cold-and-dry, WD = warm-and-dry, WH = warm-and-humid.

The most important feature in the development mechanism of the current warming is shown in Figure 2, presenting the pattern of the hundred-year rhythm model we developed based on the data of instrumental monitoring of ground air temperatures and atmospheric precipitation done at Yakutsk weather station (continuous monitoring since 1883). You can see from the figure that temperature peaks occurring in Yakutia in the 20th century came one quarter the hundred-year rhythm wave ahead of moisture content peaks. This led to successive alteration of cold-and-dry (HD), warm-and-dry (WD), cold-and-humid (CH) and warm-and-humid (WH) climatic periods.

It appears that the one-quarter rhythmic wave lag of moisture content peaks in relation to temperature peaks is universal to the global climate regime and can be explained by two facts: first, the disproportion of the areas of the dry land and oceans and the huge size of the cryo-lithozone (Pomortsev et al., 2015, Trofimtsev et al., 2017).

With the area of the ocean in the Northern hemisphere being 2.5 the size of the area of dry land accompanied by the obvious dominance of the cryolithozone in the region, water and ice absorb most of the solar heat and thus slow down the rapid development of the warming. In the case of the area of the dry land increasing on account of the expansion of the ice sheet, the interval between the temperature and moisture content peaks must become shorter. Apparently, this situation was characteristic for the beginning of the Pleistocene, when the preponderance of the land area over the area of the ocean in the Northern hemisphere led to climate continentality of

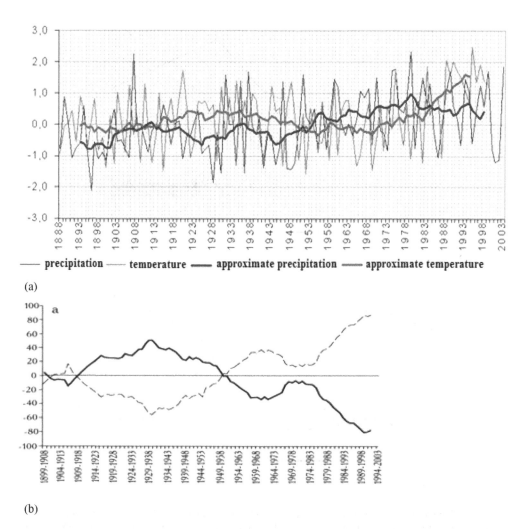

(a)

(b)

Figure 2. (a) –curves of average annual temperatures and atmospheric precipitation at Yakutsk weather station in the 20th century (natural series and 11-year fitted ones). (b) – the runs of zonal (full line) and meridional (dotted line) atmospheric circulation in the Northern hemisphere.

unprecedented proportions, thus providing the dominance of the steppe and the African form of mammals up to the coast of the Arctic Ocean (Pomortsev et al., 2015).

Today we can observe the correction of changeability of natural environments by the moisture content conditions on the example of ice caps melting at Suntar-Khayata and Chersky mountain ranges in the east of Yakutia. The data for the rates of melting are provided in the study (Ananicheva et al., 2006).

The loss of ice caps area in Chersky range, located in the more continental part of Asian North-East, in the conditions of the current warming is twice as big as the loss in Suntar-Khayata range. The Southeast end of the Suntar-Khayata reaches the Pacific ocean, so its moisture content protects the range from the effects of global warming much more effectively than the continental climate around the Chersky range.

Another example is presented by Little Ice Age (LIA). While it is separated from the modern age by just a hundred years, its moisture content conditions differ significantly from the present-day ones. With the maximum area of Arctic ice fields and mountain glaciation at the peak of the LIA (1840-1860), the global moisture content was at its minimum (Maksimov, 1972, Maksimov, 1995). It is particularly well illustrated by the "drop" of the atmospheric precipitation run at the peak of the LIA occurring in Yakutia and the Altay (Maksimov, 1972; Nekrasov et al., 1973). The key reason for this phenomenon is the increase in the area of sea ice sheets and mountain glaciers. The sea ice provided the effect of an increase in dry land area and, correspondingly, the sharp growth of climate continentality. As a result, moisture content could not curb the rise of temperatures at the peak of the LIA in 1840 to 1860 as successfully as in the course of the 1970s to 2000s warming, when ice coverage of the seas and mountain glaciation approached their minimum, while moisture content approached its maximum. The inhibiting impact of these factors nowadays effectively suppresses the extremely hazardous "outbreak" of global warming. Allowance for the ice coverage of the World ocean must be made for paleoclimatic reconstructions and the interpretations of the runs of temperatures and moisture content in rhythmic models.

On the whole, according to the rhythmic scenario we presented above, which repeatedly proved itself over the past thousands and hundreds of years, in Yakutia as well (Maksimov, 1995; Maksimov, 1977; Nekrasov et al., 1973; Pomortsev et al., 2015; Pomortsev et al., 2015, Pomortsev et al., 2018), we have entered the warm and humid climate phase developing against the background of enhancing of meridional atmospheric circulation (see Figures 1, 2, and 2a). We should expect an extended "watery" stage dominated by typhoons, floods, and mild winters. It will exceed the length of the 21st century, and cannot fail to impact the stability of the cryolithozone. If in the LIA fall of temperature in the 14th to 19th century the average annual temperatures in Yakutia reached its lowest values (-12,6°C in 1837), the 2017 warming almost halved it to -6,6°C. Alongside this, as it was already mentioned above, the multi-hundred-meter layers of permafrost have an enormous cold content accumulated as far back as in the Pleistocene and maintained by Siberian high-pressure area, which is one of the key factors curbing the ecologically hazardous advance of the warming.

4 THE RESPONSE OF THE CRYOLITHOZONE TO THE WARMING

Just like in the case of the ice cover of the Arctic Ocean, the response of the cryolithozone to the warming did not show at once: it took 20 to 30 years after the start of the active phase of the warming (late 1970s to early 1980s). The first wave of the increase in the dynamics of exogenous geological processes was registered from 2001 to 2003 in the middle Lena River valley. Here, affected by the thaw out in permafrost formations, the dynamics of superpermafrost waters increased, which caused massive caving of the sedimentary sheath on the steep sides of the valley accompanied by exposure of rock foundations, as well as emergency of thermocrust on the open areas of ice-rich accumulative plains of Central Yakutia. The phenomenon was especially prominent in Khangalas Stone area at the periphery of Khangalassky Razrez brown coal field located near Yakutsk (Figures 3). The same also happened in the Erkeny area 70 km south of the Khangalas outcrop.

At about the same time the exogenous disaster struck the automobile roads in Yakutia including federal routes "Lena" (Yakutsk to Bolshoy Never) and "Kolyma" (Yakutsk to Magadan) which provide access to gold deposits of East Siberia (Aldanskoe, Allakh-Yunskoe, Upper-Indigirka, etc.) and year-round transregional road communications of the Sakha Republic with Russian Far East, Central Russia and Magadanskiy region (Figure 4).

Later, from 2011 to 2014, the exogenous expansion reached the mountainous areas of Eastern Yakutia. It was most prominent in the vicinity of Batagayskoe cassiterite deposit (Upper Yana River). Here, in the Batagayka river basin, a huge cave-in emerged exposing the thick ice kernel of the cryolithozone which has the shape of an ice ledge, up to 20 to 25 meters high, and over 4 km long (Figure 5).

In 2013-2014, 10 years after the first response of the cryolithozone to the warming, the effect reached Oimyakon, the cold pole in the Northern hemisphere, showing in the territory of the Khangalas gold ore field in Oimyakonsky district. Here while exploring the deeply frozen adits, geologists came across snow slush that made it difficult to move along the mine workings. Besides, the quarry floor and exploration ditches were flooded by superpermafrost waters (Figure 6).

(a)   (b)   (c)

Figure 3.   Caving of the sedimentary sheath on the left side of the Lena River valley in the Khangalas Stone area. (a) – landslide blocks, view from the north; (b) – landslide slope deposit, view from the northeast; (c) – landslide surface, view from the southeast.

(a)   (b)   (c)

Figure 4.   Federal route "Lena" at the Yakutsk – Aldan section, July 2003: (a) – thawed foundation soils; (b) – waterlogged stretch. Regional route Nizhny Bestyakh - Maya: (c) – thermal erosion accompanied by ice body outcrop.

(a)   (b)   (c)

Figure 5.   The exposed ice kernel of the cryolithozone in the vicinity of Batagayskoe cassiterite deposit (photo by Innokenty Starostin, summer 2014): (a) – view from the southeast; (b) – view from the northeast; (c) – sandy hilly terrain (bed land) in the territory formerly occupied by ice body.

(a)   (b)   (c)

Figure 6.   Snow slush in the deeply frozen adits of Khangalass gold ore field in the vicinity of Oimya-kon, 2014 (photo by L. I. Polufuntikova): (a) – adit with shaftmen footprints in the snow slush; (b) – flooded quarry field; (c) – a flooded exploration ditch.

A year earlier, in the summer of 2013, and later in 2014 and 2015, there were major prob-lems on the mountain stretch of the motorway Yakutsk – Magadan in the valley of Vostoch-naya Khandyga river, Suntar-Khayata range (south of upper Yana valley). Here the abnormal atmospheric precipitation in July exceeding the long-term annual average 4 to 5 times led to hazardous engineering-geological processes that blocked the traffic movement in the motorway for a week and longer (Figure 7).

Thus the findings of our study inevitably show that the current warming poses an explicit threat to uninterrupted functioning of the major traffic arteries of Yakutia and sustainability of cryolithozone ecosystems, including those located in the territories adjacent to mining facil-ities. Considering the impact of climatic anomalies, the "product" of the warming, on the dynamics of engineering-geological conditions in the cryolithozone, today we cannot exclude the intensification of the industry-related inundation, dangerous enough as it is, of diamond-iferous deposits of Western Yakutia, as well as thawing of cryostorage disposals of hazardous drainage water brines, for example, of Levoberezhny disposal at Udachensky MPP., OOO ALROSA. All these links in a chain, exposing the growing threat of the warming. We should also emphasize once again that the impact mechanisms of the warming on the cryolithozone may vary substantially. Thus, the first wave of exogenous process expansion, which emerged at the very start of this century on the valleys of Central Yakutia, manifested itself against the background of extremely arid and hot summers that caused extremely large forest fires con-tributing to the increase in the depth of annually thawed layer in the areas affected by fire and in the dynamics of superpermafrost water runoff. The second wave that manifested itself in the upper Yana valley in 2012 to 2017, on the contrary, developed against the background of

(a)   (b)   (c)

Figure 7.   Hazardous engineering-geological processes at the "Kolyma" motorway (Yakutsk – Maga-dan) in the valley of Vostochnaya Khandyga river (Suntar-Khayata range) (July 2013): (a) – thermal ero-sion destruction of the roadbed; (b) – solution sinkhole; (c) – stormwater drainage blocked by slope detritus.

31

extremely humid summers, when the average monthly figures of atmospheric fallouts in summer months were 3 to 5 times as high as the annual mean in many years. It must be noted that the anomaly of the first exogenous wave spreading across the valleys of Central Yakutia in the first years of the current century, developed against the segment of rise of the odd 11-year solar activity (SA) cycle, while the second wave, the humid one of the second decade of the current century, which manifested itself in the mountains, developed against the even SA cycle. If the identified tendencies persist, in the coming years, considering the start of another odd 11-year SA cycle in 2017, we should expect enhancement of the hazardous exogenous geological processes dynamics in the valleys of Central Yakutia, and then in the late 2030s and in 2040s the same should be expected in the north-eastern mountainous areas against the background of the even SA cycle.

5 DISCUSSION

The contribution of this work are the systematisation of the ideas by various arguments of authors for and against the implementation of oil and gas projects in the Arctic, and the development of a system of factors and indicators. In the Russian context, a specific operator, who is the owner of the license for a given region, develops Arctic oil and gas projects. In this regard, the factors characterising the ability of the oil and gas company to realise projects are of special interest. Along with macro-level factors (political, legal, economic and other), the indicators characterising the innovative capacity of the companies strongly influence the prospects of implementation of such projects.

The research offers an integrated approach to studying the factors exerting impact on the prospects of implementation of shelf projects and an identification of the internal resources characterising the potential for such projects. This is supplemented by the classification of indicators for the assessment of prospects of implementation.

We list appropriate long-term forecasting methods that are applicable for the creation of predictions for the various elements influencing the prospects of shelf projects and highlight the need to consider a variety of alternative approaches. An area of further research is the testing of the efficacy of these forecasting methods on these tasks, and to provide guidelines for the expected uncertainties stemming from different factors. Application of these methods in combination with the developed system of factors and indicators will be presented as the results of further scientific research.

6 CONCLUSION

1. Global warming, even at its initial stage, has shown us the fragility and sensitivity of the cryolithozone ice structure. The warming presents the major threat to mining the natural resources of Yakutia and its economic activity and well-being in general;
2. According to our forecast, the warming wave is going to last throughout the 21st century, now enhancing and now mitigating the dynamics of hazardous exogenous geological processes. These are modulated by both positive temperature anomalies and the alterations of abnormally dry (arid) and abnormally humid (damp) summer seasons;
3. Yakutia is going to see warmer and shorter winters along with warmer and more humid summers. The trend has proven itself already.
   The identified climatic tendencies, as our study has shown, have the potential not only to increase the thickness of seasonally thawed layer in the locations of permafrost formations but also to modulate the anomalies of downpours and droughts, forest fires, floods, enhancement of the dynamics of land and over-permafrost runoff, which in its turn may activate the hazardous exogenic geological processes that can put a strain on mining operations in the cryolithozone (Pomortsev et al, 2015). In addition to that, in connection with the expected enhancement of zonal atmospheric circulation dynamics in the second half of the current century there is a possibility of decrease of the Arctic seas ice sheet area.

REFERENCES

Ananicheva M. D., Kapustin G. A., Koreysha M. M. 2006. Changing of the glaciers of Suntar- Khayata and Chersky ranges according to the Catalogue of Glaciers of the USSR and the 2001 to 2013 space images, *Glaciological research materials*, 101, p.161-168.

Balobaev T. V., Skachkov Yu.B., Shender N.I. 2009. The study of the Eastern Siberia cryolithozone, *Geography and natural resources*. № 2, p. 50-57.

Budyko M.I. 1980. The climate in the past and the future, Leningrad, *Hydrometeoizdat*, p.352.

Bolshiyanov D. Yu. and Verkulich S.R. 2018. Climate change in the polar regions of the Earth over the past 10,000 years, Saint-Petersburg, *Arctic and Antarctic Research Institute*, p.204.

Diskey, Jean O., Steven L. Marcus, Oliver de Viron 2011. Air Temperature and Anthropogenic Forcing: Insights from the Solid Earth, *J. Climate*, № 24, p. 563-574.

Dorofeyuk N.I., 2008. Reconstruction of the natural conditions of Inner Asia in the Late Glacial and Holocene: based on diatom and palynological analyses of lake sediments in Mongolia. *Abstract of the doctoral thesis*, Moscow, p.48.

Gavrilova M. K. 2003. Current climate change of the permafrost region in Asia, Overview of the status and trends of climate changes in Yakutia, Yakutsk, *SO RAN*, p. 13-18.

Grichuk V.P. 1961. The main features of changes in the plant cover of Siberia during the Quaternary period. *Collection of articles: Paleography of the Quaternary period in the USSR*, Moscow, p. 189-206.

Kalesnik S.V. 1975. Common geographic regularities of the Earth, Moscow, *Mysl'*, p. 283.

Kononova N. K. 2003. Study of long-term fluctuations in the atmospheric circulation of the Northern hemisphere and their application in glaciology, *Glaciological research materials*, 95, p. 45-6.

Maksimov E.V. 1972. Problems of glaciation of the Earth and rhythms in nature, Leningrad, *LGU*, p.294.

Maksimov E.V. 1995. Rhythms on the Earth and in Space, Saint-Petersburg, *SPbun-ta*, p.324.

Maksimov E.V. 1977. The rhythm of natural phenomena and its meaning, *Izvestiya VGO*, v. 109, № 5, p. 418-422.

Nekrasov I.A., Maksimov E.V., Klimovsky I.V. 1973.The last glaciation and cryolithozone of the Southern Verkhoyansk, *Yakutsk*, p.144.

Pomortsev O.A., Kashkarov E.P., Lovelius N.V. 2015. Bioclimatic chronology of the Holocene, *Vestnik SVFU* № 3 (47), p. 100-115.

Pomortsev O.A., Kashkarov E.P., Lovelius N.V., Maksimov E.V. 2015. Maksimov's theory of in nature (to the 85th anniversary), Geography: development of science and education, LXVIII *Hertzen's Readings*, St. Petersburg, p.34-39.

Pomortsev O.A., Pomortseva A.A., Rozhin S.S. 2018. Global climate warming as a hazard factor in the mining and geological development of the cryolithozone, Geology and mineral resources of the north-East of Russia, *materials of the 8th All-Russian Scientific and Practical Conference 2018*, vol.2, Yakutsk, p.267-270.

Pomortsev O.A., Pomortseva A.A. 2018. Absolute chronology of the transgressive Pleistocene cycles and their place on the scale of M. Milankovich, *Proceedings of the XXVII international coastal conference Arctic shores: the path to sustainability*, Murmansk, p. 127-130.

Shnitnikov A. V. 1957. Changeability of overall moisture content of the Northern hemisphere continents, *Zap. GO USSR*, Moscow, Leningrad, vol.16, p.337.

Trofimtsev Yu.I., Pomortsev O. A., Popov V.F., Pomortseva A. A. 2017. Numerical modeling of harmonics in meteorological time series, *American Institute of Physics Numerical modeling of harmonics in meteorological time series*, AIP Conference Proceedings 1907, 030025.

Scientific and Practical Studies of Raw Material Issues – Litvinenko (Ed)
© 2020 Taylor & Francis Group, London, ISBN 978-0-367-86153-7

Innovations in mechanical rock excavation at TU Bergakademie Freiberg

B. Grafe, T. Shepel & C. Drebenstedt
Technische Universität Bergakademie Freiberg, Germany

ABSTRACT: The following article summarizes a series of developments revolving around innovative mining technologies in the field of mechanical rock cutting that have and are being developed at the Chair of Surface Mining at TU Bergakademie Freiberg. Amongst them are undercutting, cutting of pre-weakened rock and cutting force analysis for sensing boundaries between different rock types. This research can be understood as contributions towards the knowledge base for low-impact, highly selective mining techniques of the future.

1 INTRODUCTION

Technologies for mechanical excavation are commonly used in tunnelling and to a rising extend also in production excavation – especially in softer rocks like coal, salt, and limestone but also sedimentary deposits. Intrinsic to this technology are some advantages over drilling and blasting (D+B). The material does not have to be sized as it comes out of the cutting process in a rather narrow and smaller particle distribution band. This means that boulders also do not affect the process chain and breakers can be redundant. Plinninger (2011) cites that mechanical tunnelling is beneficial to reduce overbreak in comparison to D+B by 1.5 to 2.3 times. This has an influence on roof stability and required effort towards safety measures such as scaling, bolting or more sophisticated support (Plinninger 2011, p 145).

With D+B-operations, additional time is required for venting of the blasting gases, which does not apply for mechanical excavation. Linked to the continuous operation scheme, it is possible to achieve high advance rates. Especially in development works, where only one advancing face exists, this can have a great impact. Restner (2016) cites that starting a mining operation one month earlier due to faster development work can have an impact of 0.25 – 0.5% to total NPV. This would equal ca. 100 Mio USD/month for a mine producing ca. 200 kt/day of copper ore with 1% copper content. (Restner 2016, p 12).

However, cutting excavation is limited to geotechnical conditions not exceeding a certain abrasivity and rock strength. The range of rocks that can economically be cut with point attack tools ranges from ca. 160 MPa for low Cerchar Abrasivity Indices (CAI) to ca. 80 MPa for CAI of up to 3.5. Above a CAI of 3.5 economic feasibility is usually not given (Hartlieb et al. 2017a).

Therefore the aim of numerous research and innovation activities is to overcome these barriers and enabling the technology to work efficiently also in tough conditions. These improvements can be divided into the three main groups (see also Figure 1) (Hartlieb & Grafe 2015; Sifferlinger, Hartlieb & Moser 2017; Vogt 2016; Drebenstedt & Vorona; Grafe & Drebenstedt 2017):

- development and improvement of tool materials like polycrystalline diamond picks,
- development of improved cutting techniques like activated cutting or undercutting,
- development of alternative methods to support the cutting process by weakening the rockmass like pulsed water jets, microwave irradiation or other means of inducing additional energy input into the rock mass.

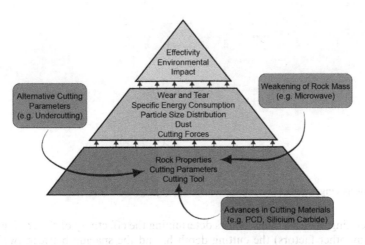

Figure 1. Pyramid of influences to the effectivity of rock cutting (modified after (Drebenstedt & Vorona (2012))).

With respect to mechanical based excavation in hard rock. The presented activities follow mainly an optimization and better understanding of the cutting parameters as well as possibilities to alter the quality of the rock mass itself.

The presented research at TU Bergakademie Freiberg focuses on point attack picks as they are widespread and common tools for rock excavation.

2 THEORETICAL BACKGROUND

During the cutting of hard rock with point attack picks, the pick is moving in a groove through the rock mass. Hereby, it is subjected to a penetration force F_z and a cutting force F_x as well as a side Force F_y. These three forces are the components of the resulting force vector F_{res}. As the pick moves, a zone of pulverized material is created and a tensor field is induced into the rock mass. To a small extent, the rock mass deformes elastically until the local strength of the rock is overcome. It results in the formation of cracks which propagate further until a chip is being separated from the rock mass. This results in a sudden release of all tensions and forces. Now the process starts again (Figure 2). The nature of this process results in highly varying force values (Kurosch).

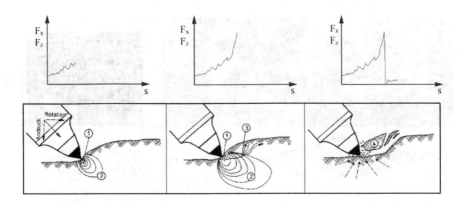

Figure 2. Schematic of the chipping process and development of the cutting forces after (Kurosch).

35

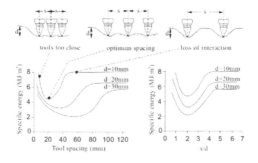

Figure 3. Tool spacing and its effect on specific energy (Speight 1997).

In regular cutting, the two main factors determining the effectivity of the cutting process are (amongst many other factors) the cutting depth h_s and the spacing between two cuts a_s. As seen in Figure 3, according to the chosen cutting depth, an optimal spacing can be found. At this spacing, the specific energy consumption W_{sp} is the lowest. As a result, the ratio between cutting depth and spacing can be calculated - Kappa: $\kappa = a_s/h_s$. For one material and cutting assembly, an optimal κ can be found regardless of the cutting depth. This means that the deeper a cut is, the bigger the spacing should be for an optimal energy consumption of the cutting process (Bilgin, Copur & Balci 2014).

Deeper cuts however always mean that the cutting forces in general rise. As such, a maximum cutting depth exists that is defined by the stability of the cutting tool. During most excavation processes, the tools are mounted on a rotating element. Due to that, the resulting cut has a crescent shape with an either in- or decreasing cutting depth during the cut. That means that optimal cutting parameters can never be achieved in reality.

3 MATERIAL AND TYPICAL WORKFLOW

For estimating the results of all cutting experiments, the linear cutting test rig HSX 1000-50 of TU Bergakademie Freiberg, specifically produced and manufactured by the company ASW GmbH, is used (Figure 4, (a)). It is used for carrying out comprehensive rock cutting tests aimed at the optimization of the rock cutting process (decreasing specific energy of rock destruction, maximizing cutting output, reducing wear and tear of cutting tools, reduction of dust and noise emissions). Programmable movement of a table with a sample in cutting direction (X-axis), of the tool holder in the respective Y and Z-axes enables cutting tests to be

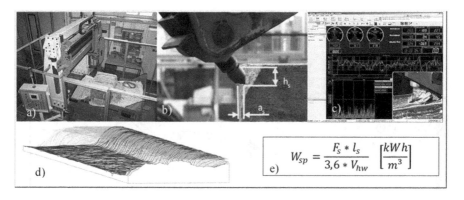

$$W_{sp} = \frac{F_s * l_s}{3,6 * V_{hw}} \quad \left[\frac{kWh}{m^3}\right]$$

Figure 4. Workflow of cutting tests with the linear cutting test rig HSX-1000-50 (Grafe et al. 2017).

carried out in a wide range of cutting parameters and trajectories. The tests are being conducted on a single tool. To simulate the relieved cut – the standard cutting scheme in excavation with point attack picks, several subsequent cuts are conducted next to each other. The relatively big size of rock samples (of up to 60x100x50 cm) provides high precision and reliability of given results as a reasonable cutting length can be provided while still avoiding edge effects.

The rig is equipped with a laser scanner for direct measurements of outbroken volumes. The power of 60 kW, flexible operating modes and modular design of the cutter head allows for different types and sizes of cutting tools to be tested (point attack, wedge type, soft rock cutting tools). In operation mode, cutting and penetration forces of up to 75 kN can be measured with piezo sensors at a maximum rate of 10 kHz while the maximum cutting speed lies at 1.67 m/s.

As shown in Figure 4, cutting measurements are conducted in the following workflow

(b) definition of the cutting parameters,

(c) measurement of the resulting cutting forces, also with visualisation via high-speed camera,

(d) calculation of the outbroken volume utilizing a laser scanner to generate rock surface before and after the cut

(e) and calculation of the specific energy consumption, as well as statistical analysis of theresulting cutting force data (Grafe et al. 2017).

4 UNDERCUTTING

With the development of the Wohlmeyer-machine in the 1960s, the idea of undercutting came up first. Until today, several attempts have been made to utilize its principles to extend the range of applicability of mechanical extraction to hard and very hard rock. The most recent generation approaches come from Sandvik and Caterpillar (Sifferlinger, Hartlieb & Moser 2017). It is reported that by using the undercutting principle, the cutting forces could be reduced to some extend as well as the specific energy consumption. Although several undercutting machines proved that they could work in very hard rock conditions, a major breakthrough on the market is missing. As the main issue, the tools high wear is mentioned to be a drawback (Grafe 2014).

In undercutting, the pick no longer attacks from the top of the face but "from the side", at a higher cutting depth in combination with a lower spacing. The geometrical conditions are shown in Figure 4. For the presented study, the calculation of κ was be switched to $\kappa = h_s/a_s$ for the undercutting manner to represent these geometrical changes.

The tests were conducted with C20/25 concrete blocks. For the tests in regular cutting, three cutting depths h were chosen: 4, 7, and 10 mm. Additionally, κ was varied from 1-3 in steps of 0.5. This means that for every cutting depth, the spacing needed to be varied according to the chosen κ. In the undercutting tests, a similar approach was taken, but instead, the spacing was defined as to be 4, 7, 10 mm. Accordingly, now, the cutting depth was to be varied. As a result, a matrix of 9 different parameter combinations resulted. Detailed information about the cutting procedure can be found in (Grafe & Drebenstedt 2017) and (Grafe 2014).

As shown in Figure 5, the following observations where made: while undercutting, the assembly allows a different crack propagation within the rock, leading to the outbreak of larger chips compared to regular cutting. This leads to:

a) A lower specific energy consumption (since fewer separation planes must be created) whereas the reduction is higher when κ is lower (up to 49%). With increasing κ, the reduction becomes less (to 99% of regular cuttings energy consumption)

b) The cutting forces vary more in undercutting, resulting in generally higher peak forces. In addition, the mean angle of attack of the resulting cutting force towards the tool is steeper (up to 50° for undercutting compared to up to 24° in regular cutting). This leads to

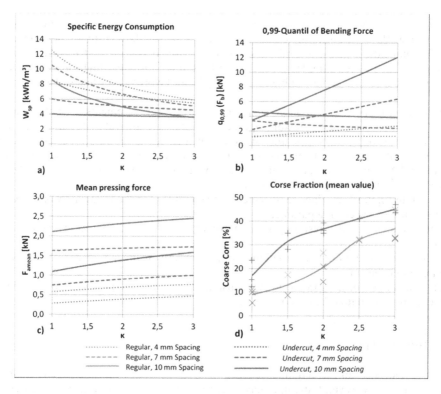

Figure 5. Diagrams of a) specific energy consumption; b) 0.99-quantile of bending force; c) Mean pressing force; d) coarse fraction; comparing regular relieved cutting with undercutting after (Grafe & Drebenstedt 2017).

a stronger bending force (defined as the force that acts on a 90° angle towards the picks axis) with higher peak forces in undercutting.

c) The mean penetration force (force in the direction perpendicular to the cutting direction and towards the face) in undercutting is generally lower (between 43 % and 75 %) than in regular cutting. The side force (not in picture) while undercutting is lower for a small κ (down to 69 %) but rises up to 305 % with a break-even point at $\kappa = 2$.

d) The grain size distribution shows that the coarse fraction is from 8 – 18 % higher while undercutting. Fine grain (< 0.5 mm, not in picture) is relatively similar (between 3.7 % less and 1.4 % higher)

What does this mean for a comparison between the two methods? The observations a), b) and c) lead to the conclusion that the different crack propagation leads to the outbreak of bigger chips. This is more efficient from an energetic point of view. However, the outbreak of such a macrochip causes a higher peak force. This high peak force could lead to a preliminary failure of the pick – especially during a hard rock application – especially in conjunction with the generally steeper angle of attack of the cutting force towards the pick.

The observation c) indicates that the force needed to penetrate the rock does come from a different angle than in regular cutting. Especially when the penetration component must be overcome by the excavation machine's weight itself. The side component could be borne by a symmetrical arrangement of the tools. This would lead to a nullification of the side forces and thus, allowing for a much lighter machine concept.

However, the machine and especially the picks and holders design must still bear the higher peak forces at a less favorable angle. Additionally due to the generally lower specific energy consumption, an undercutting machine would employ either a lighter drive or show a higher advance rate compared to a regular working machines (Grafe & Drebenstedt 2017).

Table 1. Geometry in regular and undercutting after (Grafe & Drebenstedt 2017).

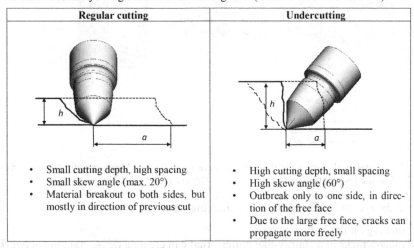

Regular cutting	Undercutting
• Small cutting depth, high spacing • Small skew angle (max. 20°) • Material breakout to both sides, but mostly in direction of previous cut	• High cutting depth, small spacing • High skew angle (60°) • Outbreak only to one side, in direction of the free face • Due to the large free face, cracks can propagate more freely

5 MICROWAVE RADIATION

When a given rock mass is being radiated by microwave radiation, a portion of the electromagnetic energy is absorbed and converted into heat energy. The process hereby is not limited to the surface but characterized by a continuous absorption within the material itself - in contrary to most other heating methods like convectional or laser-driven heating,

Due to this, the radiated areas expand. This results in the induction of stress within the rock mass. Such stress can exceed the strength of the rock, leading to the occurrence and propagation of cracks (Hartlieb et al. 2016).

For the here presented results, the microwave irradiation was conducted by Montanuniversität Leoben, Austria, in the testing facilities of Sandvik in Zeltweg. A 24 kW microwave using a frequency of 2450 MHz was applied. The radiation was focused with a rectangular waveguide. By doing so, the available power could be focused on a spot-like area of ca. 5 cm in diameter. The waveguides opening had a distance of ca. 2.5 cm from the specimen's surface.

Three granite blocks were tested with linear rock cutting experiments, one was left untreated (referred to as B0) to serve as a reference block, two were fully radiated in a chessboard-like pattern with a spacing between the radiation spots of 10 cm and 10 cm between the spot lines. Of these two Blocks, one was irradiated with 30 s radiation time for each spot (B30) and the other for 45 s (B45). To visualize the crack distribution, a regular penetration fluid was used as can be seen in (Figure 6).

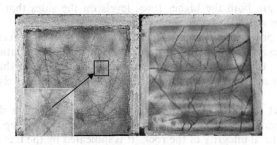

Figure 6.   Surface of 30 s radiated (left) and 45 s radiated blocks (Grafe et al. 2017).

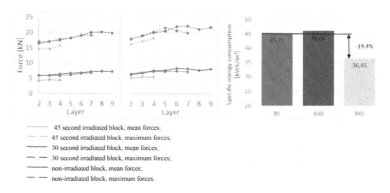

45 second irradiated block, mean forces;
45 second irradiated block, maximum forces;
30 second irradiated block, mean forces;
30 second irradiated block, maximum forces;
non-irradiated block, mean forces;
non-irradiated block, maximum forces.

Figure 7.   Mean and peak forces for the cutting force F_x (left) and Mean specific energy consumption for layers 2 – 4 of the tested granite specimen (right) after (Grafe et al. 2017).

Visible to the eye, two different kinds of cracking occur:

a) Microcracking on the radiation spot
b) Radial macrocracks in a sunray-like pattern that can create a dense cracking network

In B45, the cracks were bigger and deeper (up to 10 cm) their distribution denser. Some cracks were also visible on the sides of the block.

During the irradiation, the surface temperature of the specimen was measured with a contactless infrared measuring device. B30 showed surface temperatures of up to 300°C; B45 of up to 500°C (Hartlieb & Grafe 2017; Hartlieb et al. 2017b; Grafe et al. 2017).

The cutting experiments were executed with a cutting depth of 4 mm and two different spacings: 8 and 12 mm. From the cutting force analysis, the following observations where made (Figure 7, left):

• The average increase of the forces between each layer is ~ 3%, this can be accounted to the wear of the picks and the reduced microwave influence
• The cutting forces of the tests with 12 mm spacing have in average been 13 % higher than with 8 mm
• Between B45 and B0, a clear reduction of the cutting forces could be monitored; for F_x the reduction is 22.5 % in mean (16.5-29 %); for F_z the reduction lies at 36,9 % (29.4-43.1 %);
• Between B0 and B30, no relevant change of the cutting forces could be observed (Hartlieb & Grafe 2017)

As seen in Figure 7 right, with respect to specific energy consumption, the mean specific energy consumption with 36.45 KWh/m³ was almost 20% lower for B45 than of the other two specimens. For the comparison of E_{sp}, only the layers 2 – 4 were used – due to the failure of B45 during the cutting of layer 5. A lower specific energy consumption usually allows for a higher advance rate or lighter and more flexible machine design (Grafe et al. 2017).

Figure 8 illustrates the force distribution for the two fully irradiated blocks and the unirradiated block. Visible are both the higher force levels on the sides that have been cut with a spacing of 12 mm, as well as low force areas caused by the crackings.

It shows that high power microwave irradiation can be used to induce cracks (caused by increased thermal stresses) into massive granite. Directly on the irradiation spots, an area of small cracks is visible to the eye. Furthermore, radial macro cracks that overlap could be observed. They form a network between the radiation spots that effectively weakens the whole radiated area.

However, this effect seems to show only after a "critical treatment duration" where the crack network propagated dense enough and the cracks are opened up enough to have an influence on the structural integrity of the rock. It is indicated by the fact that a weakening of B30 was not measurable on a block-wide scale.

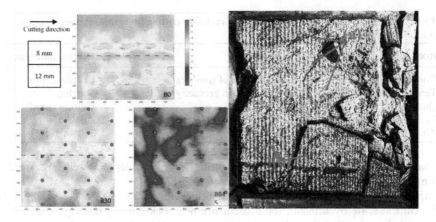

Figure 8. Cutting force maps of B0, B30, and B45, 2nd layer; red dots mark radiation points (left); foto of B45 after failure, note the overbrak (a) and the still visible penetration spray (b) (right) (Grafe et al. 2017).

For B45 however, the results show clearly that the irradiation has a considerable effect as the reduction of the cutting forces by 22.5 % and the lowering of the specific energy consumption by 19.4 % imply.

6 SENSOR CONCEPTS FOR SELECTIVITY

The cutting machine needs sensors to receive information about the boundaries of the vein and the ore quality of the cut material. Currently, multiple sensor technologies are under research and development. The range of these covers mainly electromagnetic wave sensors (IR, UV, visible light, radar, multi-/hyperspectral), for example for face mapping. Other possibilities are sensing technologies like X-ray Fluorescence, Prompt Gamma Neutron Activation Analysis and Laser-Induced Breakdown Spectroscopy (LIBS) for online material analysis. (Benndorf & Buxton 2017; Gaft et al. 2011).

The positioning of sensors in front of a possible cutting machine is very restricted. During operation, dust, vibration, water, flying debris and particles occur and disturb the measuring quality of most face sensors. Because of this, at TU Bergakademie Freiberg, research is conducted to estimate the applicability of direct force measurements during cutting. Other than taking the power consumption of the cutter drive, the data obtained have a higher information value. If the power consumption of the cutter drive is utilized, only a mean value over all tools that are currently in the face is given out. With direct force measurements at the tools, it can be distinguished spatially where high or low forces occur if the position of the individual tools is measured too. Furthermore, if specific tools generally show high force values this is an indicator for high wear. The statistical distribution of the forces can be used for a more exact analysis of the rock condition if related back spatially.

Generally, online analysis methods can already improve the exactness of the resource quality prediction as e.g. was presented by Benndorf on the example of bucket wheel excavators in coal (Benndorf 2017). There the coal quality prediction for the subsequent cut could be improved by ~65%. Such information is of great monetary value when it comes to improving blending and stockpiling efforts (Grafe et al. 2018).

The use of data on the reaction between tool and rock has already seen the application in the past, however on a more simple scale. Mainly the data were used to optimise the excavation process. Earlier approaches to the geospatial use of such information have also been undertaken by the Chair of Open Pit Mining on the example of bucket wheel excavators (Drebenstedt & Paessler 2005; Drebenstedt & Vorona). For hard rock cutting machines, the force distribution shows a greater variation than for softer rocks. Cracks and disturbances as well

Table 2. Influences towards the outcome of real-time cutting force measurements (Grafe et al. 2018).

Rock massif	Tool	Cutting parameters
Compression strength (UCS)	State of tool/wear	Cutting depth
Tensile strength (TS)	Design geometry of tool	Spacing
Quality, joints/cracks		Cutting speed
Foliation		Angle of attack
Alteration		Tilt angle

as the wearing of the picks influence the resulting forces (Galler, Stoxreiter & Wenighofer 2016; Entacher, Winter & Galler 2013; Vorona et al. 2012).

The approach is to qualify and quantify the influencing factors at first on a single pick in a semi-technical scale by utilising the aforementioned rock cutting test rig. This allows for a very detailed analysis of the influencing factors as well as for approaches to compensate their influence towards the quality of the analysis.

From literature as well as from own experiments the major factors identified are shown in Table 2. Here, the work of (Vorona et al. 2012) shall be recommended for a comprehensive review of existing studies about influences of parameters towards the cutting process (Grafe et al. 2018).

Changing these parameters now can change the following different aspects of the measured forces:

(1) general force level
(2) ratio of spatial force components, e.g. cutting force/normal force
(3) dispersion measures
(4) occurrence of zero-values

The understanding of these points with respect to the rock mass excavated can then create surplus value in the subsequent process chain.

The here presented results show the basic approach of force analysis to identify boundaries between different rock types based on the raw force values that can be obtained during cutting.

Figure 9 shows exemplary results of the cutting of a heterogeneous rock sample. The sample consisted of two lead-zinc-ore boulders (ca. 30x20x20 cm) from Reiche Zeche, the

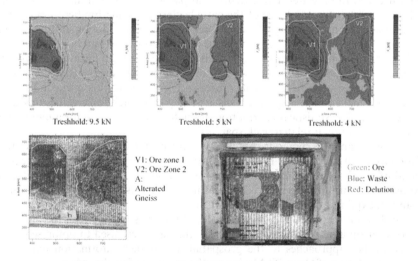

Figure 9. Exemplary results for cutting tests on a heterogeneous sample, the top row shows the overlay of the spatial cutting force analysis using different force thresholds to identify lithological boundaries. The lower row shows the identification of different zones in the rock sample (left) and overlay of the identification process with the sample (right); after (Grafe, Drebenstedt & Shepel 2018).

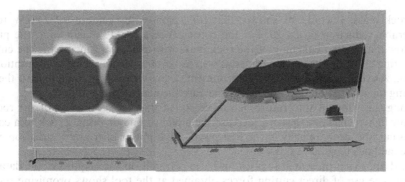

Figure 10.   3D-moving average analysis of the cutting forces showing the harder vein zones in blue.

teaching and research mine of TU Freiberg, that have been cast into a concrete block (55x80x40 cm). The block was spatially referenced and the boundaries of the two ore samples where marked. Additionally to the main vein material pieces V1 & V2, a smaller alteration zone (A) of a weathered gneiss was mapped. One of the ore samples was freshly blasted (V1), the other one was in the mine for several years and hence showed signs of weathering (V2).

Then a moving average analysis on a layer by layer basis was conducted for F_x, F_y, F_z, F_{res}. It was checked whether certain force thresholds could be formulated to identify the borders between the different rock zones as seen in Figure 8.

One issue that occurs during a layerwise analysis of the cut blocks is that an outbreak of very large chips on one layer creates a gap on the subsequent layer. A logical solution is to expand the model into the z-Dimension as well. Figure 10 shows a 3D-Force model. Compared to the two-dimensional model, it shows straighter edges and a more stable picture as peaks that occur on individual levels can be evened out.

As first results, it can be stated that the use of direct force measurement for lithology identification has the potential to become a robust tool. Of course, the lithologies that can occur must be known beforehand. This is usually given during an excavation operation. In addition, the system would have to be calibrated before use in an industrial application.

A rather simple floating average analysis posed very successful for the needs. Since the data density is very high, a complete calculation of the 3D-Model takes (depending on search radius) around 30 min. The subsequent use of further averaging filters, however, requires only a fraction of that time (10 s) and is recommended instead of the calculation of lots of individual models from the input data. However, loosing too much information during the subsequent data simplification steps should be avoided. One possibility to store the number of data points for individual blocks to retain the associated weights to the mean values.

The presented results are intermediate results. Nevertheless, it could be shown that a material identification with very simple tools can be possible.

7 CONCLUSION

The present paper shows recent work from the centre for rock cutting technology of TU Bergakademie Freiberg. As standards in mining continue to improve while the characteristics of deposits become more and more unfavorable, new technologies and approaches are needed.

One approach is the use of alternative cutting parameters which could unlock a more energy efficient way of mechanical rock excavation. This has been shown on the example of undercutting. While undercutting principles are less subjected to a working optimum with regards to the spacing ratio and are generally more energy efficient, the peak forces could pose problematic with respect to sudden breakouts of the tools.

Another approach, the alteration of the rock mass, in this example by heat-induced tensions due to high power microwave radiation shows another way of improving the excavatability of a given rock mass. It could be shown, that for a spot-wise radiation of 24 kW for 45 s, a considerable reduction of ca. 20% of the cutting forces could be achieved. The presented results show that a pre-weakening of the rock mass with the aim of increasing the cuttability has a lowering effect on the cutting forces as well as on the specific energy consumption of the excavation. As such, in an industrial environment, a pre-weakening could extend the limitations during the excavation of hard rock or increase the net cutting rate.

Following this thought into future, it can be stated that pre-weakening of the rock mass could allow lighter machine designs, increase the advance rate or bring a rock to a condition where it is cuttable economically in the first place - high power microwave radiation could be one of the methods to achieve this.

The third approach presented focuses on basic research on the acquisition of data during excavation. The use of direct cutting forces obtained at the tool shows promising results for further research with the aim to identify different rock masses or even different states of the rock mass. By doing so, near-real time information about boundary layers can be obtained and utilized for a direct update of geologic models. This could lead to higher selectivity in production. In addition, these data could be used to optimize the excavation process online as well as directly identifying altering support need due to changing rock mass qualities.

ACKNOWLEDGMENTS

We thank Dr. Hartlieb from MU Leoben, Austria for the partnership and joined research on the research field of pre-treatment of hard rock with high power microwave radiation. All microwave treatments have been conducted by MU Leoben at the Zeltweg Research Centre of Sandvik.

The project *InnoCrush - Dynamic methods of mechanical excavation and comminution for high selective production chains in Critical Raw Materials in Saxony* is financially supported by the European Union (European Social Fund) and the Saxonian Government (Grant No. 100270113).

REFERENCES

Benndorf, J 2017, 'Turning Geo-Data into Mining Intelligence – Nutzung von Online-Daten zur Echtzeitmodellierung im Gold- und Kohlebergbau', *BHM Berg- und Hüttenmännische Monatshefte*, vol. 162, no. 10, pp. 418–422. Available from: https://link.springer.com/content/pdf/10.1007%2Fs00501-017-0634-3.pdf [09 June 2017].

Benndorf, J & Buxton, M 2017, 'Proceedings of Real-Time Mining International Raw Materials Extraction Innovation Conference 10th & 11th October 2017' [02 May 2018].

Bilgin, N, Copur, H & Balci, C 2014, *Mechanical Excavation in Mining and Civil Industries*.

Drebenstedt, C & Paessler, S 2005, 'Analysis of cutting resistances for bucket wheel excavators in hard clays', *Proceedings of the 14th International Symposium on Mine Planning and Equipment Selection, MPES 2005 and the 5th International Conference on Computer Applications in the Minerals Industries, CAMI 2005*. Available from: https://www.scopus.com/inward/record.uri?eid=2-s2.0-84892650537&partnerID=40&md5=8153dc8a4562ba0ff2059df845846d52.

Drebenstedt, C & Vorona, M, 'Optimierung von Gewinnungsmaschinen im Tagebau – Vom Schneidprozess zur Virtuellen Realität' in *5. Fachtagung Baumaschinentechnik* 2012, pp. 72–86.

Entacher, M, Winter, G & Galler, R 2013, 'Cutter force measurement on tunnel boring machines – Implementation at Koralm tunnel', *Tunnelling and Underground Space Technology*, vol. 38, pp. 487–496 [09 July 2019].

Gaft, M, Nagli, L, Groisman, Y & Barishnikov, A 2011, 'Laser-Induced Breakdown Spectroscopy (LIBS) for On-line Control in Mining Industry'. *Signal recovery and synthesis. [part of] Imaging and applied optics; 10-14 July 2011, Toronto, Canada*, OSA The Optical Society, Washington, DC [28 May 2018].

Galler, R, Stoxreiter, T & Wenighofer, R 2016, *Disc cutter load monitoring and face monitoring in TBM-tunnelling - developments for detailed analysis of the cutting process*, Leoben.

Grafe, B 2014, *Potential des Hinterschneidens mit Rundschaftmeißeln zur Steigerung der Schneideffizienz*, Freiberg [01 February 2019].

Grafe, B, Ali, SE, Bongaerts, JC, Bravo, AH, Drebenstedt, C, Heide, G, Hesse, M, Kogan, I, Konietzky, H, Kühnel, L, Lieberwirth, H, Liu, J, Morgenstern, R, Popov, O, Rehkopf, A, Rosin, K, Shepel, T, Schlothauer, T, Schwarz, S, Yadav, VV, Talovina, I, Ilinov, MD & Nikiforowa, VS 2018, 'InnoCrush: New Solutions for Highly Selective Process Chains' in *Innovation-Based Development of the Mineral Resources Sector. Proceedings of the 11th Russian-German Raw Materials Conference, November 7- 8, 2018, Potsdam, Germany*, ed V Litvinenko, Chapman and Hall/CRC, Milton.

Grafe, B & Drebenstedt, C 2017, 'Laboratory Research on Alternative Cutting Concepts on the Example of Undercutting', *BHM Berg- und Hüttenmännische Monatshefte*, vol. 162, no. 2, pp. 72–76 [25 May 2018].

Grafe, B, Drebenstedt, C, Hartlieb, P & Shepel, T 2017, 'Studies on the effect of high power microwave irradiation as a means of inducing damage to very hard rock to reduce the cutting resistance during mechanical excavation', *Proceedings MPES 2017, Luleå, Sweden* [27 May 2018].

Grafe, B, Drebenstedt, C & Shepel, T 2018, 'Ways towards highly selective mechanical production chains', *Proceedings 5th International Colloquium of Non-Blasting Rock Destruction 19th - 22th November 2017*.

Hartlieb, P, Gerer, R, Grief, R & Sifferlinger, NA 2017a, *The use of diamond based tools in mining picks. A status report on advantages, challenges and the applicability*, Freiberg.

Hartlieb, P & Grafe, B 2015, *Untersuchungen zur Erhöhung der Schneidbarkeit von Gesteinen durch den Einsatz alternativer Methoden. in Tagungsband Bergbau, Energie und Rohstoffe 2015*, pp. 271–280.

Hartlieb, P & Grafe, B 2017, 'Experimental Study on Microwave Assisted Hard Rock Cutting of Granite', *BHM Berg- und Hüttenmännische Monatshefte*, vol. 162, no. 2, pp. 77–81 [25 May 2018].

Hartlieb, P, Grafe, B, Shepel, T, Malovyk, A & Akbari, B 2017b, 'Experimental study on artificially induced crack patterns and their consequences on mechanical excavation processes', *International Journal of Rock Mechanics and Mining Sciences*, vol. 100, pp. 160–169 [25 May 2018].

Hartlieb, P, Toifl, M, Kuchar, F, Meisels, R & Antretter, T 2016, 'Thermo-physical properties of selected hard rocks and their relation to microwave-assisted comminution', *Minerals Engineering*, vol. 91, pp. 34–41.

Kurosch, T, *Geologisch-felsmechanische Grundlagen der Gebirgslösung im Tunnelbau*. Habilitation, Zürich [10 November 2018].

Plinninger, RJ 2011, *Teilschnittmaschinen als alternatives Vortriebsverfahren im innerstädtischen Tunnel - und Stollenbau – Chancen und Risiken. Roadheaders as an alternative excavation method in urban tunneling - chances and risks*, Berlin, pp. 139–145. Available from: http://www.plinninger.de/images/pdfs/2011_18taging_tsm.pdf [28 April 2017].

Restner, U 2016, *Hard Rock Continuous Mining. Rapid Mine Development System*, Leoben.

Sifferlinger, NA, Hartlieb, P & Moser, P 2017, 'The Importance of Research on Alternative and Hybrid Rock Extraction Methods', *BHM Berg- und Hüttenmännische Monatshefte*, vol. 162, no. 2, pp. 58–66 [28 April 2017].

Speight, H 1997, 'Observations on Drag Tool Excavation and the Consequent Performance of Roadheaders in Strong Rock', *The AusIMM Proceedings*, no. 1, pp. 17–32.

Vogt, D 2016, 'A review of rock cutting for underground mining. Past, present, and future', *Journal of the Southern African Institute of Mining and Metallurgy*, vol. 116, no. 11, pp. 1011–1026.

Vorona, M, Drebenstedt, C, Kholodnyakov, G & Kunze, G 2012, *Optimierung des Schneidprozesses und Prognose der relevanten Arbeitsgrößen bei der Gesteinszerstörung unter Berücksichtigung des Meißelverschleißes*. Dissertation, Freiberg. Available from: http://www.qucosa.de/fileadmin/data/qucosa/documents/9636/Dissertation%20Vorona%20Maxim.pdf [28 April 2017].

Scientific and Practical Studies of Raw Material Issues – Litvinenko (Ed)
© 2020 Taylor & Francis Group, London, ISBN 978-0-367-86153-7

Technological scheme of development flooded fields of sands

D. Dzyurich & V. Ivanov
Saint-Petersburg Mining University, Saint-Petersburg, Russian Federation

ABSTRACT: The work is devoted to the consideration of the main typical technological schemes used in the development of flooded deposits of construction sand. In the paper the main equipment used, advantages and disadvantages of existing technological schemes. The graphic drawings of the considered technological schemes used in modern quarries, working off the watered reserves of deposits of construction sand are given, their advantages and disadvantages are described. A new technological scheme for the development of flooded deposits of construction sand is proposed.

1 INTRODUCTION

Currently, about 60% of sand and sand-gravel quarries have watered mineral deposits. The degree of water cut plays an important role in determining the development schemes and the choice of mechanization. Carrying out works on pumping groundwater from the bottom leads to an increase in the cost of the purchase of fixed assets and energy resources. The slightest change in water level requires consideration of a number of problems.

Development of flooded deposits of construction sands usually are carried out with the division of reserves into dry and watered parts. The excavator-automobile complex of the equipment usually makes development a dry thickness. Development of irrigated part of the inventory is done mainly for dredging method with the use of a floating suction unit, in some cases, used excavators equipment type backhoe or dragline, to implement lower digging rocks from the watered slaughter (Ivanov V. 2013).

Below are the main technological schemes that do not require pumping groundwater from the bottom.

Despite the large number of works performed in the open-pit mining of flooded deposits of non-metallic materials (Boutkevitch 2004; Harin. & Novikov 1989; Ivanov 2012; Jaltaneñ & Levanov 2008; Shpanskiy & Buyanov 1996, Guzeev 2014 and others), development of excavators watered sand deposits has disadvantages.

The problem is the use of additional loading equipment, which leads to additional costs for the purchase and maintenance, as well as the salary of the working staff.

We can solve the problem of additional loading equipment by improving the technological scheme for the development of watered sand reserves using a hydraulic excavator. This article describes a process flowchart that will reduce the cost of the above components

2 THE MAIN TECHNOLOGICAL SCHEME

2.1 *The use of dredge ship*

Development of irrigated strata deposits of construction sands by means of dredging is done using as mining machines floating suction units, which work underwater part of the ledge height to 20 m, and in the presence of surface breaking part of the ledge height up to 15 m, depending on the unit (Figure 1). The angles of the above-water and underwater part of the ledge correspond to the angles of the natural slope of rocks in a stable position and are about 25-35 degrees, depending

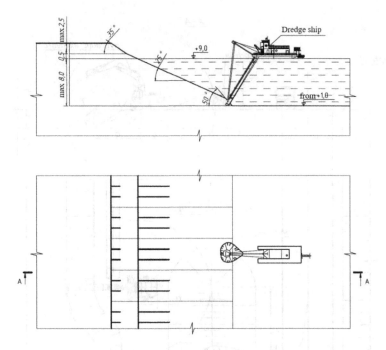

Figure 1.   Technological scheme of application of the dredge ship in the excavation of flooded sands.

on the physical and mechanical properties of rocks. The angle of the underwater part of the ledge in the place of the fence of the mountain mass by the dredge ship can exceed 45 degrees.

To move the extracted sand in the form of pulp from the dredge ship to the alluvium maps, pressure hydraulic transport is usually used. The pulp developed by the dredge ship, along the floating and onshore slurry line, is fed to an alluvium map, where sand precipitates, the water is clarified and through the spindle well into the drainage collector and returns to the water area.

To perform the annual productivity of sand extraction for each dredge ship, it is planned to create two circulating maps of alluvium, of which one is injected, the other is shipped. Alluvium maps are constructed on a sub-horizontal site within the mining allotment.

The excavator or the loader in means of transport makes shipment of sand from the finished stack of the dried map of the alluvium.

2.2 Application of a single-bucket hydraulic excavator of the «backhoe» type

Development of flooded strata, without prior dewatering, excavation of sands hydraulic excavator from the water is digging the bottom with the storage of mined rock mass in bulk for dewatering (Figure 2). The watered sands are worked out taking into account leaving of a protective layer of dry sand. The excavator or the loader in means of transport makes shipment of sand from the drained bulk.

The main advantages of the technological scheme with the use of a hydraulic excavator for excavation of flooded sands are:

1) High mobility of the equipment used;
2) Small workforce compared to the workforce in the extraction of alluviation sand;
3) Low production cost.

Disadvantages of this scheme:

1) The need to allocate additional equipment to ensure the process of loading of dehydrated rocks into transport;
2) Detailed consideration of mining face parameters is required to ensure the efficiency and safety of mining operations.

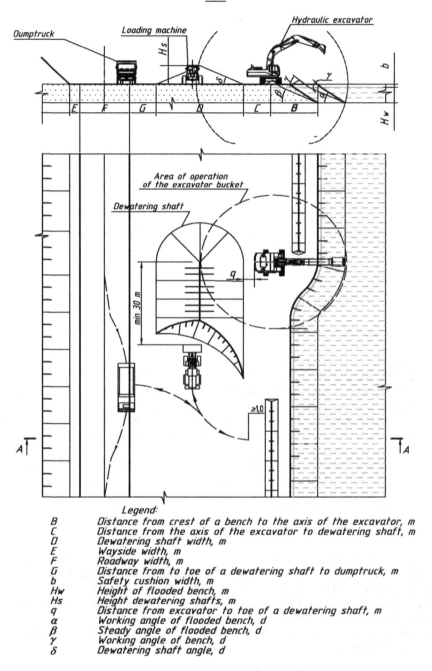

A–A

Dumptruck   Loading machine   Hydraulic excavator

Area of operation
of the excavator bucket

Dewatering shaft

min 30 m

q

≥1,0

A   A

Legend:
B	Distance from crest of a bench to the axis of the excavator, m
C	Distance from the axis of the excavator to dewatering shaft, m
D	Dewatering shaft width, m
E	Wayside width, m
F	Roadway width, m
G	Distance from to toe of a dewatering shaft to dumptruck, m
b	Safety cushion width, m
Hw	Height of flooded bench, m
Hs	Height dewatering shafts, m
q	Distance from excavator to toe of a dewatering shaft, m
α	Working angle of flooded bench, d
β	Steady angle of flooded bench, d
γ	Working angle of bench, d
δ	Dewatering shaft angle, d

Figure 2.   Technological scheme of application of hydraulic excavator for dredging of watered sands with their storage in dewatering shaft.

2.3   *Application of single-bucket excavator with dragline type equipment*

The technological scheme of working off of stocks of the watered thickness by the excavator with the equipment of dragline type (Figure 3) is similar to the above technological scheme with application of the hydraulic excavator.

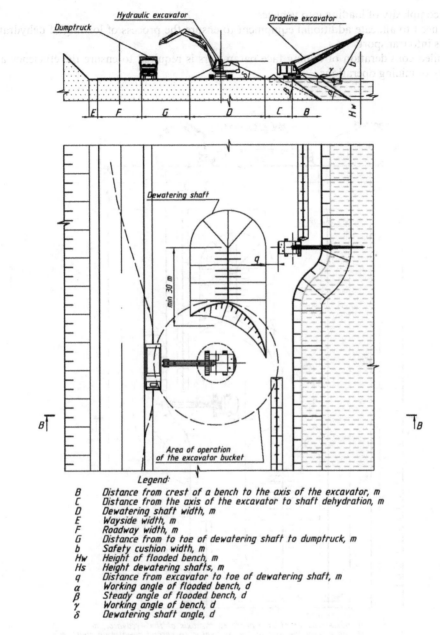

B-B

Dumptruck — Hydraulic excavator — Dragline excavator

E F G D C B Hw

Dewatering shaft

min 30 m

q

Area of operation
of the excavator bucket

B⌐   ⌐B

Legend:

B   Distance from crest of a bench to the axis of the excavator, m
C   Distance from the axis of the excavator to shaft dehydration, m
D   Dewatering shaft width, m
E   Wayside width, m
F   Roadway width, m
G   Distance from to toe of dewatering shaft to dumptruck, m
b   Safety cushion width, m
Hw   Height of flooded bench, m
Hs   Height dewatering shafts, m
q   Distance from excavator to toe of dewatering shaft, m
α   Working angle of flooded bench, d
β   Steady angle of flooded bench, d
γ   Working angle of bench, d
δ   Dewatering shaft angle, d

Figure 3. Technological scheme of application of an excavator with the equipment of the dragline excavation of sands flooded with warehousing them in bulk to dehydrate.

The main advantages of the technological scheme of working off of stocks of the watered thickness by the excavator with the equipment of dragline type are:

1) High reliability of the equipment, due to the simplicity of the design;
2) Significant parameters of the excavation approach due to the large parameters of the equipment (Butkevich G. 2014).

Disadvantages of this scheme:

1) The complexity of loading into vehicles;
2) The need to allocate additional equipment to ensure the process of loading of dehydrated rocks into transport;
3) Detailed consideration of mining face parameters is required to ensure the efficiency and safety of mining operations.

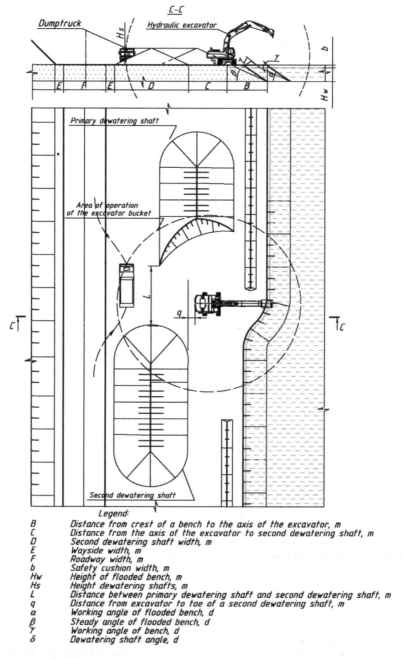

Legend:

B	Distance from crest of a bench to the axis of the excavator, m
C	Distance from the axis of the excavator to second dewatering shaft, m
D	Second dewatering shaft width, m
E	Wayside width, m
F	Roadway width, m
b	Safety cushion width, m
Hw	Height of flooded bench, m
Hs	Height dewatering shafts, m
L	Distance between primary dewatering shaft and second dewatering shaft, m
q	Distance from excavator to toe of a second dewatering shaft, m
α	Working angle of flooded bench, d
β	Steady angle of flooded bench, d
γ	Working angle of bench, d
δ	Dewatering shaft angle, d

Figure 4.   Technological scheme of application of hydraulic excavator with the formation of dewatering shaft.

3 TECHNOLOGICAL SCHEME WITH THE FORMATION OF DEWATERING SHAFTS

Development of flooded strata, without prior dewatering, excavation of sands hydraulic excavator from the water is digging the bottom with the storage of mined rock mass into the piles for dewatering (Figure 4). The watered sands are worked out taking into account leaving of a protective layer of dry sand. One excavator makes Dredging and shipment of sand from the drained bulk. In the proposed technical scheme, we can distinguish similarities with the scheme in Figure 2. But still there are fundamental differences.

3.1 *Principle of operation*

Initially, a hydraulic excavator, storing in a primary dewatering shaft, removes the useful thickness from the water.

After the first dass, the excavator is drived to the beginning of a new dass. In process of giving dump trucks under loading, the excavator carries out loading of a useful thickness in the dump truck from a primary dewatering shaft. As shipment of dehydrated sand drills next flooded dass with the formation of new shaft for dewatering.

Formation of stacks is carried out in such a way that the distance between them allowed to organize the supply of dump trucks for loading. The distance between the lower brows of the stacks is taken according to the width of the quarry road.

As can be seen in Figure 4, the main difference lies in the presence of a second dewatering shaft and in the performance of all excavation and loading operations with one excavator.
The main advantages:

1) High mobility of the equipment used;
2) Smaller staff of workers in comparison with the given technological schemes;
3) There is no need to use additional equipment;
4) Low production cost.

Disadvantages:

1) Detailed consideration of mining face parameters is required to ensure the efficiency and safety of mining operations;
2) Limited use of the technological scheme for productivity.

4 CONCLUSION

The most common technological scheme of mining for the development of watered reserves in the quarries of construction sand is the use of hydro-mechanization. The prevalence of this technological scheme is provided by the high quality of the extracted sand and as a result high demand for quarry products, but it is expensive.

The considered technological schemes have one common drawback – the need to allocate additional equipment to ensure the process of loading of dehydrated rocks into transport. In this regard, increase the cost of purchase and maintenance of equipment, thereby increasing the cost of production.

The technological scheme with the formation of dewatering shaft will reduce the amount of equipment used and thereby reduce the cost.

REFERENCES

All-Union standards of technological design of enterprises of non-metallic building materials. ONTP 18-85.//L.: Stroizdat. 1988. - 80.
Butkevich G. R. Industry of nonmetallic building materials of the USA at the present stage//Scientific, technical and production journal «Building materials». – 2014. No. 12. P. 46-47.

Boutkevitch G.R., Problems of water-bearing deposits of nonmetallic minerals. Mining Journal., 2004; 5: 27-31.

Filippova P. I. Characteristics and comparison of the quality of non-metallic building materials produced in dry and watered deposits//Proceedings of the conference «XII Prokhorov readings». – 2017. P. 130-135.

Gilev A.V., Mechanized complex for the extraction of construction materials on the watered fields. Mining Equipment and Electromechanics, 2008; 6: 18-20.

Glevickiy V.I., Hydromechanization in transport construction. M.: Transport, 1988.

Harin A.I., Novikov M.F., Hydromechanization excavation works in construction. M.: Stroyizdat, 1989.

Ivanov V. V. Analysis of technological measures to ensure the rational development of flooded deposits of sand and sand-gravel mixture/V. Ivanov, T. S. Basov//Scientific-methodical electronic journal «Concept». – 2013. – Vol. 3. – P. 2156-2160.

Ivanov V.V., Characteristics of development of water cut mass in deposits of sand and sand gravel mix. Mine mechanical engineering and machine-building, 2012; 3: 28-31.

Ivanov V.V., Basov T.S., Analysis of technological actions ensuring the rational development of water-encroached deposits of sand and sand-gravel mixture. Monthly scientific and methodological e-journal "Koncept", 2013; 4(34): 2156-2160.

Jaltañeñ I.M, Levanov N.I., Reference jetting. M.: "The world mountain book", MGGU Publisher, Publisher "Mountain Book", 2008.

Lopatnikov M.I., Tedeev T.R., Sand and gravel deposits as a possible source of local durable rubble. Construction Materials, 2007; 5: 18-19.

Melkonyan R.G., Efremov A.N., Kandaurov P.M., The technological scheme of development of deposits of quartz sand. Mining industry, 2008; 5(81),68-71.

Shpanskiy O.V., Buyanov Ju.D., Technology and complex mechanization of production of non-metallic raw materials for the production of building materials. İ., Nedra, 1996.

Scientific and Practical Studies of Raw Material Issues – Litvinenko (Ed)
© 2020 Taylor & Francis Group, London, ISBN 978-0-367-86153-7

The assessment of the roof beam stability in mining workings

M. Vilner & T.T. Nguen
Saint-Petersburg Mining University, Saint-Petersburg, Russia

P. Korchak
Kirovsk branch AO Apatit, Kirovsk, Murmansk region, Russia

ABSTRACT: The problem of reducing the danger of mining is a global one for modern industry. The article presents the problem of the stability of roof in laminated jointed rock masses. The voussoir beam scheme is considered, as it allows predicting the behavior of the roof with a large span and multiple-jointed. This scheme is not sufficiently represented in publications and is limited to horizontal bedding of the ceiling and vertical jointing. The method can be improved. The paper considers such stability factors as joint parameters and quality of contact.

1 INTRODUCTION

The development of the workings or stoped excavation causes rock discontinuity in the overburden, which can lead to the loss of stability of the excavation and inrushes. Often in coal mining there are such conditions when weak rocks lie above a layer of stiff rock, so called key strata. The stability of long shallow underground openings depends mainly on the bearing capacity of the lowest roof rock strata. Such strata are usually jointed and may be considered as discontinuous rock beam or plate structures in three-dimensional statement. In such cases a layer of stiff rock is left above the working to bear the load from the overlying rocks. The key stratum bears the overburden load in the form of a beam structure after break and when the beam consists of a single bed with vertical joints, the so called voussoir beam is obtained. They can be an alternative to pillars. Ore wasted in pillars is up to 60%, which does not correspond to the concept of rational use of mineral resources.

In this regard, it is necessary to calculate the voussoir beam correctly in order to avoid the loss of stability and inrushes. Conventional beam and plate theories cannot model the behavior of a discontinuous rock mass. Voussoir beam theory has been commonly used for stability assessment of excavations where the joints are almost perpendicular to the bedding planes. The method considers the effect of joints and lets the beam displace at the abutments or mid-span. Also, it allows investigation of a failure mode, namely failure by shear sliding along the abutments.

The voussoir beam behavior was studied analytically, empirically and with use of numerical modeling, including discontinuous analysis. Some studies focus on centrifuge model (Talesnick & Bar Yaacov 2007). All in all, little experimental work has been performed on the mechanical strength of a voussoir beam. Passaris et al. have studied the crushing strength of the beam. The mechanism of shear sliding alongside walls has been investigated by Ran et al. Both used nonlinear finite element analysis. Evans developed a design procedure for voussoir beam geometry, a method which was later extended by Beer and Meek (Beer & Meek, 1982) and is reviewed in detail by Brady and Brown (Brady & Brown, 1993). To account for the influence of jointing more directly, several investigators have used discontinuum analysis (Hatzor et al., 1998), but these studies have been limited to simple contact models between rock joints. The design of support pressure for a laminated roof with beds of varying thickness is discussed by Obert and Duvall (Obert & Duvall, 1967) and Goodman (Goodman, 1989), with the use of beam theory principles.

To predict deformations within a jointed rock mass, it is necessary to adopt accurate contact models for the rock joints. Some authors (Passaris et al.) studied the case of multiple mid-joints and the spacing between joints, but friction along the discontinuities was not modelled. Others presented solutions in which only a single layer is modeled (Evans et al.), and the influence of spacing and friction between the vertical joints are ignored.

The paper presents the comparison of the following factors on the stability of voussoir beam: stress-strain state of the rock mass; the beam span; the number of joints and joint spacing; contact quality (roughness). To sum up, the voussoir beam method presented in the literature is still incomplete and there is a lack of validated approaches for jointed rock masses.

2 APPROACH DESCRIPTION

The main task of this study is to determine the influence quantity of various factors on the stability of voussoir beam. Studies are conducted by numerical modeling using Abaqus CAE; the results are compared with the analytical classical solution.

In contrast with classic beam theories, voussoir beam method considers joint effect. After experiment series (Bucky, 2001), it was concluded that the vertical joints at mid-span control the deformation and stability of the beam. In this condition, rock mass can be considered as consisting of individual blocks. These blocks are called voussoirs and the beam composed of voussoirs is called voussoir beam (Figure 1).

The maximum stress values and deflection for a simple fixed end beam can be calculated as following:

$$\sigma_{max} = \frac{\gamma s^2}{2d} \qquad\qquad \delta = \frac{\gamma s^4}{32Ed^2} \qquad (1)$$

where E – Young's modulus of the rock; γ – specific weight of the beam; s – horizontal span; d – thickness.

The principal geometric parameters are beam span, beam height, individual layer thickness and joint spacing. An experiment conducted by Hatzor, was reproduced to test the model. The joints are considered planar with zero tensile strength. The model is described by the elastic-ideal-plastic model of Mohr-Coulomb. The input material parameters are stated in Table 1.

The layer thickness is considered to be 0.25, 0.5, 1 m; the beam span – 4 and 8 m. Joint spacing values depend on the number of blocks. The maximum deflection at mid-section for given parameters is noted in each run.

In addition to previous experiments, not only smooth vertical contact is considered; the behavior of the voussoir beam is studied at different roughness. Roughness is set explicitly on

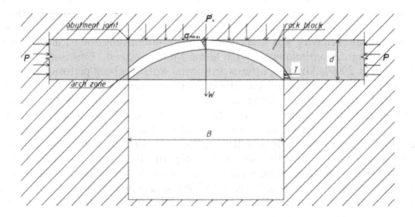

Figure 1. Voussoir beam scheme and load distribution.

Table 1. Input parameters of the model.

Parameter	Voussoir beam	Overburden
Mass per unit area, kg/m^3	2500	2000
Young's modulus, GPa	5	2
Poisson's ratio	0,3	0,1
Cohesion, kPa	6	3,1
Internal friction angle	40	32

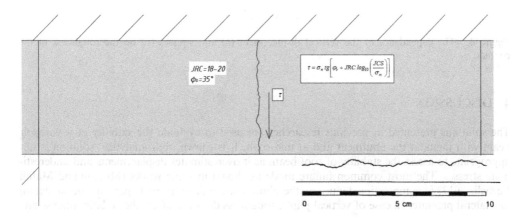

Figure 2. Roughness profile by Barton and modelled profile.

the base of Barton's classification. Barton diagram was used to determine the roughness coefficient of the joint. The roughness profile for the block was set by scaling the profile at JRC = 18-20. The maximum roughness depth is 10 cm (Figure 2).

3 EXPERIMENTAL RESULTS

Numerical modelling showed results, presented in fig. 3-4.

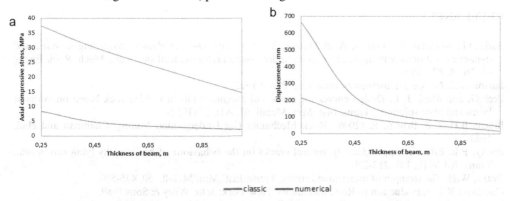

Figure 3. The comparison of numerical results with analytical solution: stresses (a) and displacements (b).

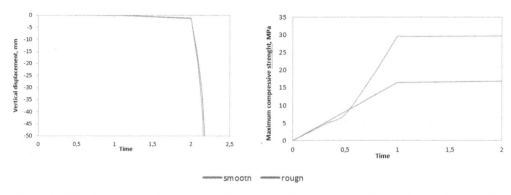

Figure 4. The dependence of the vertical displacement (a) and stresses (b) on the roughness of the contact.

4 DISCUSSION

The solutions presented in previous researches are used to evaluate the stability of a voussoir beam with joints at the abutment and at mid-span. It is shown, that analytical solution is not appropriate to assess the stability of roof beam as it overestimates displacements and underestimate stresses. The most common failure mode as shown in some works (Mohammad Mahdi Fazaeli, 2017) is compression failure at the abutments. Failure at mid-span can occur within low lateral pressure. In case of vertical joints, roughness does not affect the stability intensively.

5 CONCLUSIONS AND FURTHER RESEARCH

All in all, important conditions for sustainable work of voussoir beam are sufficient horizontal stress and layer thickness. Classic voussoir beam analysis ignore the influence of joint spacing and their roughness, cannot explain the failure.

The use of numerical models is necessary for verification of the results of the voussoir beam method

Further research will focus on derivation of the constrains of the roughness effect and joint angles on the stability of blocks, as well as the search for additional parameters of stability.

REFERENCES

Barla, G., Monacis, G., Perino, A. & Hatzor, Y. H. (2010). Distinct element modelling in static and dynamic conditions with application to an underground archaeological site. Rock Mech. Rock Engng 43, No. 6, 877–890.

Barton et al. Norwegian method of tunneling. Part 2 1992.

Beer G. and Meek J. L. Design curves for roofs and hanging walls in bedded rock based on voussoir beam and plate solutions. Trans. Inst. Min. Metall. 91, A18-22 (1982).

Brady, B.H.G. & Brown, E. (2006). Rock Mechanics For Underground Mining. Chapman and Hall. 1993.

Bucky, P.B. Effect of approximately vertical cracks on the behaviour of horizontally lying roof strata. Trans. A.I.M.E.. 109. 212-229.

Evans, W.H. The strength of undermined strata. Trans. Inst. Min. Metall.. 50. 475-500.

Goodman R.E. Introduction to Rock Mechanics. New York: John Wiley & Sons 1989.

Hatzor, Y. H., Tsesarsky M. Continuous and discontinuous stability analysis of the bell-shaped caverns at Bet Guvrin, Israel, International Journal of Rock Mechanics and Mining Sciences 39 (7), 867-886.

Hatzor, Y. H., Benary, R. 1998. The stability of a laminated Voussoir beam: back analysis of a historic roof collapse using DDA. Int. J. Rock Mech. Min. Sci. & Geomech. Abstr. 35(2):165-181.

Hatzor Y.H. The voussoir beam reaction curve, Ben-Gurion University, Department of Geological and Environmental Sciences Beer-Sheva,.

Mohammad Mahdi Fazaeli Stability assessment of salt cavern roof beam for compressed air energy storage in South-Western Ontario Canada 2017.

Obert, L. and Duvall, W. I. 1967. Rock Mechanics and the Design of Structures in Rock. New York: John Wiley & Sons.

Passaris E. K. S., Ran J. Q. Shear sliding failure of the jointed roof in laminated rock mass, Rock Mechanics and Rock Engineering, Volume 27, Issue 4, 1994.

Ran, J. Q., Passaris, E. K. S. and Mottahed, P. 1991. Shear sliding failure of the jointed roof in laminated rock mass. Rock Mech. Rock Engng.27 (4): 235-251.

Talesnick M. L., Bar Yaacov N. Modeling of a multiply jointed Voussoir beam in the centrifuge, Rock Mechasnics and Rock Engineering, The Netherlands, 2007.

Scientific and Practical Studies of Raw Material Issues – Litvinenko (Ed)
© 2020 Taylor & Francis Group, London, ISBN 978-0-367-86153-7

Modern mine survey techniques in the process of mining operations in open pit mines (quarries)

A. Blishchenko
Saint-Petersburg Mining University, Saint-Petersburg, Russian Federation

ABSTRACT: The possibility of cooperative usage of electronic tacheometers and GNSS-receivers for mine surveys of openpit mines (quarries) has been reviewed. One of the ways of the combined mine surveying was represented. The advantages, disadvantages and factors influencing on the error of the proposed hybrid survey are determined.

1 INTRODUCTION

The 20th century is the century of introducing electronic technologies into various areas of production, and this large-scale process could not help but touch upon such an unimportant field as surveying.

Surveying is the definition of rectangular spatial coordinates of various points on the earth's surface and within the bulk contours of mineral deposits for the preparation of drawings of mining graphic documentation.

The objects of surveying are the relief and the situation of the earth's surface, natural and artificial outcrops of rocks, mouths of mining and exploration workings, conducted mining (during exploration or development), elements of the geological structure of deposits, sampling points, boundaries of hazardous areas, structures and various communications in mining.

The basic principle of surveying — a gradual transition from General, more accurate geometric constructions to private, less accurate constructions, in accordance with this survey process includes the construction of planned and high-altitude surveying support networks on the earth's surface and in mine workings, construction of survey networks and survey work (actually determining the coordinates of individual points).

The development of land and underground construction in the mining industry requires improving the performance and accuracy of surveying and geodetic works based on the use of new generation devices and the latest measurement methods. Modern devices used in mine surveying provide high accuracy of measurements, but their use requires consideration of the influence of influencing factors. In addition, there is a need to take into account instrumental errors (Surveying and Subsoil Use).

The development and improvement of electronic geodesy and mine surveying measurement systems cause a necessity of reviewing survey methods that are based on traditional measurement techniques. In this way, it is possible to share use various electronic devices in order to simplify the fieldwork and improve the quality of both office processing of survey data and the final result. As an example of such symbiosis we can outline the joint use of electronic total stations (tacheometers) which are characterized by their versatility for surveying both underground and openpit mines and GNSS-receivers which have the ability to quickly and accurately perform surveying (The history of the development of a new geodetic instrument "Total Station").

The main methods of modern surveying: stereophotogrammetric (air, ground and underground); location (sonar, light, radar); total station; theodolite (polar and orthogonal); leveling areas and combined.

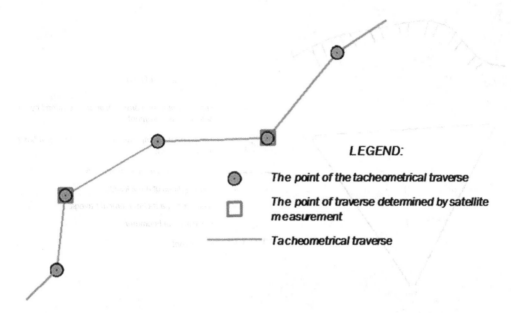

Figure 1.   A scheme of devices shared use during the construction of a tacheometrical traverse.

2   PROTOTYPE OF THIS METHOD

There is a method of combined use of electronic total stations (Sharing GPS receivers and electronic total stations) (tacheometers) and GNSS-receivers (Klabukov & Koretskaya, 2017) based on measurements where the points (which coordinates are determined by satellite technology) are change points or temporary points for radiation method or the construction of a tacheometrical traverse (Figure 1) (Markov, 2016). GNSS-measurements in this example are often made using static methods, although sometimes kinematic methods can also be used. It depends on the required accuracy in determining the coordinates, the time resource, the type of receiver, the distance between the defined points etc (Bases shooting with the use of GNSS).

The combinations of electronic total stations (tacheometers) and GNSS-receivers are diverse. One of the versions of their joint or shared use is the hybrid survey integrated during the survey of sand quarries in the Leningrad Region, Russia.

3   METHOD DESCRIPTION

The principle of this technique consists of two usual surveying processes: creation of a survey justification by satellite method and back sight by electronic tacheometer. For example, it can be necessary when surveying the new position of the quarry banks (Figure 2). Let's describe the survey methodology by stages.

1. Setting up an electronic tacheometer at an arbitrary point, exactly where it is possible to fully measure the required section of the quarry bank.
2. Following the rules of the cross-bearing we create three (or more) temporary survey justification points around the tacheometer station point. Thus, the process of traditional cross-bearing is reversed, i.e. the tacheometer is not set relative to the survey justification points but the justification points themselves are set relative to the convenient location of the device.
3. Performing satellite measurements at temporary survey stations. Observation data mode can be different, the choice is influenced by the time resource factor, technical capabilities of the receiver etc.

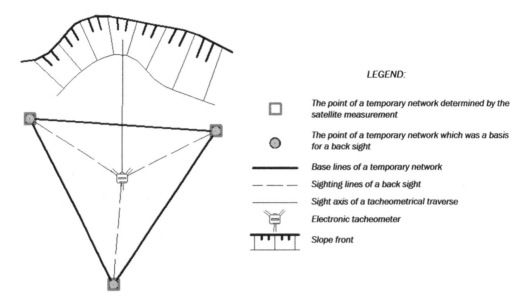

LEGEND:

☐   The point of a temporary network determined by the satellite measurement

◉   The point of a temporary network which was a basis for a back sight

━━━   Base lines of a temporary network

— — —   Sighting lines of a back sight

────   Sight axis of a tacheometrical traverse

Electronic tacheometer

Slope front

Figure 2.   A hybrid survey scheme.

4. When the coordinates of the temporary geodetic control points are received the back sight from them is performed with the help of the electronic tacheometer.
5. The final step is the performing of a tacheometrical survey of the necessary section of the quarry.

This method is notable for its simplicity and speed in comparison with other combinatorial types of surveying and is adapted to the specific conditions of solving the problem of surveying. In addition, the main advantage of this technique is the possibility of simultaneous use of electronic tacheometer and GNSS-receiver for the intended purpose after the stage of receiving the coordinates of the points of the temporary network using satellite measurements, thereby increasing the productivity of field work.

4 CONDITION OF THE METHOD

However, the final accuracy of the point coordinates survey decreases. This trend is influenced by the error in determining the location of the temporary survey network using satellite measurements, back sight and survey stake coordinates setting being determined by the tacheometer. Each of these errors has its own factors that cumulatively can have both positive and negative effects on the error of determining the coordinates of each final measurement point.

The main attention is paid to the consideration of systematic errors, causing the appearance of offsets of measurement results. In their study and the creation of methods for weakening their influence, a modeling method has become widespread, in order to develop it, one has to carefully study the mechanism of the effect of such sources of error on measurement results so that, on the basis of such a study, develop effective methods for minimizing the observed effect. Based on the analysis of the measuring process characteristic of GPS and GLONASS systems, all the main sources of errors can be divided into three main groups.

The main errors of the hybrid method are shown in Table 2.
Factors influencing on the accuracy of the back sight:

– selection of a cross-bearing type (single, multiple);
– geometry of the a cross-bearing;

Table 1.   The three main groups of sources of error.

1. Errors associated with inaccuracy of knowledge of the source data, of which the decisive role belongs to the errors of knowledge of satellite ephemeris, the values of which should be known at the time of measurement.
2. Errors caused by the influence of the external environment, among which there are such sources as the influence of the atmosphere (ionosphere and troposphere) on the results of satellite measurements, as well as radio signals reflected from surrounding objects.
3. Instrumental sources of errors, which, as a rule, include inaccuracy of knowledge of the position of the phase center of the receiver antenna, unaccounted time delays during the passage of information signals through the equipment, as well as errors associated with the operation of recording devices of satellite receivers. Along with the error groups listed above, it is necessary to take into account individual factors that cause the appearance of errors that are not characteristic of any of the above listed groups. In particular, such errors can be attributed to errors resulting from the non-optimal relative position of the observed satellites (geometric factor). In addition, a number of errors may occur in the process of transition from one coordinate system to another.

Table 2.   Error factors of satellite measurements.

1. Quality of satellite geometry;
2. Quality of satellite geometry;
3. Gravitational influences;
4. The influence of the ionosphere and troposphere;
5. Signals reflection;
6. Relativity of time measurement;
7. Computational errors;
8. Type of equipment used;
9. Selection of a network of base stations to provide coordinates;
10. Geometry of receiver and base station location etc.

– device type and its technical capabilities;
– the number of points from which the cross-bearing is being carried out etc.

The following reasons may affect the error in the tacheometrical (total stationery) survey:

– device type and its technical capabilities;
– the distance at which the object is located etc (The concept of information technology and information systems).

These are not the only factors of influence – we did not mention the measurement errors resulting from changes in weather conditions and measurement inaccuracy by a specialist.

During measurements, an important factor is the environment in which they are carried out. The reasons for the negative effect are presented in the Table 3.

The specialist contributes an element of subjectivity to the measurement process, which, if possible, should be reduced. The subjectivity of the operator depends on the following factors:

Table 3.   The reasons for the negative effect.

1. Bad weather conditions (precipitation, wind, fog, high temperature);
2. Non-standard location of the object (swamps, dams, high mountains);
3. The presence of technical means generating vibrations (proximity to railways, subway, hydroelectric power plants, etc.);
4. The presence of malicious animals;
5. Winter time when the temperature is about zero degrees.

– qualifications;
– psychophysiological state;
– comfort;
– sanitary and hygienic working conditions and more.

The specialist can have a significant impact on the measurement accuracy (Factors affecting measurement accuracy).

5 CONCLUSION

Such variety and variability of the quantitative and qualitative component of the reasons for the impact on the final result of hybrid survey leads to the fact that the error of the coordinates of the point is a floating value and its calculation is a sophisticated process. Therefore, it is necessary to further investigate the degree of influence of each presented factor on the accuracy of measurements and considering the processes of such errors' formation obtain the integral accuracy characteristics of hybrid survey. Taking into account the results of these studies, adjustments to improve the hybrid survey should be made. However, due to its simplicity, speed and performance, and in some cases its efficiency, hybrid surveying can already be of practical interest for such technology-related objects as quarries.

REFERENCES

Markov S.Y. Practical use of GPS. Using GPS for inventory purposes. KNUSA, 2016.
Genike A.A., Pobedinsky G.G. Global satellite positioning systems and their application in geodesy. Publication 2, revised and updated. - M.: Kartgeocenter, 2004. - 355 p.
The challenge of choice between GNSS and total stations. Klabukov I.V., Koretskaya G.A., Kuzbass State Technical University, 2017.
Bases shooting with the use of GNSS. https://www.aspector.ru/osnovy-sputnikovoy-semki-/
The concept of information technology and information systems. Modern concepts, ideas and problems of information technology development. The role and tasks of information technology in the development of society. Siberian State University of Geosystems and Technologies.
Sharing GPS receivers and electronic total stations. https://megalektsii.ru/s45258t6.html
The history of the development of a new geodetic instrument "Total Station". State university for land management.
Accuracy of geodetic measurements. https://domzem.su/tochnost-geodezicheskih-izmerenij.html
Factors affecting measurement accuracy. https://studopedia.ru/19_385152_faktori-vliyayushchie-na-tochnost-izmereniy.html
"Surveying and Subsoil Use" - http://geomar.ru/articles/mine-surveying/379-new-generation-tools-at-mining-enterprises.html

Scientific and Practical Studies of Raw Material Issues – Litvinenko (Ed)
© 2020 Taylor & Francis Group, London, ISBN 978-0-367-86153-7

Output prediction of Semi Mobile In Pit Crushing and Conveying Systems (SMIPPC)

R. Ritter & C. Drebenstedt
Technische Universität Bergakademie Freiberg, Germany

ABSTRACT: In this paper, a structured method for the capacity determination of SMIPCC systems under consideration of the random behaviour of system elements and their interaction was developed, which can be applied for future projection of these systems. The developed method is based on a structured time usage model specific to SMIPCC systems and empirical data of the operational and the disturbance behaviour of each system element. The method is used in a case study based on a hypothetical mine environment to analyse the system behaviour with regards to time usage model component, system capacity, and cost as a function of truck quantity and stockpile capacity. Furthermore, a comparison between a conventional truck & shovel system and SMIPCC system is provided. Results show that the capacity of a SMIPCC system reaches an optimum in terms of cost per tonne, which is 24% lower than a truck and shovel system. In addition, the developed method is found to be effective in providing a significantly higher level of information, which can be used in the mining industry to accurately project the economic viability of implementing a SMIPCC system.

1 INTRODUCTION

As ore grades decline, waste rock to ore ratios increase and mines become progressively deeper mining operations face challenges in more complex scenarios. Today´s predominant means of material transport in hard-rock surface mines are conventional mining trucks however despite rationalisation efforts material transportation cost increased significantly over the last decades and currently reach up to 60% of overall mining cost (Lieberwirth, 1994; Fabian, 1989). Thus, considerations and efforts to reduce overall mining costs, promises highest success when focusing on the development of more economic material transport methods.

SMIPCC systems represent a viable, safer and less fossil fuel dependent alternative however its viability is still highly argued as inadequate methods for the long term projection of system capacity leads to high uncertainty and consequently higher risk.

Therefore, in this paper presents a structured method for the determination of SMIPCC system that incorporates the random behaviour of system elements and their interaction. The method is based on a structured time usage model specific to SMIPCC system and is supported by a stochastic simulation. The method is used in a case study based on a hypothetical mine environment to analyse the system behaviour with regards to time usage model component, system capacity, and cost as a function of truck quantity and stockpile capacity. Furthermore, a comparison between a conventional truck & shovel system and SMIPCC system is provided.

2 SMIPCC CAPACITY CALCULATIONS

In a simplified SMIPCC system (refer to Figure 1) a truck fleet (consisting of multiple trucks) is loaded by a single loader inside the pit. The trucks discontinuously transport the material to a semi-mobile crusher station inside the pit where it is crushed to a conveyable size. The

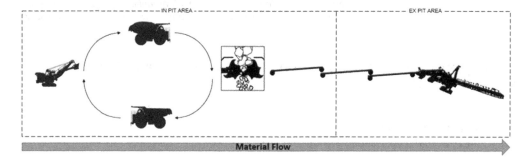

Figure 1. Illustration of simplified SMIPCC system.

material is transported out of the pit by a conveyor system (consisting of multiple conveyors flights) to a single spreader where it is discharged onto a waste dump.

The capacity of such SMIPCC systems for longer time periods is influenced by process and element specific characteristics as well as by the overall system layout. The operation of a SMIPCC systems is characterized by a high level of mechanization and automation as the majority of the transport distance is realized by conveyors. This creates the requirement for high utilisation of the machine system. The logical consequence is that capacity planning is carried out in relation to the machine system. Whereby the winning element (loader) acts as the capacity determining element under consideration of their technological connections. Once the *average hourly capacity of the winning element* C_L is known, the *capacity of the entire SMIPCC system* C_S can be determined based on the *effective operating time* t_{O_e} of winning element. It holds

$$C_S = t_{Oe} C_L \qquad \text{in t/a.} \qquad (1)$$

The problem of capacity calculation of discontinuous loaders is closely related to the general equipment selection problem, which is a wide research field in itself and has been extensively studied in the past by many researchers and shall not be the focus of this paper. Burt and Caccetta (Burt & Caccetta, 2014) outlined various modelling and solution approaches for this problem in their review paper. Further references are Hardy (Hardy, 2007) and Kühn (Kühn, 1955).

The prerequisite for the calculation of t_{O_e} is the investigation of different operational and downtime states of system elements by a time usage model. For the calculation of the effective operating time t_{Oe} each time component described in (refer to Figure 2) needs to be determined. Some time components can be simply approximated as constants. This includes the time components $t_{Dp}^{(1)}$, $t_{Dp}^{(2)}$, $t_{Dp}^{(3)}$ and $t_{Dp}^{(5)''}$.

Planned shift delays represent the sum of individual planned shift delays $t_{Dp_i}^{(4)}$ in h/shift multiplied by the available shifts per annum. To approximate the annual planned shift delays the following holds

$$t_{Dp}^{(4)} = \frac{t_C - t_{Dp}^{(1)} - t_{Dp}^{(2)} - t_{Dp}^{(3)} - t_{Dp}^{(5)''}}{t_{Shift}} \sum_{i=1}^{n} t_{Dp_i}^{(4)} \qquad \text{in h} \qquad (2)$$

The majority of technological downtimes are proportional to t_{Oe}. This includes but is not limited to times for blasting and conveyor trackshifts. The factor τ is referred to *technological downtime ratio* and is dimensionless. It holds

$$t_{Dp}^{(5)'} = \tau t_{Oe} \qquad \text{in h.} \qquad (3)$$

64

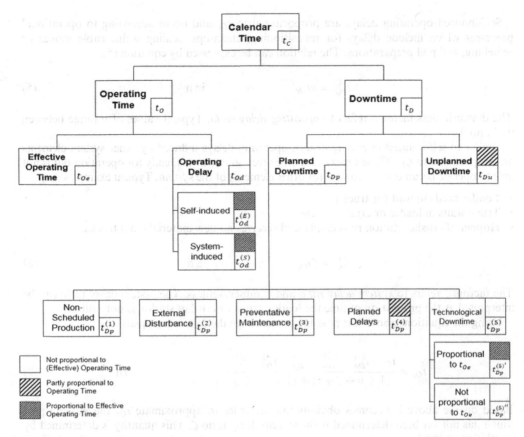

Figure 2.   SMIPCC time usage model.

Unplanned downtimes proportional to t_{Oe}. Which means that with increasing operating time unplanned downtimes increase as well. Unplanned downtimes can be estimated by the following equation:

$$t_{Du} = \varkappa t_O \qquad \text{in h.} \qquad (4)$$

The factor $\varkappa$ is referred to *repair ratio,* is dimensionless and represent the ratio between operation time $t_O(\tau)$ and the unplanned downtime $t_{Du}(\tau)$. Table 1 lists typical $\varkappa$ values which were statistically derived for different system elements.

Table 1.   Summary of repair ratio values of SMIPCC system elements.

Equipment type	Sample size	Repair Ratio			
		Min	Mean	Max	CV
Loader					
cable shovel	11	0.083	0.170	0.270	0.35
hydraulic excavator	13	0.091	0.157	0.299	0.40
Trucks	20	0.0310	0.1280	0.2300	0.50
Crusher	10	0.035	0.117	0.238	0.62
Spreader	19	0.022	0.059	0.118	0.48
Conveyor					
shiftable	29	0.004	0.019	0.043	0.77
relocatable	21	0.001	0.012	0.050	1.06
fix	26	0.000	0.007	0.035	1.41

Self-induced operating delays are proportional to t_{Oe} and occur according to operational processes which include delays for repositioning, clean-ups, scaling walls, cable moves or refuelling, and pad preparations. The relation can be expressed by equation (5).

$$t_{Od}^{(E)} = \nu t_{Oe} \qquad \text{in h} \qquad (5)$$

The dimensionless factor ν refers to *operating delay ratio*. Typical values of ν range between 0.05 and 0.1.

Similar to self-induced operating delays, operating delays induced by other system elements are proportional to t_{Oe}. These times occur whenever an element is ready for operation but is not able to operate because it has to wait for other elements of the system. Typical examples are:

- Loader needs to wait for trucks
- Truck waits in loader or crusher queue
- Hopper of crusher station runs empty and receives no new material from trucks.

$$t_{Od}^{(S)} = \zeta t_{Oe} \qquad \text{in h} \qquad (6)$$

The factors ζ refers to *system delay ratio* and is dimensionless. The system delay ratio can be interpreted as the proportion of time the loader; truck or crusher station is not utilized.

Using the equations above it is possible to rearrange the following equation

$$t_{O_e} = \frac{t_C - t_{Dp}^{(1)} - t_{Dp}^{(2)} - t_{Dp}^{(3)} - t_{Dp}^{(4)} - t_{Dp}^{(5)''}}{(1 + \nu + \zeta + \tau) + (1 + \nu + \zeta)\chi} \qquad \text{in h.} \qquad (7)$$

Based on the above it becomes obvious that in order to approximate t_{Oe} the missing part which has not yet been determined is the system delay ratio ζ. This quantity is determined by simulation methods.

3 SIMULATION RESULTS BASED ON HYPOTHETICAL COAL DEPOSIT

A hypothetical coal deposit was created which is loosely based on the characteristics found at the Clermont coal mine in Queensland, Australia (refer to Figure 3). For the overburden layer

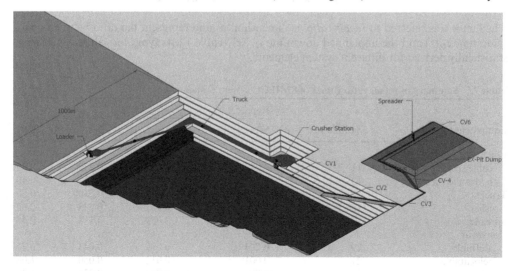

Figure 3. Hypothetical coal mine layout.

Table 2. Disturbance parameters of SMIPCC system elements.

Element Disturbance Parameters	Mean repair time [min]	Repair ratio	Mean Work Time [min]
Loader	132.7	0.170	782
Truck	288.1	0.128	2251
Crusher Station	33.1	0.117	282
Conveyor			
CV1	31.8	0.012	2661
CV2	21.0	0.007	3205
CV3	21.0	0.007	3205
CV4	21.0	0.007	3205
CV5	31.8	0.012	2661
CV6	32.7	0.019	1722
Spreader	52.1	0.059	878

a SMIPCC system is utilised in which a P&H4100 electric rope shovel excavates the overburden material with a mean hourly capacity of $C_L = 9,398$t/h. The shovel loads a homogeneous truck fleet consisting of Komatsu 930-4SE trucks which transport the material along the indicated truck travel path (in blue) to the semi-mobile in-pit crusher station located at the permanent wall. The crusher station, with a nominal capacity of 9,400 t/h and a hopper capacity of 725 t, has 3 truck bridges which allows the trucks to discharge material into the hopper of the crusher station. The conveyor system has the same nominal capacity as the crusher station and consist of a series of 6 conveyors (CV1 - wall conveyor; CV2 - ramp conveyor; CV3 – overland conveyor; CV4 - dump ramp conveyor; CV5 - extendable dump conveyor; CV6 – trackshiftable dump conveyor). The conveyor system transports the material out of the pit to an ex-pit dump where it is discharged by a spreader. The overburden layer consists of consolidated sandstone with an average insitu density of 2.37 t/m³. After blasting, the loose density of the material amounts to 1.78 t/m³ applying a swell factor of 1.33.

The hypothetical mine is planned to operate 362 days per annum, allowing 3 days for non-worked holidays, in two 12 hour shifts per day. It is estimated that the mine stops operation due to bad weather conditions and other external downtimes for a total of 5 days per year. Preventative maintenance $t_{Dp}^{(3)}$ is planned to amount to 896 hours per annum. Annual planned shift delays $t_{Dp}^{(4)}$ are approximated to 640 hours. The operating delay ratio ν of the loader was estimated to 0.08 to account for minor short-term delays such as minor pad preparations, face clean-ups, tramming, etc. The technological downtime ratio τ is set to 0.074. The mean truck travel time unloaded t_{T_U} and truck travel time loaded t_{T_L} were estimated to 190 s and 290 s, respectively. The coefficient of variation of loaded and unloaded travel time was estimated to 0.15. In addition, a 45 s manoeuvre and spot time at the loader t_S and a 60 s manoeuvre and dump time at the crusher t_D was projected. The estimated disturbance parameter for the system elements are listed below.

Results of the simulation indicate that the annual capacity of a SMIPCC system increases as more trucks are introduced to the system (Figure 4). However, the increase of SMIPCC

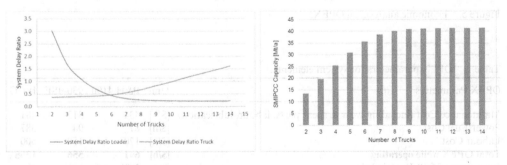

Figure 4. System delay ratios for loader and truck (left) and SMIPCC system capacity (right) for various number of trucks.

capacity shows diminishing marginal returns as the number of trucks in the system increases. Furthermore, the results indicate that annual SMIPCC capacity approaches a limit. In this particular case study, the limit of the annual SMIPCC system capacity was approximately 41.5 Mt/a.

The progression of the cost per tonne curve (Figure 5) of a SMIPCC system over an increasing number of trucks indicates two stages; one in which the cost per tonne decreases until they reach a minimum and one in which the cost per tonne increases. The positively sloped portion of the cost per tonne curve is directly attributable to the diminishing marginal returns of the annual SMIPCC system capacity. In this particular case study, the minimum cost per tonne of the system was found at 6 trucks and 0.69 $/t. The corresponding system capacity for that minimum was 35.6 Mt/a. Refer to input OPEX parameter in Table 3.

A sensitivity analysis shows that the annual SMIPCC capacity increases as the mean repair time of the system elements decreases. However, in this particular case study for 6 trucks the reduction of the mean repair time of the continuous part of the SMIPCC system indicated the highest increase of SMIPCC system capacity. For example, by reducing the mean repair time of the continuous part of the SMIPCC system by 10% the annual capacity of the system increased by 3.6%, while for the same change of the mean repair time for the loader or the trucks the system capacity increases only by 1.4% and 1.1%, respectively.

The introduction of a small stockpile in front of the crusher station increases the annual SMIPCC system capacity. The annual SMIPCC system capacity increases as the stockpile capacity increases. However, the SMIPCC system capacity shows diminishing marginal returns as the stockpile capacity increases. In this particular case study, an increase of the

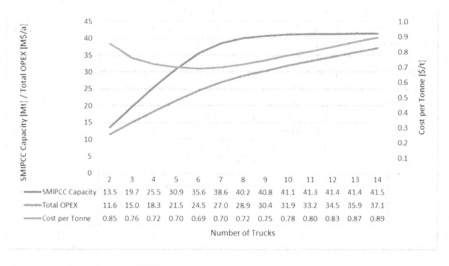

Figure 5.   Economic analysis on OPEX.

Table 3.   OPEX parameters for system elements.

OPEX Parameters	Unit	P&H 4100	930-4SE	IPCC
Maintenance Cost (including wear & spear parts, labour, lubrication)	[$/h]	434	312	481
Power/Fuel Cost	[$/h]	87	94	387
Labour Cost	[$/h]	170	150	500
Total OPEX while operating	**[$/h]**	**691**	**556**	**1368**
Percentage of OPEX while Idle	[%]	30%	32%	42%
Total OPEX while Idle	**[$/h]**	**205**	**178**	**568**

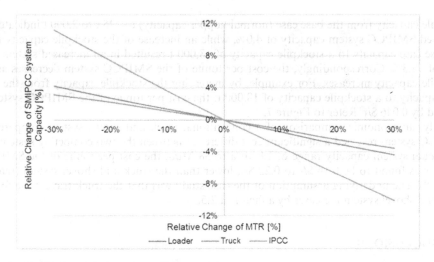

Figure 6.   Sensitivity analysis on mean time to repair.

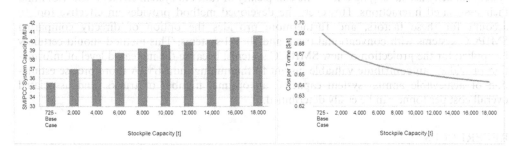

Figure 7.   SMIPCC system capacity (left) and cost per tonne of SMIPCC system for various stockpile capacities.

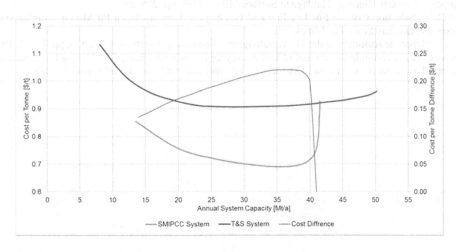

Figure 8.   Annual system capacity vs. cost per tonne.

stockpile capacity from the base case (normal hopper capacity) of 725 t to 2,000 t indicated an increased SMIPCC system capacity of 4.0%, while an increase of the stockpile capacity from the base case capacity to a stockpile capacity of 18,000 t resulted in an increased system capacity of 14.3%. Correspondingly, the cost per tonne of the SMIPCC system decreases as the stockpile capacity increases. For example, by increasing the stockpile capacity from the base case capacity to a stockpile capacity of 18,000 t, the cost per tonne of the SMIPCC system is reduced by 0.046 $/t. Refer to Figure 7.

Lastly, an economic comparison of a conventional truck and shovel system compared to SMIPCC system revealed a significant cost difference between the two competing systems. In the annual system capacity range of 14 Mt/a to 38 Mt/a, the cost per tonne of the SMIPCC system was found to be 0.14 $/t to 0.22 $/t, lower than the truck and shovel system (refer to Figure 8). The underlying assumption of the case study was that the truck travel time for the truck and shovel system increases by a factor of 2.5.

4 CONCLUSIONS

As an overall conclusion, it can be said that the accurate determination of annual SMIPCC system capacity is challenging due to the complexity of random system element behaviour and their associated interactions. However, the developed method provides an effective tool to account for these factors, and furthermore provides the option of directly comparing SMIPCC systems with conventional truck and shovel systems. This method should certainly be applied for the projection of new SMIPCC systems, because the increased level of information provided can contribute valuable insight to the mining industry. A more precise estimation of achievable annual system capacity, an optimal number of trucks, and associated overall cost per tonne can be easily determined.

REFERENCES

Burt C. and Caccetta L., "Equipment Selection for Surface Mining: A Review," Interfaces (Providence)., vol. 44, no. 2, pp. 143–162, 2014.
Fabian J., "Cyclic mining systems in Czechoslovakian surface mines," in Proceedings of International symposium on Off-Highway Haulage in Surface Mines, 1989, pp. 205–209.
Hardy R.J., "Selection Criteria For Loading and Hauling Equipment - Open Pit Mining Applications," Western Australian School of Mines, 2007.
Kühn G., "Über die Ausnutzung der Universalbagger," Fördern und Heb., vol. 1, no. 5, pp. 16–21, 1955.
Lieberwirth H., "Economic advantages of belt conveying in open-pit mining," in Mining Latin America/ Minería Latinoamericana, 1st ed., D.C. Bailey, Ed. Springer Netherlands, 1994, pp. 279–295.

Scientific and Practical Studies of Raw Material Issues – Litvinenko (Ed)
© *2020 Taylor & Francis Group, London, ISBN 978-0-367-86153-7*

Arrangements for increase the efficiency of mining operations on the deep ore mines

A.V. Kholmskiy & D.V. Sidorov
Saint-Petersburg Mining University, Saint-Petersburg, Russian Federation

ABSTRACT: The article is dedicated to the problem of decreased efficiency of mining operations on the deep ore mines, particularly, on the North Ural bauxite basin. The world researches in this area are reviewed. The most suitable type of mining equipment is chosen. The question of applicability of the cyclical-and-continuous method is discussed and the methodology of the determination of hydraulic breaker in given conditions is proposed.

1 INTRODUCTION

At the present moment, the change from the cyclical mining method to the cyclical-and-continuous (continuous) mining method is the topic of especial importance, considering the demand of the high extraction volumes of ore. At the same time, the extraction volumes are directly connected with the efficiency of mining operations. There are several factors, related with the deep mining (Galchenko, Yu.P. et al. 2018) which is affects negatively on the possibility to increase the efficiency of mining operations, such as: increased amount of air for the mining ventilation; increased probability of dynamic events of the rock pressure (Kosukhin, N.I. et al. 2019). Usage of the drilling-and-blasting method in such conditions is limiting the possibility of the change to the cyclical-and-continuous (continuous) method.

More detailed look on this problem is connected with the mining of the North Ural bauxite basin. North Ural bauxite basin includes Cheryomukhovskoye, Novo-Kalyinskoye, Kalyinskoye and Krasnaya Shapochka ore deposits (Karapetyan, E.A. 2009). All of these deposits are marked by the following features: meridional strike of ore bodies; monoclinal dip of ore bodies from West to East, the inclination is 20-35 degrees; block structure of ore bodies; the same basis of lithologic structure of overburden; different types of emplacement on the Cheryomukhovskoye deposit, two types on the Novo-Kalyinskoye, Kalyinskoye and Krasnaya Shapochka ore deposits. The sheetlike bauxite ore deposit is divided on two sublevels. The upper sublevel consists variegated pyritized bauxites, lower sublevel consists red bauxites. Mining of the bauxites is situated on the depth of 1200-1400 meters. The current mining method is room and pillar with drilling and blasting operations. The parameters of mining method are:

- equipment – drillrig Rocket Boomer L2, load-and-haul dump TORO 400, shot loader ZMK 1–A;
- room and pillar parameters – size of excavation block is 120x120 m, width of rib pillar is 3.5 m, cross-cut of the face is 30 m^2, length of room is 60 m;
- drilling and blasting parameters – length of drillholes is 2.5 m, use ratio of drillholes is 0.9, charge density is 1100 kg/m^3, specific charge is 1.8 kg/m^3, diameter of drillholes is 45 mm.

For the purpose of optimization of the technological parameters, for preparation and extraction of bauxites the milling cutter can be used (Titov, K.S. 2006). Thus, the combination of drilling-and-blasting method with the milling cutter is recommended. The results of trials of non-explosive bauxite mining revealed the reliability of the milling cutter. Usage of milling cutter allows to decrease the costs for drilling and blasting.

In addition to milling cutters, hydraulic breakers are applicable for such mining operations as drifting and stoping. The analysis of world practice of tunnel drifting shows that high impact energy breakers are compete successfully with the drilling-and-blasting method. Usage of drilling and blasting is connected with the crack formation in overburden and seismic effect on the surface buildings. Usage of hydraulic breakers makes the change to continuous mining method more possible, visual control of the quality of ore is assured, selective extraction is possible. According to the researches, hydraulic breakers is suitable on the hard rocks. If the rock jointing affects on the drilling and blasting operations negatively, so for hydraulic breakers jointing contributes the effective breakage. The producing costs (Sidorov, D.V. et al. 2018) for drilling-and-blasting method are 0.5-2.3 USD per m^3. For hydraulic breakers producing costs is 0.3-0.5 USD per m^3.

Thus, hydraulic breakers were successfully used for the drifting of tunnel between Messina and Palermo (Sicily) (Makarov, A.B. & Romanov, A.N. 1996). The drifting was performed from both ends, first of them was drifted using drilling-and-blasting method, second of them was drifted with the hydraulic breaker. As the result, hydraulic breaker assured the decrease of producing costs for 30%. Drifting of two tunnels in Hyogo Prefecture (Japan) with hydraulic breakers has shown the face advance of 20 m per month (Lysikov, B.A et al. 2004). There is experience of usage of the hydraulic breakers on the iron mine in Khromtau (Kazakhstan).

2 METHODOLOGY

For the determination of capacity of the hydraulic breaker the mechanism of impact effect is analyzed. Generally, the value of impact force depends on the pressure of liquid in the hydraulic system. The stress, that appears in the plane of face after the impact not only parallel to the direction of impact, but also perpendicular. Thus, it concluded that the maximum (peak) of stress corresponds with cracking and slacking of rock mass.

In the peak of stress the cracking and shearing is the most probable. That means that the peak of stress is the exact point in the plane of face where the next impact should be performed.

For the determination of the stress in the plane of face caused by the impact, the correspondence with the pressure of liquid in hydraulic system, considering the dynamic coefficient is used (Table 1).

Using these values, the stress in the plane of face is 15 MPa.

The determination of distance to the peak of stress is based on the correspondence of distance with the diameter of anvil and compressive strength of rock (Table 2).

The diameter of anvil is taken from the characteristics of Atlas Copco HB 3000 hydraulic breaker. As the result, the distance to the peak of stress in plane of face is about 20 cm. In this manner, the potential zone of crack formation and shearing of rock is at the distance of 20 cm

Table 1. The values that are taken for example.

Pressure of liquid in hydraulic system, MPa	Dynamic coefficient
25	2

Table 2. The values for the correspondence of distance with the diameter of anvil and compressive strength of rock.

Diameter of anvil, mm	Approximate compressive strength of bauxites, MPa
165	60

Table 3. The values for the determination of required number of impacts.

Cross-cut of the face, m²	Distance to the peak of stress, cm
30	20

Table 4. The values for the determination of duration of stoping.

Duration of move to the point of impact, sec	Thickness of layer, cm
3	10

Table 5. The values for the determination of capacity of hydraulic breaker.

Duration of shift, h	Duration of stoping of a layer, min
7	30

from the performed impact. A possibility to design the scheme of stoping for the hydraulic breaker appears.

Knowing the distance to the peak and the cross-cut of the face it is possible to determine the required number of impacts (Table 3)

Thus, the required number of impacts is 567 impacts.

Also, it is possible to determine the duration of stoping of one layer. The correspondence is based on the duration of the move of the breaker (Table 4).

In this manner, the duration of stoping a layer 10 cm thick is 30 minutes.

Knowing the duration of stoping a layer 10 cm thick it is possible to determine the capacity of hydraulic breaker. The capacity of hydraulic breaker is defined by the duration of shift on the mines of the North Ural bauxite basin (Table 5).

Consequently, the capacity of hydraulic breaker in such conditions is 1.4 m per shift (4.2 m per day).

The comparison of the efficiency of mining operations is carried out according to productivity of drilling-and-blasting method and capacity of hydraulic breaker. Productivities are measured in tons per day. Determination of daily productivity of drilling and blasting is based on classical approach to the comparison of technical and economical parameters of mining methods. At the data mentioned above, the productivity of drilling and blasting is 225 tons per day. Capacity of hydraulic breaker, considering the face advance of 4.2 m per day, is 353 tons per day.

Besides the results of comparison it is important to notice the other advantages of hydraulic breaker usage: applicability of the cyclical-and-continuous mining method; the significant timeout for ventilation after the blasting is excluded.

3 CONCLUSION

In the course of research done the problem of increase the efficiency of mining operations is discussed and the applicability of cyclic-and-continuous mining method in conditions of the North Ural bauxite basin is analyzed. The world experience in this area is reviewed, different types of mining equipment is examined. The choice of hydraulic breaker is based on analysis of world experience and by reason of workability in given conditions and possibility to use the cyclic-and-continuous mining method. The usage of hydraulic breaker answers the following requirements: the amount of air for mining ventilation is decreased; the negative seismic effect is excluded; the application of cyclic-and-continuous mining method is possible. The methodology for the determination of capacity of hydraulic breaker is purposed and the comparison of productivities of drilling-and-blasting method and hydraulic breaker is performed. The comparison views the advantage of hydraulic breaker at more, than 50%.

REFERENCES

Galchenko, Yu.P. et al. 2018, Solution of geoecological problems in underground mining of deep iron ore deposits, *Eurasian Mining*. № 1. P. 35–40.
Kosukhin, N.I. et al. 2019. Assessment of Stress-Strain and Shock Bump Hazard of Rock Mass in the Zones of High-Amplitude Tectonic Dislocations. *IOP Conference Series: Earth and En-vironmental Science*. 224 (1), DOI: 10.1088/1755-1315/224/1/012014.
Louchnikov, V. et al. 2014. Ground support liners for underground mines: energy absorption capacities and costs. *Eurasian Mining No 1*. P. 54–62.
Lysikov, B.A et al. 2004. Non-explosive environmentally friendly tunneling with hydraulic hammers *Ecological problems*. Donetsk, p. 172.
Makarov, A.B. & Romanov, A.N. 1996. Non-explosive excavation of mine workings with KR and PP hydraulic hammers. *Underground space of the world*. p. 51–53.
Sidorov, D.V. & Ponomarenko, T. 2017. The development of a software suite for predicting rock bursts within the framework of a system for ensuring geodinamic safety of mining operations. *International Multidisciplinary Scientific Geo-Conference Surveying Geology and Mining Ecology Management, SGEM*. 17 (22), pp. 633-638, DOI: 10.5593/sgem2017/22/S09.079.
Sidorov, D.V. et al. 2018. Economic justification of innovative solutions on loss reduction in the aluminium sector of Russia. *Gornyi Zhurnal*. P. 65-68, DOI: 10.17580/gzh.
Sidorov, D.V. et al. 2018. Forecasting Rock Burst Hazard of Tectonically Disturbed Ore Massif at the Deep Horizons of Nikolaevskoe Polymetallis Deposit. *Journal of Mining Institute*. Vol. 234, p. 604-611. DOI: 10.31897/PMI.2018.6.604
Karapetyan, E.A. 2009. Optimization of the parameters of the process of extraction of bauxite in the development of open pit deposits in difficult mining and geological conditions. 25.00.22. Ekaterinburg, 173 p.
Titov, K.S. 2006. Experience of using non-explosive thin-layer technology for the mining of Middle Timan bauxites. *Materials of Ural Mining Decade*.

Scientific and Practical Studies of Raw Material Issues – Litvinenko (Ed)
© 2020 Taylor & Francis Group, London, ISBN 978-0-367-86153-7

Influence of technological factors on the formation of spontaneous combustion centers in underground mining

D.D. Golubev
Saint-Petersburg Mining University, Saint-Petersburg, Russia

ABSTRACT: Analysis of existing coal mines in Russia shows that currently mainly coal deposits of the Kuznetsk basin are mined, the coal seams of which are prone to spontaneous combustion and dangerous by gas factor. In most cases mining system by longwalls with the preparation of the dual mine workings is used. The research presents the results of studies of the influence of technological factors in the development of coal seams prone to spontaneous combustion on the formation of centers of spontaneous combustion of coal in underground mining. In particular, the conclusions about the influence of coal pillars, the insulated air exhaust from working face through goaf, the size of the extraction area, the depth of mining and the combination of these factors on the endogenous fire hazard, are made.

1 INTRODUCTION

Currently, Russia is one of the world's largest exporters of fossil fuels and minerals, and remains an important part of the global mineral commodity market (Tarazanov 2019). This is important to accelerate the economic growth of the country (Zubov 2018). In recent years, of particular importance is the increase in coal exports, which stimulates the development of the entire coal industry in Russia today and will remain its driver in the future.

In such circumstances, it is necessary to maintain a stable growth rate of coal production. Also, the quality of fossil fuels is of great importance. Taking into account these factors, the development of the coal industry in Russia is largely connected with the coal deposits of the Kuznetsk basin. In 2018, more than 58% of the total coal produced in Russia was produced on its territory (Figure 1).

In addition, the enterprises of the Kuznetsk basin annually increase their productivity by 4-6%. Half of them mine coal by underground method. This is due to the high quality of coal, the conditions of bedding of coal seams and their thickness. These factors make it possible to use high-performance mining technologies and the most efficient modern equipment.

However, when mining coal seams of the Kuznetsk coal basin, it must be considered that about 70% of them are prone to spontaneous combustion, and more than 60% of them are dangerous by gas factor. The main danger is connected with the fact that occurred center of spontaneous combustion can cause an explosion of methane. Such accidents can have disastrous consequences. There are known several such cases at the enterprises of the Kuznetsk basin. For example, methane explosions at the Ulyanovskaya mine on 19 March 2007 (110 people died), methane explosion at the Yubileynaya mine on 24 May 2007 (38 people died) and two methane explosions at the Raspadskaya mine on 8 and 9 May 2010 (91 people died) (Meshkov & Sidorenko 2017).

Today, there are a large number of special means to reduce the risk of spontaneous combustion of coal. In recent years, the mines of the Kuznetsk coal basin often use nitrogen for these purposes. It and other means are intended to reduce the chemical activity of coal. Coal with low chemical activity is not able to absorb the necessary amount of oxygen to heat to a critical temperature and spontaneous combustion does not occur (Golubev 2018).

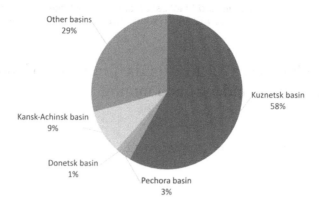

Figure 1.   The amount of coal mining in the main basins of Russia in 2018.

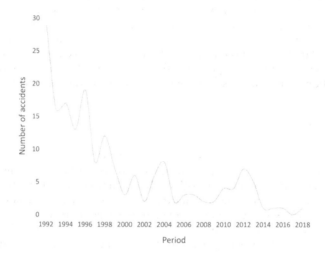

Figure 2.   Statistics of accidents connected with spontaneous combustion of coal.

However, accidents connected with spontaneous combustion of coal continue to occur every year (Figure 2). For example, the center of spontaneous combustion was registered at the mine named after Lenin on April 7, 2018. This case shows that the existing means of preventing spontaneous combustion of coal is not effective enough. The significant consequences of the accidents connected with the spontaneous combustion of coal indicate the need to develop new ways to reduce the risk of endogenous fires in the conditions of constant growth of productivity of coal mining enterprises (Golubev 2019).

Thus, of particular interest is the study of the influence of modern high-performance mining technologies on the risk of formation of centers of spontaneous combustion of coal.

2   METHODS

To assess the influence of modern high-performance mining technologies on the risk of formation of centers of spontaneous combustion of coal, the experience of mines that mine coal seams prone to spontaneous combustion in the Kuznetsk basin of Russia for the last 30 years was studied.

Modern scientific, regulatory and technical literature related to the mining of coal seams, prone to spontaneous combustion, by the underground method was studied.

Official data on the accidents connected with spontaneous combustion of coal in mines in the territory of the Kuznetsk coal basin and their reasons were taken into account, and materials with alternative opinion were also considered.

3 RESULTS

As a result of the research, it was concluded that longwall mining system with leaving of the coal pillars in goaf used for underground mining of coal seams prone to spontaneous combustion in the Kuznetsk coal basin, and currently it has no alternative (Figure 3).

Today, regulatory documents (such as safety rules for coal mines, etc.) require to leave coal pillars when mining coal seams prone to spontaneous combustion. This is due to the need for reliable isolation of previously mined areas and the prevention of air leaks to them (Skritsky 2017).

Coal pillars are important for high-performance technologies (Sankovsky et al. 2018). They help to maintain entries during the entire period of mining of the extraction area (Gromtsev & Kovalsky 2018). This allows to use a simple roof bolting in all entries (Kovalski et al. 2018b). As a result, the speed of construction of entries increases and their cost decreases. At the mines of the Kuznetsk coal basin, they can reduce the influence of the gas factor, as they exclude the influence of goaf of the neighboring area, which is one of the main sources of methane (Vinogradov et al. 2017). In addition, coal pillars allow to use the insulated air exhaust from the working face to the entry with the fresh air flow through the goaf. This also allows to significantly reduce the influence of the gas factor. Several regulatory documents on the possibility of using insulated air exhaust through the goaf were published in 2016.

Reducing the gas factor allows to use high-performance equipment at modern mines. Firstly, such equipment allows to mine more than 1.5 million tons of coal per month from one working face. For example, in September 2018 at the mine named after Yalevsky a world coal mining record was set. It amounted to 1.627 million tons. Mine named after Yalevsky located on the territory of the Kuznetsk coal basin and is mining coal seams dangerous by gas factor and spontaneous combustion, therefore, this example reflects the effectiveness of the technology of leaving coal pillars.

Secondly, the use of modern equipment can significantly increase the length of both working faces and extraction areas (Figure 4 & Figure 5). In some cases, in the high-performance mines of the Kuznetsk coal basin, the length of the working face reaches 400m, and the length of the

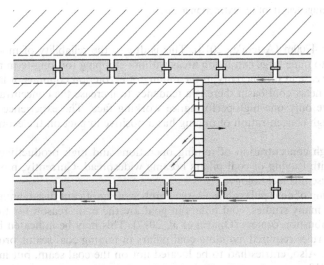

Figure 3.   Longwall mining system with leaving of the coal pillars in goaf.

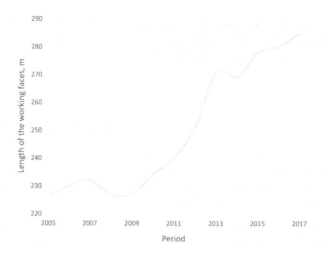

Figure 4.   The average length of the working faces, m.

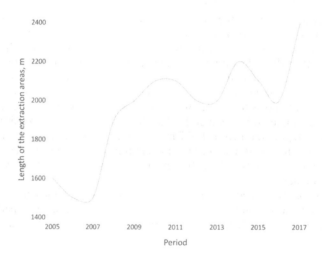

Figure 5.   The average length of the extraction areas, m.

extraction area is 4700m (Artem'ev 2018). This is due to the reliability of modern equipment and its potential to mine large extraction areas without stopping for long-term maintenance.

Due to the increase in productivity in the mining of extraction columns of large sizes in the mines of the Kuznetsk coal basin there is a tendency to increase the concentration of mining. Many mines have only one high-performance working face. The experience of these mines shows that the high concentration of mining has a positive impact on the economic efficiency of enterprises.

Due to the high concentration of mining operations and high productivity, the longwall mining system with leaving of coal pillars became widespread. Currently, more than 95% of mines in Russia use this technology.

However, leaving of coal pillars has a large number of significant disadvantages. In particular, according to many studies, coal pillars in goaf are the main reason for the formation of spontaneous combustion centers (Oparin et al. 2017). This may be indicated by the fact that until 2015 safety rules required avoiding coal pillars in mining coal seams prone to spontaneous combustion. Also, entries had to be located not on the coal seam, but in the rock mass (Kazanin et al. 2019).

In the mines, spontaneous combustion centers are often formed in goaf, or in zones of geological disturbances (Zubov & Nikiforov 2017). This is due to the fact that these areas can form clusters of shattered coal. In zones of geological disturbances they occur naturally, but occurrence of clusters of shattered coal in goaf is the lack of the technology (Zubov et al. 2017). Abutment rock pressure in goaf destroys the boundary parts of coal pillars (Karpov & Leisle 2017). As it shown in Figure 6, increase in length of the working face (distance between coal pillars) increases the value of the abutment rock pressure (Golubev et al. 2019).

Based on this, it can be concluded that the current trend towards an increase in the size of the extraction columns leads to the fact that more clusters of shattered coal are formed in goaf.

Spontaneous combustion of coal occurs as a result of its oxidation. In the process of oxidation, the temperature of coal rises. When the temperature of coal reaches a critical value, there is spontaneous combustion of coal. Taking into account that modern high-performance mines use the insulated air exhaust from the working face through goaf, the air is filtered and delivers oxygen to the active centers through the clusters of shattered coal (destroyed boundary parts of coal pillars). Shattered coal has a small coefficient of thermal conductivity. Consequently, coal loses little heat, and the temperature gradually increases.

It is very important that the depth of mining operations at high-performance mines increases almost every year (Figure 7).

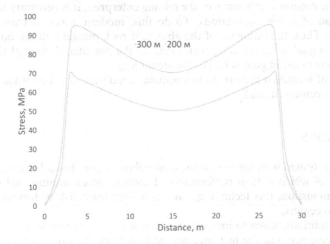

Figure 6. Stress distribution diagram for coal pillar.

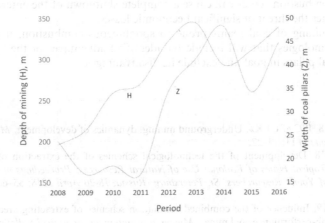

Figure 7. Average depth of mining and width of coal pillars.

79

With increasing depth of mining the size of coal pillars also increases (Zubov 2017). This is due to the increase in rock pressure at great depth and the need to preserve the undamaged middle part of coal pillar to ensure the isolation of goaf (Kovalski et al. 2018a). Thus, the amount of coal in goaf increases annually, as well as the influence of the abutment rock pressure on the boundary parts of the pillars increases (Sidorenko et al. 2019).

4 DISCUSSIONS

The research has shown that with the increase in productivity when using the longwall mining system with leaving of coal pillars in goaf, the danger of the formation of spontaneous combustion centers also increases.

It is necessary to leave pillars of coal to isolate goaf when mining coal seams prone to spontaneous combustion. Thus, there is always coal in goaf, i.e. the potential danger of the formation of spontaneous combustion centers is laid in the technology.

To increase the productivity of the mine, it is necessary to reduce the influence of the gas factor. Modern mines use the insulated air exhaust from the working face through the goaf. Thus, air is always supplied to goaf, i.e. the technology has potential opportunity for oxidation of coal clusters in goaf.

To increase the economic efficiency of the mining enterprise, it is necessary to try to increase the concentration of mining operations. To do this, modern mines increase the size of the extraction areas. Thus, the influence of the abutment rock pressure on the boundary parts of the coal pillars in goaf is significantly increased, i.e. the potential danger of the formation of clusters of shattered coal in goaf is laid in the technology.

The complex of identified factors creates optimal conditions for the formation of spontaneous combustion centers in goaf.

5 CONCLUSIONS

Longwall mining system with leaving of the coal pillars in goaf has a large number of advantages, the main of which is high performance. However, when mining coal seams prone to spontaneous combustion, this technology creates a significant risk of formation of spontaneous combustion centers.

In modern conditions, when to increase the productivity of the mine the size of the extraction areas are increased and the insulated air exhaust from the working face through goaf is used, the danger of the formation of spontaneous combustion centers increases significantly.

With the current level of concentration of mining operations in mines, the emergence of a spontaneous combustion center can cause a complete shutdown of the enterprise for a long period. This creates the threat of significant economic losses.

For the safe mining of coal seams prone to spontaneous combustion, it is necessary to develop new technologies that will include the identified advantages of the technology with leaving of the coal pillars in goaf and exclude its disadvantages.

REFERENCES

Artem'ev, V.B. 2018. JSC «SUEK». Underground mining, dynamics of development. *Mining information and analytical Bulletin* 11: 13–22.

Golubev, D.D. 2018. Development of the technological schemes of the extraction of coal seams for modern mines. *Topical Issues of Rational Use of Natural Resources: Proceedings of the International Forum-Contest of Young Researchers, St. Petersburg, Russia, 18-20 April 2018*: 55–60. London: CRC Press.

Golubev, D.D. 2019. Influence of the combined ventilation schemes of extracting area on endogenous fire hazard of high-performance coal mines. *Mining information and analytical Bulletin* 4: 66–74.

Golubev, D.D., Dmitriyev, P.N. & Sankovsky, A.A. 2019. Influence of a state of interpanel pillars on spontaneous combustion of coal. *International Journal of Civil Engineering and Technology (IJCIET)* 10(4): 2074–2082.

Gromtsev, K.V. & Kovalsky, E.R. 2018. Justification of stowing methods of the mined-out space during potasil deposits mining using longwall method. *Topical Issues of Rational Use of Natural Resources: Proceedings of the International Forum-Contest of Young Researchers, St. Petersburg, Russia, 18-20 April 2018*: 61–66. London: CRC Press.

Karpov, G.N. & Leisle, A.V. 2017. Qualitative assessment of strain stress distribution of rock massif in the vicinity of pre-driven recovery room. *Journal of Industrial Pollution Control* 33: 840–846.

Kazanin, O.I., Sidorenko, A.A. & Sirenko, Y.G. 2019. Analysis of the methods of calculating the main roofcaving increment in mining shallow coal seams with long breaking faces. *ARPN Journal of Engineering and Applied Sciences* 3(14): 732–736.

Kovalski, E.R., Karpov, G.N. & Leisle, A.V. 2018a. Geomechanic models of jointed rock mass. *International Journal of Civil Engineering and Technology (IJCIET)* 9(13): 440–448.

Kovalski, E.R., Karpov, G.N. & Leisle, A.V. 2018b. Investigation of underground entries deformation mechanisms within zones of high stresses. *International Journal of Civil Engineering and Technology (IJCIET)* 9(6): 534–543.

Meshkov, S. & Sidorenko, A. 2017. Numerical simulation of aerogasdynamics processes in a longwall panel for estimation of spontaneous combustion hazards. *E3S Web of Conferences* 21: 1–6.

Oparin, V.N., Ordin, A.A. & Nikolskiy, A.M. 2013. About negative consequences of selective mining of coal seams in Kuzbass. *Proceedings of the all-Russian forum with international participation, Tomsk, Russia, 24-27 September 2013*: 622–626. Tomsk: Tomsk Polytechnic University.

Sankovsky, A.A., Aleksenko, A.G. & Nikiforov, A.V. 2018. Practical experience analysis: Superimposed seams series mining at the Verkhnekamsk potassium-magnesium salts deposit applying room-and-pillar mining method. *International Journal of Civil Engineering and Technology (IJCIET)* 6(9): 715–728.

Sidorenko, A.A., Ivanov, V.V. & Sidorenko, S.A. 2019. Numerical simulation of rock massif stress state at normal fault at underground longwall coal mining. *International Journal of Civil Engineering and Technology (IJCIET)* 1(10): 844–851.

Skritsky, V.A. 2017. Causes of methane explosions in high-performance coal mines of Kuzbass. *Innovation and expertise* 2: 171–180.

Tarazanov, I.G. 2019. Russia's coal industry performance for January-December, 2018. *Coal* 3(1116): 64–79.

Vinogradov, E.A., Yaroshenko, V.V. & Kislicyn, M.S. 2017. Method of gas emission control for safe working of flat gassy coal seams. *IOP Conference Series: Earth and Environmental Science* 87(2): 1–7.

Zubov, V.P. & Nikiforov, A.V. 2017. Features of development of superimposed coal seams in zones of disjunctive geological disturbances. *International Journal of Applied Engineering Research* 5(12): 765–768.

Zubov, V.P. 2017. The state and directions of perfection of coal seams system on the advanced coal mines of the Kuznetsk coal basin. *Zapiski Gornogo instituta* 225: 292–297.

Zubov, V.P. 2018. Applied technologies and current problems of resource-saving in underground mining of stratified deposits. *Mining Journal* 6: 77–83.

Zubov, V.P., Nikiforov, A.V. & Kovalsky, E.R. 2017. Influence of geological faults on planning mining operations in contiguous seams. *Ecology, Environment and Conservation* 2 (23): 1176–1180.

Scientific and Practical Studies of Raw Material Issues – Litvinenko (Ed)
© 2020 Taylor & Francis Group, London, ISBN 978-0-367-86153-7

Evaluation problem of harmful effects in mining works during construction of subway escalator tunnels with the help of soils freezing method

D. Mukminova & E. Volohov
Saint-Petersburg Mining University, Saint-Petersburg, Russian Federation

ABSTRACT: This research deals with the technological features of escalator tunnels construction using soils freezing technique. The nature responsible for harmful effects of such mining on the earth's surface was analyzed. The characteristic of existing methods of displacements estimation and deformations are given, their shortcomings are revealed. Methods and techniques for solving assessment problem of harmful mining effects in the escalator tunnels construction by freezing were proposed with the purpose of buildings and structures protection.

1 INTRODUCTION

One of the most important elements in the underground transport infrastructure in modern megacities is the subway. In terms of its deep-laid for the stations linkage with the surface mainly escalator tunnels are used. Usually up to 4 escalator systems are placed at the station, which are responsible for passenger delivery. This scheme is characterized by a large cross-section (the tunnel diameter (.5 ÷ 10.5 meters), relatively small length (up to 100 ÷ 120 meters), significant tilt of its axis (the standard tilt angle is 30°) and special mining and geological conditions of excavation.

Engineering and geological conditions of the upper (usually Quaternary) rock strata are estimated as extremely unfavorable, due to the presence of weak, unstable rocks and due to the presence of aquifers (sometimes several). To ensure the rocks stability in the conduct of mining operations, exclude falls and collapses possibility, the elimination of significant water flows, as well as reduction rock mass deformation and the subsidence of the earth's surface, special methods are used. During the construction of escalator tunnels in such conditions, it is mainly envisaged to create a temporary ice-ground fence.

An important rational aspect of the protective measures mentioned above is the reliability of predictive displacements and deformations assessment. Here the situation looks safer, engineering forecasting methods are developed and used for the conditions of the escalator tunnels penetration by freezing. For example, in St. Petersburg to date, the impact assessment method described in the "Manual on the measures to protect operated buildings and structures from the mining operations influence in the construction of the subway" Podakov V.F. (Podakov et al., 1973). The forecasting method is focused on the application of typical curves method and is associated with the primary assessment of the main parameters for displacements muld: the maximum subsidence value, the maximum subsidence point position. The magnitude and position of maximum subsidence for the main longitudinal cross-section calculated according to the methodology proposed Ph. D. by Silvestrov S. N., Ovsyanko E. A. and the V. F. Podatkowym (Silvestrov, 1987):

$$\eta_0 = \frac{\left[\delta_p^2 + \delta_p(0{,}64D_B + 0{,}36D_\Gamma)\right]l_{cp.63.}\sqrt{m_o h}}{0{,}23K_p L_y}, \tag{1}$$

where δp - wall thickness fencing, m; D - the vertical diameter, m; DH is the horizontal diameter, m; CA.ot. - the weighted average value of the frozen thickness relative compaction; h - power, m; Lu - length of the shear muld in the main section, m; m0 - parameter determining the slopes steepness, 1/m2; Kr - loosening coefficient (for watered soils take Kr = 1.0; for unwater soils - Kr = 1.08).

The maximum subsidence point position in the main section of the displacement muld, passing along the axis of escalator tunnel, is determined by the formula:

$$d = 0{,}26L_{\ni}, \tag{2}$$

where L_Э - length of an escalator tunnel.

As can be seen from the formula, the technique is replete with parameters and coefficients, the values of which should be determined by individual studies. These parameters are not given in the Manual (Podakov et al., 1973). They are largely dependent on geological conditions and time factor and are associated only with the stage of thawing.

Another approach to the assessment of movements and deformations can be considered to obtain dependences based on the processing of field surveying data. So, Ph. D. Long MV in his Thesis (Dolgikh, 1999), according to the results of field studies on the objects of the St. Petersburg metro in the first approximation, suggested that the maximum subsidence of the earth's surface is directly proportional to the average subsidence of the lining arches and is determined by the linear dependence:

$$\eta_m = 2{,}7\Delta Z,$$

$$\Delta Z = \Delta Z_m \left(0.317 \lg \frac{t}{t_0}\right)^3 \tag{3}$$

where η_m - subsidence on the earth's surface, ΔZ - the average subsidence of the arch lining, as a function of time t (time since the installation of the lining), ΔZ_m - the expected total subsidence of the arch, t0 = 1 day.

However, these equations, according to the author, require further confirmation, for which it is necessary to lay special observation stations with soil reference points on the surface and deformation points in the arch and the lining tray. In addition, this dependence is poorly suited for the stage of thawing, when subsidence does not develop due to deformations of the support, but due to changes in the stress-strain state and transformation of the rocks properties after disabling freezing.

More and more information in modern works often devoted to the assessment of mining impact and construction works on the stress-strain state, where researchers rely on mathematical modeling by the finite element method. For example, in the works of Potemkin Da (Potemkin, 1999) and Belyakov N. Ah. (Belyakov, 2012) an attempt was made to estimate the displacements of rocks and the earth's surface by the results of such modeling. The authors concluded that the displacements on the earth's surface are directly related to the thickness of the ice barrier. Comparison of the data obtained in this simulation with the data on the method from "Manual..." (Podakov et al., 1973) and by nature showed a significant difference. This is partly due, often taking place in practice, a significant freezing of the soil in excess of the design (calculated under the conditions of the ice barrier strength) volume, as well as the fact that the total value of subsidence consists of two component groups: due to freezing and rocks thawing and the actual mining operations during tunneling, which are difficult to take into account at the same time. The main drawback of such model studies that should be recognized is not sufficient usage of data from the surveying measurements results, which in principle allows to verify such finite element models.

2 APPROACH DESCRIPTION

The greatest distribution in practice, the metro got way brine freezing. The essence of the soil freezing method is resulted around a quality when future construction artificially creates a temporary ice-ground fence. At first rotary drilling machines are used to pipes casing, afterwards they install a freezing column, then mount it to the collector and the brine distribution pipe network (see Figure 1). After brine freezing wells are equipped with freezing columns, they continuously circulate the coolant. When passing through the freezing column between the column and the surrounding rocks, heat exchange occurs (the brine is heated and the rocks are cooled). The brine circulates through the freezing columns in a closed loop and is cooled again at the freezing station. As a result, the water contained in the rocks gradually turns into ice, which leads to a significant change in their physical, mechanical and filtration properties: significantly increases the compressive and tensile strength. Adhesion, increases the deformation modulus, dramatically reduces the filtration characteristics of water-saturated rocks (Trupak, 1983).

During the escalator tunnel excavation with the help of soils freezing method is necessary to allocate following basic stages:
– The process of active freezing – the process of ice barrier formation. It is characterized by minimum coolant temperatures and maximum capacity of the freezing station. Active freezing period depends heavily on the depth of works and other geological conditions and can reach 50 ÷ 60 days. The tunneling of the escalator tunnel can be started when the desired temperature is reached in the thermometric wells;
– The process of passive freezing – the process of maintaining the soil in a frozen state for the entire period of construction. Here the coolant temperature is slightly higher. The penetration itself with the maintenance of the ice-ground fence can last up to 8 ÷ 10 months;

Figure 1. The escalator tunnel excavation with the help of soils freezing method.

– Immediately after the penetration completion in the frozen thickness, refrigeration units are switched off, equipment in the freezing and hydro-observation wells is dismantled, the wells themselves are plugged. The process of natural thawing of rocks continues and after the completion of construction and commissioning of the tunnel, it can last up to $3 \div 5$ years.

3 ANALYSIS

The first stage (the stage of the ice fence formation) is characterized by the manifestation of uneven rocks heaving. They occur on the surface and lead to the development of the most dangerous tensile deformation for existing objects. In tunnels (for example, durng re-freezing), this can lead to significant deformation of the lining. The maximum beams values beams here can reach $100 \div 200$ mm (according to the available data from surveying observations). Displacements and deformations at this stage are the most intensively developed, are very significant and can be dangerous for objects in the zone of influence. However, they should be considered the least studied and therefore poorly predictable.

For the second phase passive freeze phase) the values of displacements and deformations, as a rule, are insignificant, because the stage is related to the array maintenance in a given temperature regime.

The third stage (the rock thawing period) is characterized by the gradual development of subsidence and deformation in the muld by restoring the volume of the soil liquid phase, secondary stresses redistribution, strains and strength reduction and deformation rocks properties. The maximum subsidence at the end of observations can reach $400 \div 500$ mm, and the length of the muld displacement in the main longitudinal direction can reach $200 \div 250$ m, in the main cross-section $150 \div 200$ m. An example of the displacement development of muld during the escalator tunnel construction of the Sennaya station in the main longitudinal and cross sections is shown in Figure 2.

In this case (during the escalator tunnel of Sennaya subway station construction in St. Petersburg) the size of the muld displacement at the end of observations reached 205 m and 185 m in the longitudinal and transverse directions, respectively (Figure 3). Maximum subsidence on the ground surface reached up to 418 mm, slopes deformation i = 13.6·10-3, Curvature deformation k = 6 ·10-4 1/m (Figure 4) (Mukminova & Novozhenin, 2015).

As it can be seen from presented data and data (Mukminova & Novozhenin, 2015) about effects (muld displacement) of works on escalator tunnels construction by freezing method is extensive and surface deformations are very significant, here they are much higher compare with the level of limit deformations according to the classical criteria of Protection Rregulations. The mold displacements, as a rule, are characterized by the asymmetry of bolumunden in the cross sections, abrupt (not smooth) subsidence distribution in the mulde (in graphs mainly represented by smooth curves), a significant difference between values of maximum soil subsidence and mould profiles from tunnel to tunnel. That is why escalator tunnels are considered the most difficult, in terms of building and structural protection, subway development.

However, the dangerous deformation zone for buildings and structures in muld is still limited and isometric in General, which allows researchers to consider the space-planning solution for the underground complex, when the designer uses the areas of streets and areas free from development, placing production axis with the least damage as the main protection measure of buildings and structures. Mountain protection measures for obvious reasons are practically not considered here (due to the lack of effective methods of influence on the negative effects of rocks freezing), and constructive (construction) measures can be relevant only in the marginal muld parts.

So, main disadvantages of methods and approaches used so far for displacement and deformation evaluation in the construction of escalator tunnels in the way of freezing should be classified:

– the usage of one class of methods in the assessment of displacements and deformations, the lack of an integrated approach to identify patterns;

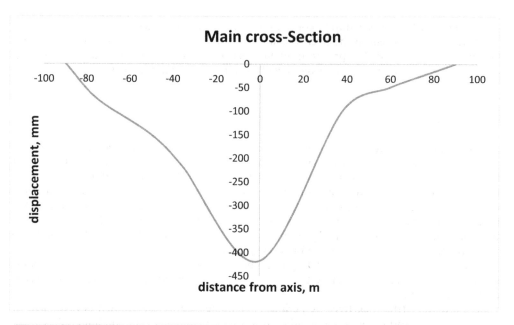

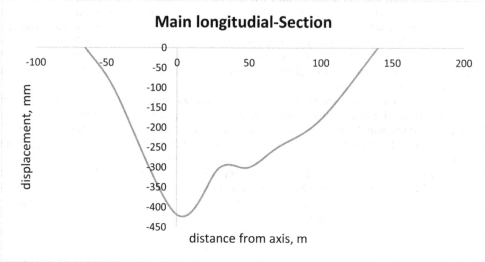

Figure 2.   Muld of the main longitudinal and cross-section of "Sennaya". metro station in Saint-Petersburg.

– averaging settings leopardsnow fence by the lithological differences and the cross section generation;
– ignorance in the majority of works the stage of formation of the ice-ground fencing at the phase of active freezing and evaluation of tensile deformations associated with temporary frost heave of rocks;
– lack of direct indications and parameters for specific mining and geological conditions, which are traditionally present in engineering methods;
– not taking into account in the calculations of complex effects of destruction, transformation of rocks during freezing and a significant change in their original properties after defrosting (Dashko,1987), leading to the manifestation of complex hydro-geomechanical rheological processes in the thawing rock mass, significantly stretched in time.

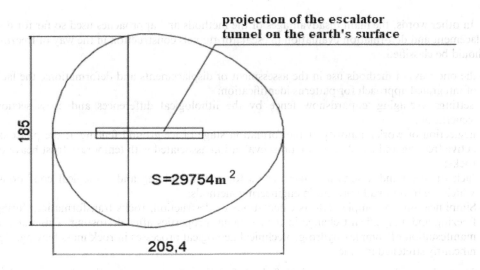

Figure 3. The muld Boundary of vertical displacements (subsidence) on the earth's surface during tunneling the escalator tunnel of "Sennaya" subway station (plan view).

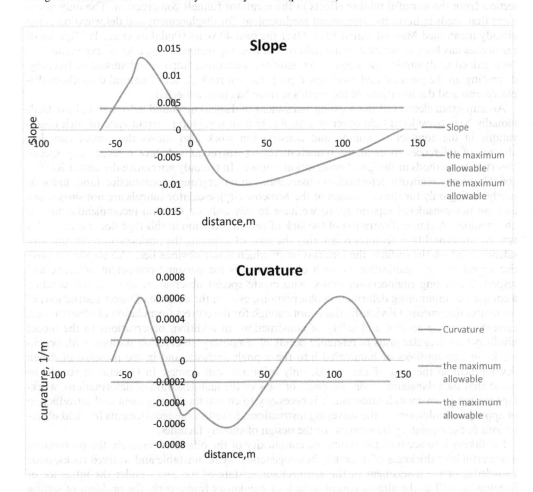

Figure 4. Slope values and curvature in the main longitudinal section for "Sennaya" subway station.

In other words, the main disadvantages of the methods and approaches used so far for displacement and deformation evaluation in escalator tunnels construction in the way of freezing should be classified:

– the one class of methods use in the assessment of displacements and deformations, the lack of integrated approach for patterns identification;
– settings averaging leopardsnow fence by the lithological differences and cross section generation;
– neglection of works majority at the formation stage of ice-ground fencing at the phase or active freezing and tensile deformations evaluation associated with temporary frost heave of rocks;
– lack of direct indications and parameters for specific mining and geological conditions, which are traditionally present in engineering methods;
– Simplifications in complex effects calculations of destruction, rocks transformation during freezing and a significant change in their original properties after defrosting. This leads to manifestation of complex hydro-geotechnical geological processes in rock mass hawing, significantly stretched in time.

From the regulatory support point of view of, it is necessary to recognize almost complete absence of regulatory framework in the shifts assessment field, buildings and structures protection from the harmful mining effects in the escalator tunnels construction. The only document that sheds light on the assessment methodology for displacement and deformation is the already mentioned Manual, dated 1971. Over the past 40 years (Podakov et al., 1973), a lot of researches has been conducted in the field of soil freezing technologies, a lot of experience has been gained in dynamics assession of ice barriers formation, impactmechanisms of freezing-defrosting on the physical and mechanical properties of rocks, a lot of natural data about displacements and deformations of the earth's surface has been collected.

An important element of the ongoing assessment of displacement and deformation have traditionally been considered field observations. Analysis of regulation current state of such observations in the conduct of mining and construction work also shows the almost complete absence of references to escalator tunnelsSurveying instructions do not consider any special observation methods or design of observation stations. In the only normative document RD 07-166-98, directly relatedto deformation observations in underground construction (and, unfortunately, acting only for the conditions of the Moscow city), escalator tunnels are not singled out and are not considered separately, so we have to rely only on General recommendations for observations. A simple illustration of the lack of proper attention in this type documents to this type of works and their features is ignoring the issue of assessing the position of maximum subsidence point on the surface, the localization of which is not obvious here. An important positive aspect in the organization of such observations is the personal competent initiative and responsible mining engineers-surveyors, who create special observation stations and conduct a complex of monitoring deformation observations, even in the absence of direct instructions of normative documents. Obviously, this is not enough for the correct formulation of observations, current surveying control and safety of construction. In addition, observations to the model niveliruya on irregular grids of reference points of temporary observation stations, with benchmarks in lighting poles or hammered into the asphalt anchors, and in the presence of urban development in the area of the mould, only primitive wall frames. In General, it should be noted that such dangerous, from the point of view of the manifestation of deformations, workings are not given enough attention, it is necessary to ensure the development and introduction of appropriate additions to the surveying instructions, to reflect the requirements for field observations in the regulatory framework for the design of subway facilities.

So, taking into account the geometric complexity of the object of research, the presence of a powerful host thickness of discretely heterogeneous, weak, unstable and watered rocks, poor knowledge of the assessment of the geomechanical state of the array under the influence of freezing, as well as the almost complete lack of regulatory framework, the problem of setting the assessment of displacements and deformations of rocks and protection of workable structures in the construction of escalator tunnels by freezing should be recognized as still relevant.

Our research and studies of other authors show a broad perspective of application of methods of mathematical modeling of displacements and deformations on the basis of the finite element method. The most difficult aspect of such modeling is to take into account the effects of freezing, for which special modeling tools are rarely offered, so you have to rely on the available Arsenal of tools and their compilation. The main thing is that this approach allows for consideration of geomechanical processes on the basis of taking into account the real physical picture of the phenomena. A key element in ensuring the reliability of the simulation here is the use of field surveying data obtained on escalator tunnels for the so-called calibration of models.

Studies of the possibilities of organization of field observations have shown the need for: input in the typical observation station additional profile lines in the cross stretch of the axis of the tunnel (at least in the zone of maximum subsidence and in the zone of maximum width of the muld), input to the station group of reference points on the contour of the mouth of the development, the allocation of a separate period of observations in the phase of active freezing to control the time heave and evaluation of strain stretching, extension of the standard periods of periodic observations at the stage of subsidence (when the array is defrosted). Taking into account the essential features of the considered development, freezing technology and capabilities of modern measuring systems, require a separate justification of the basic elements of the standard methodology of field observations: methods and means of ground and underground deformation monitoring; schemes of deformation networks and observation stations; evaluation of the required accuracy of measurements; periods, frequency and timing of observations.

4 CONCLUSION

After the end of this research the following conclusions were made:

– at the present time the freezing method remains the main methodology of deep underground escalator tunnel constriction;
– this underground construction is characterized by complex geometry and geological conditions, multi-stage physical impact on weak rocks, the manifestation of the largest deformations on the earth's surface;
– development of reliable predictive assessment of the harmful effects of such mining and protection of buildings and structures should be based on the integrated use of dependencies describing the dynamics of the formation of ice-ground fences, mathematical modeling of displacements by numerical methods and field surveying data;
– for the correct formulation of field observation systems on the escalator tunnels under construction, it is necessary to take into account not only the complex asymmetric geometry of the muld of displacement and large deformations, but also the multistage operation of the freezing system, which determines the multidirectional displacement and a significant change in deformation stages, as well as initiating the manifestation of complex, highly stretched in time, rheological processes in the thawing rock mass.

REFERENCES

Belyakov N. A. 2012. Development of a method for predicting the stress-strain state of the lining of transport tunnels in the disturbed array. Autoref. Diss.PhD. SPb.: Spggi.
Dashko R. E. 1987. Rock Mechanics. Textbook. for universities. Moskow: Nedra.
Dolgikh M.V. 1999. Displacement of the earth's surface in the construction of metro facilities in St. Petersburg. Autoref. Diss.PhD. SPb.: Spggi.
Mukminova D.Z., Novozhenin S.Yu. 2015. Analysis of field data in the construction of escalator tunnels in St.-Petersburg. Mine Surveying Bulletin №6.

Polakov V.F., Solovyov Y.F., Kapustin V.M. 1973. A manual for design of measures for the protection of exploited buildings and structures from the effects of mining operations in the subway construction. Leningr.: Stroyizdat (Leningr.).

Potemkin D. A. 1999. Rationale thickness leopardsnow fencing in layered array taking into account the thermo-physical rock properties and process parameters freeze. PhD.kand.tech.sciences. SPb.: Spggi.

Silvestrov S. N. 1987. Experience of research by the method of equivalent materials surface sediment in the construction of underground tunnels with artificial freezing of rocks. The Leningrad Institute of transport engineers: Leningr.

Trupak N.G. 1983. The birth of the method of artificial freezing of soils. We build the subway. Collected papers.

Vasiukov P.A. 1983. Experience in the application of artificial freezing of soils. We build the subway. Collected papers.

Scientific and Practical Studies of Raw Material Issues – Litvinenko (Ed)
© 2020 Taylor & Francis Group, London, ISBN 978-0-367-86153-7

Optimization of gas pipeline operation modes considering the condition of gas compressor units

A.V. Kokorin & A.M. Schipachev
Saint-Petersburg Mining University, Saint Petersburg, Russian Federation

ABSTRACT: One of the leading directions of natural gas saving in PJSC "Gazprom" is the optimization of operation modes of compressor stations. This method is performed without additional capital costs for reconstruction and modernization. The research is dedicated to optimization taking into account the current condition of compressing facilities by entering values of condition rates. To increase productivity and accuracy of calculations, a program was created. The program is able to make calculations at the design stage as well as at the operational stage of the gas pipeline life cycle. The program could be developed by adding parameters of modern equipment, namely gas turbines, centrifugal compressors, and gas air cooling units. The effect of reducing the condition of gas turbines and centrifugal compressors on costs in a compressor station is investigated. It is demonstrated that the fuel gas condition rate of gas turbines has a greater effect than the other rates.

1 INTRODUCTION

The development of energy conservation and the increase of efficiency in energy-intensive sectors of the economy, which include the gas industry, is one of the key areas of (Russia's energy strategy for the period up to 2030). The relevance of the energy saving problem in PJSC "Gazprom" is determined by the significant volume of the exhaustible resource's consumption; reduction of facilities' condition, which leads to the loss of gas during extraction, transportation, processing and storage; the increase of gas production costs due to the continuing decline in reservoir pressure in the main fields of Western Siberia: Medvezhye, Urengoyskoye, Yamburgskoye (Khvorov 2017).

Gas pipelines constructed in the 1970s-80s make up 70 % of all Russian pipelines (Kryukov 2017). The specific energy intensity of these pipelines and adjoining facilities is 30-40 % higher than the rate of modern ones (Khvorov 2017). The maximum contribution to pot.ential energy savings refers to gas transportation with 73 % or 28.2 TOE (The concept of energy saving... 2011). For the period from 2011 to 2016, one of the main directions of gas savings in transportation was optimization of operation modes of compressor stations with 24.3 % (Aksyutin 2017). Moreover, this method is performed without any additional capital costs for reconstruction and modernization. In general, the energy component of gas pipelines operating costs (which also include salaries, maintenance, overhaul, spare parts and materials) is close to 20 % (Machula, 2013).

It should be noted that any energy-saving technologies would be introduced into operation only if there are economic benefits from implementing these measures. But the price of gas in the Russian Federation is artificially low: the gas is almost 2.5 times cheaper than oil and 1.5 times cheaper than coal (in tons of oil equivalent) (Pystina & Yacenko 2011). Such a ratio of energy prices entails an inefficient use of gas and does not encourage both individual gas consumers and the whole PJSC "Gazprom" to save gas and to introduce modern technologies for mitigation of gas overuse.

During operation, the condition of gas compressors inevitably decreases, which leads to economic losses. The main reasons for the decrease in gas compressor condition are pollution and an increase in clearance of the equipment flow section, blade defects, and heated air at the entrance to an axial compressor (Kalinin 2011). All this leads to deterioration in the output of energy-technological indicators of the gas-pumping unit: effective efficiency, available power, fuel gas consumption. To take into account the current condition of the equipment, rates of condition are introduced. These rates are a power condition rate of a turbine (K_{Ne}), a fuel gas condition rate of a turbine (K_{fg}), and a condition rate of a compressor (K_H).

This work is devoted to optimization of gas pipelines operation modes bearing in mind the current condition of gas compressor units.

2 METHODOLOGY

We observe here two compressor stations adjoining a pipeline section. In each station there are a compressor system and a cooler system. (After pressurising the gas, the temperature is quite high so it is arbitrary to reduce the gas temperature to prevent deterioration of pipe coating.)

The total cost of gas transportation is equal to the sum of costs of a compressor system and a cooler system. The total costs C_Σ are an optimization criterion and it must tend to minimum. Figure 1 depicts the change of the optimization criterion with the operation modes of air coolers. By switching the air coolers on and off one by one, we maintain the optimum discharge temperature T_1 to ensure the minimal total costs. Iterative calculations are provided until total costs are the lowest or boundary condition of gas temperature regime is reached.

The boundary conditions of gas temperature regime of gas transportation consist of two points:

- gas cooling should be carried out to a value not lower than the ground temperature, otherwise, overcooled gas can provoke changes in the aggregation state of water contained in the soil which surrounds buried gas pipelines;
- the maximum gas discharge temperature should not exceed the regulated maximum operating temperature: from 313 to 353 K (National Standard GOST R 51164-98 1999).

As far as condition rates are concerned, they all are estimated as a ratio of a current parameter to a nominal value. To put this in greater detail, first, the condition rate of a turbine equals actual power of a turbine divided by a nominal power; second, the fuel gas condition rate of a turbine is a ratio between actual fuel gas consumption and a nominal one; third, a condition rate of a compressor is equal to the actual polytropic efficiency divided by a nominal value (Kalinin 2011).

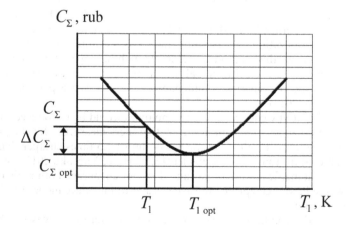

Figure 1.   Change of optimization criterion with operation modes of air coolers.

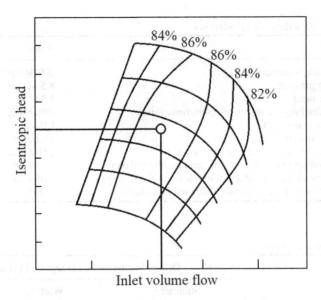

Figure 2. Layout of a compressor map.

3 PROGRAM

To conduct the research a program has been built. The program is based on a technological algorithm of estimating gas transportation costs at a pipeline section between two compressor stations. It also calculates the savings from operation mode optimization. The program has built-in protection against the introduction of incorrect raw data.

The dynamic characteristics of centrifugal compressors are depicted on compressor maps. The layout of compressor maps is shown in Figure 2. The operating point is located at the intersection between the inlet volume flow and the isentropic head. From there, values of specific rotor speed, efficiency, and available power (the distribution of the values of available power is shown at the top of Figure 1) are retrieved. With each new iteration (switching air coolers on and off), a new pair of the inlet volume flow and the isentropic head is formed due to changing of the discharge temperature which leads to the revaluation of the inlet volume flow and identifying the new operating point.

The program is able to compute at the design stage and the operational stage of a life cycle of gas transmission pipelines. It is able to estimate a number of pipelines, choose types and a number of gas compressor equipment, calculate and optimize a number of air coolers. The program is flexible and it is possible to extend a variety of gas compressors and cooling facilities by adding performance data of state-of-the-art equipment.

4 CALCULATIONS

In a such technological algorithm, where gas transportation is observed, it is crucial to determine different operational regimes during different seasons. Two seasons are estimated because of the summer valley: it is an ordinary situation when performance of gas system in winter is 10 % or much higher than in summer. The main raw data shown in Table 1 include average parameters of gas pipelines constructed in the 1970s-80s and operated until now which we are to optimize.

Table 2 demonstrates detailed results of calculations. This table is shown at the final step in the program and it contains all the necessary information of gas temperature and pressure regime, types of chosen equipment, load of gas compressor units and cooling system, and fuel gas and electricity (in a cooling system) savings as a result of the optimization. Since we

Table 1.   Raw data for calculations.

Parameter	Value
Performance (summer/winter)	$300/340 \cdot 10^6$ m^3/day
Operating pressure	8.5 MPa
Isentropic head	1.44
Max gas discharge temperature (summer/winter)	298/294 K
Pipeline section length	125 km
Pipeline diameter	1220 mm
Average soil temperature (summer/winter)	279/275 K
Power condition rate of a turbine (nominal/boundary)	0.95/0.80
Fuel gas condition rate of a turbine (nominal/boundary)	1.05/1.20
Condition rate of a compressor (nominal/boundary)	1.00/0.85

Table 2.   Results of calculations.

Parameter	Value	Optimal value	Value	Optimal value	Dimension
	Summer		Winter		
Number of pipeline strings	5				pcs
Performance per string		60.0		68.0	10^6 m^3/day
Discharge pressure	8.50				MPa
Suction pressure	6.84	6.83	6.34	6.33	MPa
Discharge temperature	298.0	299.7	294.0	294.8	K
Suction temperature	286.2	287.3	282.0	282.6	K
Number of shops	5				pcs
Number of units per shop	2				pcs
Type of gas tubine	GPA-C-16				
Type of gas compressor	NC-16/100-1.44				
Rotor frequency	4080	4080	4820	4820	rpm
Polytrophic efficiency	0.820	0.820	0.815	0.815	
Discharge temperature before cooling	304.7	306.1	307.1	307.8	K
Available capacity per shop	28.98	28.98	37.79	37.79	MW
Effective power per shop	16.07	16.03	24.68	24.64	MW
Fuel gas consumption per shop	8368	8357	10898	10886	m^3/hour
Costs of gas compressing per shop	843.5	842.4	1098.5	1097.3	10^3 rub/day
Number of air coolers per shop	9				pcs
Type of air coolers	Nuovo Pignone				
Two working fans per air cooler	0	0	0	0	pcs
One working fan per air cooler	9	7	7	6	pcs
Natural convection per air cooler	0	2	2	3	pcs
Costs of cooling system per shop	25.9	20.2	20.2	17.3	10^3 rub/day
Total costs *	4347.5	4313.0	5593.5	5573.0	10^3 rub/day
Energy savings due to optimization		34.5		20.5	10^3 rub/day

* For all shops in a compressor station.

consider two seasons, four values of regime's parameters are shown: before and after optimization for each season.

Owing to a slight increase in the gas discharge temperature (1.7 K and 0.8 K for summer and winter respectively), we could save a small amount of fuel gas on compressors as well as electricity because of switching off fans which lead to considerable energy savings (in the last line of Table 2). As there is a strong correlation between the performance of a gas system and energy savings due to optimization, we can estimate the minimal savings per year utilising raw data during summer valley. Consequently, these savings are more than 7 million roubles for one particular compressor station.

To analyse the effect of reducing condition of the equipment on gas transportation costs (costs of a compressor station), the costs were calculated for various combinations of condition rates of turbines and compressors (within their boundary values).

The study of the effect of each condition rate on gas transportation costs was conducted. Three combinations of the condition rates were considered, under which one was changed, and the other two were constant. Furthermore, the increase of gas transportation costs was calculated between these triplets. Table 3 depicts the relative increase of gas transportation costs due to deterioration of condition of gas compressor units. The analysis shows that the effect of condition rate changing on gas transportation costs is diverse, although the same changing step of condition rate values (0.05) was utilised.

According to the results, the fuel gas condition rate of a turbine has the greatest effect on the increase in gas transportation costs. The condition rate of a compressor has a lesser impact on the costs. The power condition rate of a turbine has the least impact on the costs, which is 3.3 times lower than the influence of the fuel gas condition rate of a turbine. But at the same time, the power condition rate of a turbine narrows the regulation area of compressor stations (Porshakov et al. 2014).

We also compared pairs of values of the gas transportation costs. The first costs were estimated with nominal value of condition rates. The second ones were calculated with the worst (boundary) value of one rate and the two remaining rates are nominal. Figure 3 illustrates the results of the calculations in winter and summer seasons.

Table 3. Increase of gas transportation costs due to changing condition rates of turbines and compressors.

Condition rates (CR)	Changing step	Relative increase of gas transportation costs, %	
		Summer	Winter
Fuel gas CR of turbine	0.05	4.2–4.4	4.3–4.5
CR of compressor	0.05	3.2–3.6	2.7–3.4
Power CR of turbine	0.05	1.2–1.3	1.3

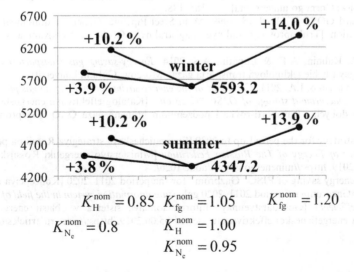

Figure 3. Increase of gas transportation costs due to reducing condition of gas compressor equipment.

95

As a result of these calculations, it was revealed that the gas transportation costs could dramatically increase up to 14 % due to reducing condition of a gas turbine and a gas compressor. This situation can occur until maintenance and overhaul are underway. That is why the proper timing of maintenance and overhaul is pivotal in gas transportation.

5 CONCLUSION

Optimization of gas pipelines operation modes is an actual approach in the field of energy savings. In this research it was revealed that these measures could save a considerable amount of money. But the methods would be implemented if only the gas price was set at a reasonable level. Furthermore, the cheapness of gas is one of the reasons for its ineffectual utilisation and it provokes a growth in gas demand and excess consumption.

Since we observed here one pipeline sector between two compressor stations, this work needs further study in the modelling of extended pipelines. Nowadays digitalization of the gas industry is a significant direction and the created program is a good tool in providing calculations during the design stage and the operational stage of a life cycle of gas pipelines and considering the condition of gas compressor equipment. Moreover, the program could be developed by adding performance data of state-of-the-art equipment. Regarding condition rates, the fuel gas condition rate of a gas turbine affects more than the others.

REFERENCES

Aksyutin, O.E. 2017. Realization of the energy saving potential in the gas transportation in PJSC "Gazprom" [Realizaciya potenciala energosberezheniya v magistral'nom transporte gaza PAO «Gazprom»]. *Gas Industry [Gazovaya promyshlennost]* 750(1): 52–58.
Kalinin, A.F. 2011. Calculation, regulation, and optimization of operating modes of gas compressor units [Raschet, regulirovanie i optimizaciya rezhimov raboty gazoperekachivayushchih agregatov]. Moscow: MPA-Press.
Khvorov, G.A. 2017. Assessment of energy saving potential in the main gas transportation: problems, implementation, prospects [Ocenka potenciala energosberezheniya v magistral'nom transporte gaza: problemy, realizaciya, perspektivy]. *Gas Industry [Gazovaya promyshlennost]* 750(1): 76–85.
Kryukov, O.V. 2017. Analytical models of gas transportation [Analiticheskie modeli transporta gaza]. *Theses of NGTU R.E. Alekseev* [Trudy NGTU im. R.E. Alekseeva] 1: 161–173.
Machula I.A. 2013. Assessment of the energy efficiency of the gas main in Komi Republic [Ocenka povysheniya energoeffektivnosti magistral'nogo transporta gaza v respublikie Komi]. *Journal of Mining Institute* [Zapiski Gornogo universiteta] 206: 185–188.
National Standard GOST R 51164–98. 1999. Main Steel Pipeline. General requirements for the corrosion protection [Truboprovody stal'nye magistral'nye. Obshchie trebovaniya k zashchite ot korrozii].
Porshakov, B.P., Kalinin, A.F. & Lopatin, A.S. 2014. *Energy-saving gas transportation technologies* [Energosberegayushchie tekhnologii transporta gaza]. Moscow: Gubkin University.
Pystina, N.B. & Yacenko, I.A. 2011. *Catalogue of the energy-saving technologies in gas production, transportation and underground storage of OJSC "Gazprom"* [Katalog effektivnyh energosberegayushchih tekhnologij v dobyche, transporti-rovke i podzemnom hranenii gaza OAO «Gazprom»]. Moscow: Gazprom.
Russian energy strategy for the period up to 2030 [Energeticheskaya strategiya Rossii na period do 2030 goda]. *Ministry of Energy of The Russian Federation* [Ministerstvo energetiki Rossijskoj Federacii], viewed 05.06.2019. https://minenergo.gov.ru/node/1026/.
The concept of energy saving of PJSC "Gazprom" for the period 2011–2020 [Koncepciya energosberezheniya PAO «Gazprom» na period 2011–2020 gg]. *State information system in the field of energy saving and energy efficiency* [Gosudarstvennaya informacionnaya sistema v oblasti energosberezheniya i povysheniya energeticheskoj effektivnosti] viewed 05.06.2019. https://gisee.ru/articles/organizations/19316/.

Scientific and Practical Studies of Raw Material Issues – Litvinenko (Ed)
© 2020 Taylor & Francis Group, London, ISBN 978-0-367-86153-7

Studies of mixing high viscosity petroleum and pyrolysis resin to improve quality indicators

R. Sultanbekov & M. Nazarova
Saint-Petersburg Mining University, Saint-Petersburg, Russian Federation

ABSTRACT: This paper analyzes the problem of recycling tires and rubber products. Environmentally friendly recycling is pyrolysis. The resulting pyrolysis fuel can be used for furnace boilers. However, the quality of this fuel is low due to the high sulfur content and ash content. The solution to this problem can be a mixture of pyrolysis fuel with fuel oil. This will allow to achieve energy saving, environmental friendliness and a positive economic effect. The mixture of fuels allows to improve the quality indicators, namely to reduce the viscosity and sulfur content. Laboratory studies on the mixing of pyrolysis resin with a highly viscous petroleum product were carried out, various ratios were analyzed and the quality indicators of the mixture were determined. Completed the rationale for the proposed solution.

1 INTRODUCTION

At present time the waste disposal and recycling is the one of the actual problem. Millions industrial wastetons, which are annually taken to landfills, are surrounded on all sides by large industrial cities. The mass of secondary resources has been decomposed over years in landfills, damaging the environment, the processing of which would significantly reduce the cost of producing various products.

The number of cars is growing steadily, and with them the amount of waste rubber, which makes world leaders seriously think about the problem of recycling and recycling of this type of waste. In Europe alone, over 2 million tons of used tires are dumped annually (SHuldyakova 2016).

One of the known solutions is tire recycling through pyrolysis. The resulting product can be used for mixing with highly viscous petroleum products, this will reduce viscosity and expand the possibility of using, for example, for ship installations or for boiler rooms.

Recently, the world has established a tendency for deeper processing of oil, namely the use of residual products of the cracking processes. However, along with the deepening of refining, the quality of petroleum products is reduced due to the increase in the proportion of asphalt-resinous products in the composition of heavy fuels. Therefore, the stability and compatibility of petroleum products is an important parameter. Since when mixing different types of dark oil there is a risk of precipitation (Mitusova & Averina 2005).

A high total sediment in petroleum products adversely affects the operation of engines and fuel systems, contributes to their wear and disruption, and also leads to clogging of filters and separators (Karimov & Mastobaev 2012). Also, precipitation of the total sediment contributes to the active accumulation of the "dead" residue in the tanks, which can lead to a deterioration in the quality of the oil being drained and reduce the useful volume of the tanks. The permissible total sediment content is governed by standards, in Russia it is (GOST R 33360-2015), which complies with the international standards ASTM D4870-IP 375, IP 390 and ISO 10307. According to ISO 8217, the total sediment content in fuels should not exceed 0.1%.

There are several actual problems that are caused by the incompatibility of petroleum products during technological operations on tanks:

- Deterioration in the quality of petroleum products;
- Promotes wear and clogging of process systems;
- An increase in the intensity of accumulation of "dead" residue on the tanks, which leads to a decrease in the useful volume of the tanks;
- Economic losses due to lower prices for petroleum products.

2 THEORY AND METHODS

Mixing fuels is a common way to improve the quality of petroleum products (Mitusova et al. 2018; Kondrasheva et al. 2018). New requirements for sulfur content in ship's highly viscous fuels according to ISO 8217 will be 0.5% from 2020. Therefore, the relevance of developing ways to reduce the sulfur content in fuels is increasing. However, when mixed, there is a risk of static electricity and precipitation due to fuel incompatibility (Sultanbekov & Nazarova 2019). Research is underway to determine and improve the stability of marine fuels (Kondrasheva et al. 2019; Sultanbekov & Nazarova 2019).

Manifestations of "incompatibility" when mixing petroleum products are associated with the emergence of strong intermolecular interactions caused by changes in the structural group composition and the relative ratio of concentrations of high molecular compounds of petroleum products, which leads to the formation of associates of molecules, bulk colloidal particles of various shapes and structures.

There are several methods for assessing the stability and compatibility of the fuel. The simplest is the method of determining the stability and compatibility of the spot according to GOST R 50837.7. The essence of the method is a visual assessment of the core and stain stains on the paper filter, formed by a drop of the sample under test conditions, and comparing the spot with standard stains. However, this method does not have high accuracy and has a large error.

There are known methods for determining toluene and xylene equivalent according to GOST R 50837.3 and GOST R 50837.4, respectively, which make it possible to judge the stability of petroleum products.

Toluene and xylene equivalents are criteria for the stability of a dispersed structure, showing the degree of aromaticity of the fuel, which is necessary to preserve asphaltenes in a dispersed state.

The value of the xylene equivalent, less than or equal to 25/30, is a criterion for straightness (i.e., the absence of secondary residual products of oil refining in the studied mixed fuel).

A common drawback of both methods is low accuracy due to the subjectivity of visual assessment of the presence of a darker spot inside the entire (also darkened) spot from a drop of a dilute portion of fuel with a mixture of toluene or xylene with n-heptane.

Existing methods for assessing the stability and compatibility of petroleum products do not have high accuracy and have a large error.

To increase the stability of petroleum products, it is possible to use dispersing additives or, thanks to an accurate method for determining compatibility before mixing fuels, it is possible to select the necessary components of petroleum products (Mitusova et al. 2017).

To determine the compatibility of petroleum products, an algorithm has been developed that allows obtaining results accurately and quickly.

The principal difference of the method used is that to determine the compatibility of several types of petroleum products, it is first necessary to perform tests to determine the total sediment with preliminary chemical aging (Total Sediments Accelerated - TSA) according to the method according to GOST R 33360-2015 for each component of the mixture, which are planned for mixing. Further, after determining the total sediment for each of the components, these petroleum products should be thoroughly mixed in the required proportion and then per-form tests to determine the total sediment of the prepared mixture of petroleum products.

The experiments were carried out in a laboratory installation of Total Sediment Tester of Seta Clean company, to which a vacuum pump is connected, an oil bath was also used for aging the sample and scales with an accuracy of 0.01 and 0.0001 g. All laboratory glassware and reagents meet the necessary requirements.

3 EXPERIMENTAL RESULTS

Studies of the effect of the mixing ratio of petroleum product and pyrolysis resin on compatibility and key quality indicators, such as density, viscosity and sulfur, have been carried out. Component №1. Fuel oil 100, component number 2. Heavy pyrolysis resin, grade A.

In the experiments (Table 1), compatibility analyzes were carried out with different ratios of the mixture of petroleum product and pyrolysis resin (Figure 1). As can be seen from the results, total sediments (TSA) values correspond to the required rate with different mixing ratios correspond to the required rate and are fully compatible. Maximum precipitation occurred at a mixing ratio of 50% to 50%.

Table 1. Determination of total sediment at different ratios of the mixture of components.

Component №1. Fuel oil 100, low-ash (STO 05747181-2013).
Quality indicators:
$\rho_{15} = 977,0\,kg/m^3$; $\upsilon_{50} = 689,5\,\frac{mm^2}{s}$; $water = 0,1\%$; $TSA = 0,03\%$; sulfur $= 2,32\%$
Component №2. Heavy pyrolysis resin, grade A (TU2451-183-72042240-2013).
Quality indicators:
$\rho_{15} = 1025,0\,kg/m^3$; $\upsilon_{50} = 17,9\,\frac{mm^2}{s}$ c; $water = 0,05\%$; $TSA = 0,04\%$, sulfur $= 0,21\%$.

№	Component mixing ratio, %	TSA, %
1	№1 – 90, №2 – 10	0,03
2	№1 – 80, №2 – 20	0,04
3	№1 – 70, №2 – 30	0,04
4	№1 – 60, №2 – 40	0,06
5	№1 – 50, №2 – 50	0,07
6	№1 – 40, №2 – 60	0,05
7	№1 – 30, №2 – 70	0,04
8	№1 – 20, №2 – 80	0,04
9	№1 – 10, №2 – 90	0,04

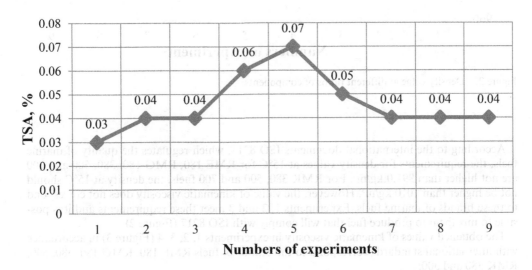

Figure 1. Analysis of petroleum products for compatibility with different ratios of components.

Table 2. Determination of density, kinematic viscosity and sulfur at different ratios of the mixture of components.

Component №1. Fuel oil 100, low - ash (STO 05747181-2013).
Quality indicators:
$\rho_{15} = 977,0\,kg/m^3$; $\upsilon_{50} = 689,5\,\frac{mm^2}{s}$; $water = 0,1\%$; $TSA = 0,03\%$, sulfur $= 2,32\%$

Component №2. Heavy pyrolysis resin, grade A (TU2451-183-72042240-2013).
Quality indicators:
$\rho_{15} = 1025,0\,kg/m^3$; $\upsilon_{50} = 17,9\,\frac{mm^2}{s}\,c$; $water = 0,05\%$; $TSA = 0,04\%$, sulfur $= 0,21\%$.

№	Component mixing ratio, %	Density (15°C), kg/m³	Kinematic viscosity (50°C), mm²/s	Sulfur %
1	№1 – 90, №2 – 10	981,5	434,7	2,11
2	№1 – 80, №2 – 20	986,0	275,1	1,92
3	№1 – 70, №2 – 30	990,6	169,8	1,68
4	№1 – 60, №2 – 40	995,5	121,7	1,48
5	№1 – 50, №2 – 50	1000,2	84,55	2,11
6	№1 – 40, №2 – 60	1005,0	58,37	2,11
7	№1 – 30, №2 – 70	1010,3	42,01	2,11
8	№1 – 20, №2 – 80	1014,9	31,49	2,11
9	№1 – 10, №2 – 90	1020,7	23,77	2,11

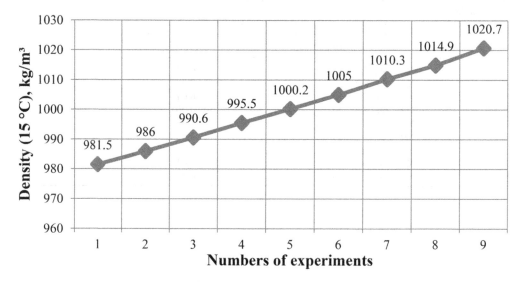

Figure 2. Density value at different ratios of components.

According to the international documents ISO 8217, which regulates the quality of marine fuels, the requirements for density values at 15°C for RME 180, RMG 180, 380, 500 and 700 are not higher than 991.0 kg/m³. For RMK 380, 500 and 700 fuels, the density at 15°C should not be higher than 1010 kg/m³. However, the value of kinematic viscosity does not correspond to these brands of marine fuels. Experiments 1, 2 and 3 meet these requirements and it is possible to mix them to produce fuel that will comply with ISO 8217 (Figure 2).

The obtained values of kinematic viscosity in experiments 1, 2, 3, 4 (Figure 3) in accordance with international standards correspond to the brands of fuels RME 180, RMG 180, 380, 500, RMK 380 and 500.

100

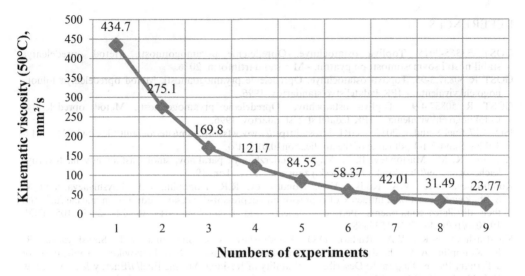

Figure 3.   The value of kinematic viscosity at different ratios of components.

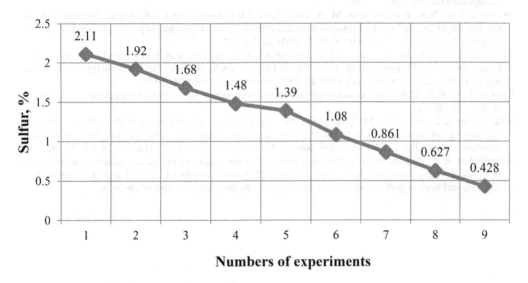

Figure 4.   The value of sulfur at different ratios of components.

4   CONCLUSION

The resulting product when disposing of old tires and rubber products with the help of pyrolysis, it is possible to mix with highly viscous petroleum products. This allows you to expand the use of pyrolysis resin and improve energy efficiency. According to the results of laboratory experiments, mixing up to 30% (number of experiments 1, 2, 3) pyrolysis resin produced fuels, where the density, viscosity and total sediment fully meet the requirements of the standards, and a decrease in sulfur in the mixture (Figure 4), which is also a positive result.

REFERENCES

GOST 33365-2015. Topliva ostatochnye. Opredelenie pryamogonnosti. Metod opredeleniya stabil'nosti i sovmestimosti po pyatnu. - M.: Standartinform, 2016.

GOST R 50837.3-95. Topliva ostatochnye. Opredelenie pryamogonnosti. Metod opredeleniya toluol'-nogo ehkvivalenta. – IPK Izdatel'stvo standartov, 1996.

GOST R 50837.4-95. Topliva ostatochnye. Opredelenie pryamogonnosti. Metod opredele-niya ksilol'nogo ehkvivalenta. – IPK Izdatel'stvo standartov, 1996.

ISO 8217 Fuel Standar. 2017. - URL https://https://www.wfscorp.com/sites/default/files/ISO-8217-2017-Tables-1-and-2-1-1.pdf (date of the application10.01.2019).

Karimov, R. M., Mastobaev, B. N. Vliyanie soderzhaniya parafinov, smol i asfal'tenov na tovarnye kachestva neftej//Bashkirskij himicheskij zhurnal. - 2012.-Tom 19.

Kondrasheva N.K., V.A. Rudko, D.O. Kondrashev, R.R. Konoplin, K.I. Smyshlyaeva, and V. S. Shakleina. Functional influence of depressor and depressor-dispersant additives on marine fuels and their distillates components. Petroleum Science and Technology 2018 36 (24), 2099-2105. DOI: 10.1080/10916466.2018.1533858.

Kondrasheva N.K., V.A. Rudko, D.O. Kondrashev, V.S. Shakleina, K.I. Smyshlyaeva, R. R. Konoplin, A.A. Shaidulina, A.S. Ivkin, I.O. Derkunskii, O.A. Dubovikov. Application of a Ternary Phase Diagram to Describe the Stability of Residual Marine Fuel/ //Energy & Fuels. 2019. Vol. 33, № 5. P. 4671–4675.

Mitusova T.N., Averina N.P., Pugach, I.A.. Ocenka stabil'nosti i sovmestimosti ostatochnyh topliv. Mir nefteproduktov №1, 2005, 33-35.

Mitusova T.N., N.K. Kondrasheva, M.M. Lobashova, M.A. Ershov, and V.A. Rudko. Influence of dispersing additives and blend composition on stability of marine high-viscosity fuels. Journal of Mining Institute 2017 228, 722-725. DOI: 10.25515/PMI.2017.6.722

Mitusova N.N., N.K. Kondrasheva, M.M. Lobashova, M.A. Ershov, V.A. Rudko, and M.A. Titarenko. Determination and Improvement of Stability of High-Viscosity Marine Fuels. Chemistry and Technology of Fuels and Oils 2018 53 (6), 842-845. DOI: 10.1007/s10553-018-0870-6.

SHuldyakova K.A. Utilizaciya iznoshennyh avtomobil'nyh shin v Rossii//Molodoj uchenyj. — 2016. — №26. — S. 739-742.

Sultanbekov R.R., Nazarova M.N. Research effect of humidity in vapor space of the vertical steel tank for storing oil and oil products on the generation of static electricity. Gornyy informatsionno-analiticheskiy byulleten'. (2019);4/7:498-506. [In Russ] DOI: 10.25018/0236-1493-2019-4-7-498-506.

Sultanbekov R.R., Nazarova M.N. Determination of compatibility of petroleum products when mixed in tanks. EAGE., Tyumen, (2019), DOI: 10.3997/2214-4609.201900614. Available at: http://earthdoc.eage.org/publication/publicationdetails/?publication=96369 (Accessed 30 March 2019).

Scientific and Practical Studies of Raw Material Issues – Litvinenko (Ed)
© 2020 Taylor & Francis Group, London, ISBN 978-0-367-86153-7

The use of secondary polymers to ensure environmental safety storage of mineral waste ore-dressing and processing enterprise in the Southern Urals

D. Babenko

Saint-Petersburg Mining University, Saint-Petersburg, Russia

ABSTRACT: The article contains results of the research in the field of reducing the negative impact of technogenic massifs' hazardous wastes on the quality of environmental components. The waterproofing of the technogenic massifs foundation with the help of the material, based on the mixture of secondary high-density and low-density polyethylenes is offered as a technology of reducing the negative impact. Research methodology includes experimental methods in field and laboratory conditions. Samples of surface water and ground water in the area of the study object were selected during the research, chemical analysis was performed, quality of natural waters was estimated, the physical and mechanical properties of polymers, as well as resistance to the aggressive environments impact were investigated.

1 INTRODUCTION

The studied enterprise performs underground mining of copper ore and its flotation concentration. Flotation is based on the differences in physicochemical properties of mineral surfaces (Abramov 1984, Krasotkina 2017). The studied tailings of the Gaisky Mining and Processing Plant (Gaisky GOK) include a tailings pond, a pond of clarified water and a pond of acid mine waters. Depleted pits of the Gai deposit are considered too, as one of the quarries is being reclaimed by filling with beneficiation waste (Aleksandrova 2019). Enrichment tailings are mixed with acid mine waters in the main building of the factory and then pumped to the abandoned quarry and partly to the tailings storage. The tailings storage of the Gaisky GOK is a hydraulic-fill hill-type dam, put into operation in 1966 (Recommendations... 1986). Its total area is approximately 190 hectares; the maximum storage capability is 52.5 million m^3.

In the article (Babenko 2018) the results of the investigation on the definition of the state of surface water and groundwater in the vicinity of tailing ore-dressing and processing enterprise in the southern Urals are presented. On Figure 1 it is shown the results of determination of natural waters chemical composition.

As a result of ore beneficiation by flotation process waste products are produced for the factory, which are called enrichment tailings. The tailings are mixed with acidic mine water in the main building of the factory, then via the slurry pumping station are passed in tailings pond of hydraulic type and are used for mine reclamation of abandoned open-cut minings (coffins) and stowage of underground mine.

The mineral composition of tailings is presented by pyrite, chalcopyrite, sphalerite, quartz, feldspar, chlorite and sericite. Chemical composition is FeS_2, $CuFeS_2$, ZnS, SiO_2, mixtures of aluminosilicates, chlorous acid salts $KAl (AlSi_3O_8)(OH)_2$.

Currently the project is developed for mine reclamation of abandoned open-cut minings (coffins) with the use of wastes which are formed during the process of ore beneficiation.

In order to increase the degree of protection from filtering pollutants of industrial wastes during the storage operation of tailings into abandoned open-cut minings (coffins), it is possible to use a mixture of secondary polyethylenes qua impervious coating. This coating

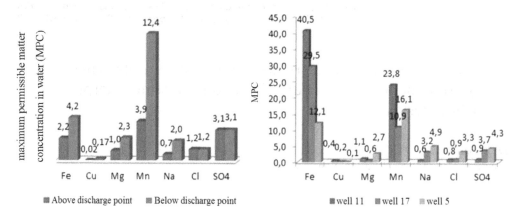

Figure 1. The analysis of surface waters (left) and underground waters (right).

should have high degree durability and be resistant to aggressive environment. In this connection it was necessary to determine the physicomechanical properties of polymers, definition of impervious material resistance to the effects of aggressive environment (Bolshunova 2017, Gałuszka 2018, Yusupov 2016).

2 METHODS

Secondary high-density and low-density polyethylenes and also polypropylene were selected as object of our investigation.

Low-density polyethylene (LDPE) is a waxy material of unexpressed colour, obtained industrially by the polymerisation of gaseous polyethylene. LDPE is a thermoplastic polymer with a density of 910–930 kg/m^3. The literature review shows that this polymer has relatively high reliability at the break, resistance to multiple bending, impact and low temperatures.

High-density polyethylene (HDPE) is a less waxy polymer than LDPE, resistant to fats and oils, but subjected to impact and bending. Values of resistance to compression and stretching of LDPE and HDPE are comparable. The density of HDPE is 940–960 kg/m^3.

Polypropylene (PP) is a linear polymer obtained during the propylene polymerisation in the presence of catalysts. The density of polypropylene is 900–920 kg/m^3. It is highly resistant to the impact but can be destroyed by multiple bending and under low temperatures.

The samples were made in press molds which were formed in training repair shop of St. Petersburg Mining university in the form of sheets of 60×130 mm by heating in a muffle furnace and subsequent cooling with speeds reflecting real field conditions. According to reference data melting points of polymers vary and are in the range 105-175°C (Adrianova 2008). Practically from highly-elastic state to viscous-flow state the transition temperature of studied polymers for high-density polyethylene is 120°C, for low-density polyethylene is 135°C, for polypropylene is 170°C (Lyamkin 2000).

Definition of indicators characterizing the durability of polymeric materials was carried out on a universal testing machine with servo-electromechanical drive H75K-S for materials static testing in tension, compression and bending.

Samples testing of durability to tearing was performed in accordance with GOST 11262-80. Tear speed was 200 mm per minute by GOST.

The general trend for all investigated samples is going up of tensile strength with increasing temperature of polymers processing with peaks at temperatures in the range 185-195°C and subsequent sharp decline of strength characteristics. Figure 2 reflect the dependence of changing polymer strength on the temperature.

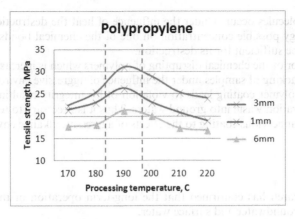

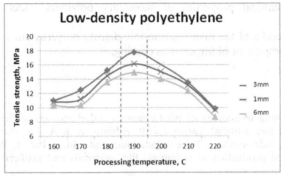

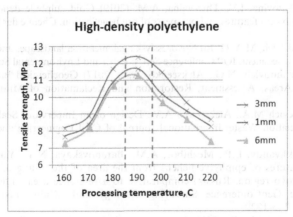

Figure 2. The dependence of strength characteristics changes of polymer samples by temperature.

The strength of samples processed at different temperatures does not change linearly. The value of the tensile strength of polymers at temperatures commensurate with the melting point is not the greatest. This is due to incomplete homogenisation of the material, the presence of excessive moisture and volatile impurities. The temperature range of 185-195°C for polymeric material processing is the optimal. The decrease in the strength of the test samples at temperatures above 200°C is determined in connection with thermal and thermo-oxidative destructive processes.

The sharp decline in tensile value is due to the thermal destruction of the polymer material. Under the action of high temperatures in the polymer at the supramolecular level tearing of macromolecules or separation of the side groups from them often accompanied by cross-

105

linking of macromolecules occur. Under the influence of heat the destruction is caused by an increase in the energy possible concentration on one of the chemical bonds in the macromolecule of which will be sufficient for its destruction.

The main indicator of the chemical disrupting of polymers which is increasing the mechanical destruction is the cracking of samples under the influence of aggressive environment. The cracking of the formed polymer coating leads to violation of his integrity resulting in a going up in the amount of infiltration waste into ground waters. Thus it is necessary to research the sustainability of proposed impervious coating to the effects of aggressive alkaline environment.

3 CONCLUSION

The conducted research has confirmed that the long-term operation of the tailings leads to contamination of groundwater and surface water.

Physical and mechanical properties of secondary polymers under tensile loads were determined.

According to the results of the investigation low-density polyethylene is the most appropriate for use as main component of impervious coating.

REFERENCES

Adrianova G.P. Processing technology of plastic masses and elastomers in the production of polymeric film materials and artificial leather / G.P. Adrianova, K.A. Polyakova, Yu.S. Matveyev; under total ed. G.P. Adrianova. − 3rd ed. reclaiming and add. − Part 1. Physical and chemical bases of creation and production of polymeric film materials and artificial leather. − Moscow: Kolos, 2008.

Aleksandrova, T.N., Talovina, I.V., Duryagina, A.M. (2019) Gold–sulphide deposits of the Russian Arctic zone: Mineralogical features and prospects of ore benefication. Chemie der Erde DOI:10.1016/j.chemer.2019.04.006

Alekseenko, A.V., Pashkevich, M.A. (2016) Novorossiysk agglomeration landscapes and cement production: Geochemical impact assessment. IOP Conference Series: Earth and Environmental Science 43(1),012050.

Alekseenko, V.A., Maximovich, N.G., Alekseenko, A.V. (2017) Geochemical Barriers for Soil Protection in Mining Areas. Assessment, Restoration and Reclamation of Mining Influenced Soils pp. 255-274.

Babenko D.A., Pashkevich M.A., Alekseenko A.V., Design of an Impervious Seal for the Tailings of Copper Ore Beneficiation Waste, International Journal of Civil Engineering and Technology 10(5), 2019, pp. 797-805.

Bolshunova, T.S., Rikhvanov, L.P., Mezhibor, A.M., Baranovskaya, N.V., Yusupov, D.V. (2017) Biogeochemical features of epiphytyc lichens from the area of the tailing of a gold- polymetallic deposit (Kemerovo region, Russia) comparative to a reference area. International Multidisciplinary Scientific GeoConference Surveying Geology and Mining Ecology Management, SGEM 17(51), pp. 165-172.

Bonotto, D.M., Garcia-Tenorio, R. (2019) Investigating the migration of pollutants at Barreiro area, Minas Gerais State, Brazil, by the 210 Pb chronological method. Journal of Geochemical Exploration 196, pp. 219-234.

Bushuev, Y.Y., Leontev, V.I., Machevariani, M.M. (2018) Geochemical features of Au- Te epithermal ores of the Samolazovskoye deposit (Central Aldan ore District, Yakutia). Key Engineering Materials 769 KEM, pp. 207-212.

Coulombe, V., Bussière, B., Côté, J., Garneau, P. (2012) Performance of insulation covers to control acid mine drainage in cold environment Proceedings of the International Conference on Cold Regions. Engineering pp. 789-799.

Dos Santos, L.S., Gardoni, M.D.G.A. (2014) Study of the durability of geomembranes for waterproofing of reservoirs of gold tailings dams in Brazil. 10th International Conference on Geosynthetics, ICG.

Kasimov, N.S., Kosheleva, N.E., Timofeev, I.V. (2016) Ecological and Geochemical Assessment of Woody Vegetation in Tungsten-Molybdenum Mining Area (Buryat Republic, Russia). IOP Conference Series: Earth and Environmental Science 41(1),012026.

Kosheleva N.E., Timofeev I.V., Kasimov N.S., Kisselyova T.M., Alekseenko A.V., Sorokina O.I. (2016) Trace element composition of poplar in Mongolian cities. Lecture Notes in Earth System Sciences pp. 165-178.

Krasotkina, A.O., Machevariani, M.M., Korolev, N.M., Makeyev, A.B., Skublov, S.G. (2017) Typomorphic features of niobium rutile from the polymineral occurrence Ichetju (the Middle Timan). Zapiski Rossiiskogo Mineralogicheskogo Obshchestva 146(2), pp. 88-100.

Krzaklewski, W., Pietrzykowski, M. (2002) Selected physico-chemical properties of zinc and lead ore tailings and their biological stabilisation. Water, Air, and Soil Pollution 141(1-4), pp. 125-142.

Lee, J.K., Shang, J.Q. (2013) Thermal properties of mine tailings and tire crumbs mixtures. Construction and Building Materials 48, pp. 636-646.

Lyamkin, D.I. (2000) Mechanical properties of polymers: Textbook. Moscow, 64 p.

Makarov, A.B. (2000) Technogenic deposits of mineral raw materials. Soros educational journal, 6(8).

Nagornov, D.O., Kremcheev, E.A., Kremcheeva, D.A. (2019) Research of the condition of regional parts of massif at longwall mining of prone to spontaneous ignition coal seams. International Journal of Civil Engineering and Technology 10(1), pp. 876-883.

Pochechun, V.A., Melchakov, Yu.L. Babenko, D.A. (2014) The use of a systematic approach in the study of natural and man-made geosystems. Bulletin of Tambov University. Series: Natural and technical Sciences. 19(5), pp. 1551-1554.

Semyachkov, A.I., Drebenstedt, C, Vorobiev, A.E. (2012) Geoecology. Textbook for higher mining and geological educational institutions. Yekaterinburg, 289 p.

SK 2.1.5.1315-03 Maximum allowable concentrations (MACs) of chemicals in waters of drinking and recreational water resources managements.

Sliti, N., Abdelkrim, C., Ayed, L. (2019) Assessment of tailings stability and soil contamination of Kef Ettout (NW Tunisia) abandoned mine. Arabian Journal of Geosciences 12(3),73.

Weishi, L., Guoyuan, L., Ya, X., Qifei, H. (2018) The properties and formation mechanisms of eco-friendly brick building materials fabricated from low-silicon iron ore tailings. Journal of Cleaner Production 204, pp. 685-692.

Yusupov, D.V., Bolshunova, T.S., Mezhibor, A.M., Rikhvanov, L.P., Baranovskaya, N.V. (2017) The use of Betula Pendula R. Leaves for the assessment of environmental pollution by metals around tailings from a gold deposit (Western Siberia, Russia). International Multidisciplinary Scientific GeoConference Surveying Geology and Mining Ecology Management, SGEM 17(41), pp. 665-672.

Yusupov, D.V., Karpenko, Yu.A. (2016) REE, Uranium (U) and Thorium (Th) contents in Betula pendula leaf growing around Komsomolsk gold concentration plant tailing (Kemerovo region, Western Siberia, Russia). IOP Conference Series: Earth and Environmental Science 43(1),012053.

Zhao, F.-Q., Li, H., Liu, S.-J., Chen, J.-B. (2011) Preparation and properties of an environment friendly polymer-modified waterproof mortar. Construction and Building Materials 25(5), pp. 2635-2638.

Zhu, P., Zheng, M., Zhao, S., Wu, J., Xu, H. (2015) Synthesis and thermal insulation performance of silica aerogel from recycled coal gangue by means of ambient pressure drying. Journal Wuhan University of Technology, Materials Science Edition 30(5), pp. 908-913.

Scientific and Practical Studies of Raw Material Issues – Litvinenko (Ed)
© 2020 Taylor & Francis Group, London, ISBN 978-0-367-86153-7

Studying the possibility of improving the properties of environmentally friendly diesel fuels

A. Eremeeva & N. Kondrasheva
Saint-Petersburg Mining University, Saint-Petersburg, Russian Federation

K. Nelkenbaum
Institute of Petrochemistry and Catalysis RAS, Ufa, Russian Federation

ABSTRACT: Samples of biodiesel fuel were obtained by the transesterification method, the main characteristics, especially the low-temperature properties of the synthesized biodiesel fuel, were investigated. Developed formulations of environmentally friendly diesel fuel based on synthesized biodiesel fuel and hydrotreated oil diesel. The influence of domestic and foreign depressant-dispersant additives on the low-temperature properties of pure biodiesel, mixed ecologically friendly fuel, as well as petroleum-treated diesel fuel is considered.

1 INTRODUCTION

The production of biofuels is a relatively new industry, which has great prospects. The production technology is not new, but due to certain circumstances it is not properly developed in all countries. Today, many governments understand that the issue of ecology cannot be ignored. Harmful emissions from cars, boiler stations, factories, and so on have a strong environmental impact.

In contrast to European countries, there is still no unified state program for the development of biodiesel fuel in Russia (Sidracheva 2009). For several years now, special regional programs have been created that support the production of biodiesel in several cities (Eremeeva 2014).

Owing to the following factors, biodiesel is advantageous.

Good lubricating properties of fuel. Petroleum diesel loses its lubricity when sulfur compounds are removed from it. Biodiesel, despite having a significantly lower sulfur content, is characterized by good lubricating properties, which prolongs the life of the engine. This is due to its chemical composition and oxygen content. Various tests have shown that when an engine is running on biodiesel, its moving parts are lubricated at the same time. As a result, the service life of the engine and the fuel pump is increased by 60% on average. It is important to note that there is no need to upgrade the engine (Kozin & Bashkireva 2009).

Higher cetane number. The cetane number for most diesel fuel derived from petroleum is 42-45. However, that of biodiesel is at least 51 (according to the European Standard EN 14214).

High flash point. Biodiesels have flash point exceeding 150 ° C, making bio-fuels relatively safe substances.

Due to the high cost of biodiesels, they are usually mixed with diesel fuels. Another reason for using a mixture of biodiesel and diesel fuel is the poor starting properties of an engine operating on biodiesel at low temperature (Vasiliev 2005).

One of the main characteristics of diesel fuel is its low-temperature properties. Despite this, refineries produce mainly summer diesel fuel amounting to 89% of total production, winter - 10% and arctic - 1% (Kozin & Bashkireva 2009, Iovleva et al. 2013).

Table 1. GOST 305-82 diesel fuel and specification.

Name of the indicator	Standard		
	Summer	Winter	Arctic
Pour point, °C, not higher for climate zone: Temperate	-10	-35	–
Cold	–	-45	-55

The most effective and cost-effective way to improve the low-temperature properties of the fuels is the use of pour point depressant. A significant temperature depression and improvement of fluidity is achieved at low temperatures when they are introduced in small doses, most often 0.05-0.10 % mass (Kemalov et al. 2010, Bormann & Stocker 1975).

Fuel pour point is not standardized in the European standard EN 14214, but according to GOST 305-82 "Diesel fuel and Specifications" 3 types of diesel fuel are distinguished: summer, winter and arctic (Table 1).

2 EXPERIMENTAL

2.1 Biodiesel fuels

For the production of biodiesel fuel, any vegetable oils, solid oils of animal origin, waste oil and fat production and slaughterhouses can be used. Any vegetable oil is a mixture of triglycerides, i.e., esters connected to the glycerol molecule (Viter & Zuboko 2008). Glycerin gives viscosity and density to vegetable oil. The problem in obtaining biodiesel is removing glycerin and replacing it with alcohol. For this purpose, the transesterification of vegetable oil with a monohydric alcohol was carried out. The exchange between alcohols and esters is called alcoholysis. Glycerin flows well in the presence of a catalyst. Otherwise, the reaction proceeds very slowly, even when heated to 250 °C. With an increase in the molecular weight of the alcohol, the reaction slows down. Thus, for sunflower oil, when using ethyl alcohol, the conversion depth is approximately equal to 35.3%, and when using amyl alcohol, it is 11.5%. However, the equilibrium position can be shifted by changing the ratio of triglycerides and alcohol or withdrawing from the reaction zone one of the resulting products, for example, glycerol (Smirnova & Podgaetsky 2007, Tovbin et al. 1981).

Esters were obtained from camelina oil and n-butanol (Sample 1) (Kondrasheva et al. 2016), and camelina oil and isopropanol (Sample 2). The transesterification was carried out using sulfuric acid as a catalyst.

Reaction conditions are shown in Table 2.

The composition of the products obtained is shown in Figure 1 and Figure 2.

Then the samples were cleaned from glycerol and other impurities. Excess alcohol and acids are absorbed by the glycerol. The part of the acid remaining in the ether phase was neutralized with sodium carbonate (Na_2CO_3) (Kondrasheva et al. 2016).

After carrying out the experiment and obtaining the ester, the physicochemical properties of biodiesel were investigated, which are presented in Table 3.

The properties of sample 2, precisely the flash point in a closed crucible and viscosity do not comply with the European standard EN 14214 "Automotive fuels - methyl esters of fatty

Table 2. Conditions for carrying out transesterification reactions.

	Synthesis temperature, °C	synthesis time, min	Mixing speed, rpm
n-butanol	115	240	250
Isopropyl alcohol	80	240	250

The composition of biodiesel fuel (sample 1)

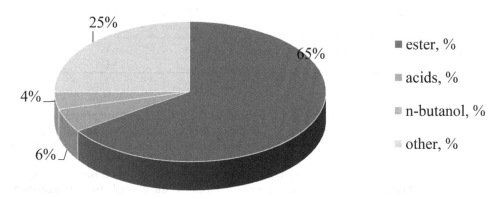

Figure 1.   Composition of biodiesel fuel synthesized from camelina oil and n-butyl alcohol.

The composition of biodiesel fuel (sample 2)

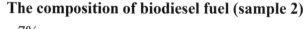
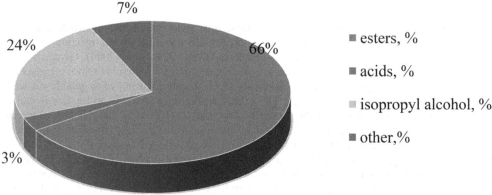

Figure 2.   Composition of biodiesel fuel synthesized from camelina oil and isopropyl alcohol.

Table 3.   Physical and chemical properties of the esters.

Property	EN 14214	Sample 1	Sample 2
Density at 20°C, kg/m^3	860-900	876.87	879.87
Viscosity at 40°C, mm^2/s	3.5-5	4.9535	9.1997
Fractional composition, °C	-	275 - 300	269 - 306
Flash point in closed crucible, °C	≥101	123	71
Pour point, °C	-	-10	-16

acids (FattyAcidMethylEsters, FAME) for diesel engines - requirements and testing methods". Hence, biodiesel fuel derived from isopropanol is not of good quality and unsuitable. For this reason, sample 1was selected for further research.

Sample 1 does not satisfy GOST 305-82, which justifies the reason it was choosen for further research into improving its low-temperature properties.

2.2 Depressants and depressant-dispersant additives

To improve the low-temperature properties of diesel fuels, pour point depressants are used.

Currently known pour point depressants, according to their chemical nature, can be classified as follows:

- copolymers of ethylene with polar monomers (copolymers of ethylene-vinyl acetate and their compositions, ternary copolymers based on ethylene and vinyl acetate, copolymers of ethylene with other polar monomers);
- polyolefin-type products (copolymers of ethylene-propylene, ethylene-propylene-diene and their decomposition products, copolymers of □-olefins, modified polyolefins);
- polymethacrylate additives (poly alkyl (meth) acrylates, copolymers of alkyl (meth) acrylates);

Almost all modern practically significant additives contain copolymers of olefins (ethylene) and vinyl acetate as active components. They effectively reduce the pour point of fuels, but do not prevent their separation during cold storage, when the fuel is stratified, forming an upper transparent layer and turbid layer below, enriched with paraffins. Both layers are mobile, but when the fuel is withdrawn from the lower layer, the engine works intermittently. This problem is solved by using special additives - dispersants or paraffin anti-precipitants. The effect of their use is the formation of very small paraffin crystals with high sedimentation stability.

Paraffin dispersants are of particular importance in countries with a long cold season. Therefore, in Russia, the use of depressants and dispersants of paraffins is highly recommended. Despite numerous attempts to develop domestic depressors, there has been little success in this area. The additives approved for use after their launch date were developed and used in very small volume, and by now they are morally obsolete. The following additives are currently used in Russia: Deprolux STD, Dodiflow 5416, Dodiflow 4851, Dodiflow 4273 (manufacturer Clariant), Keroflux 3501, Keroflux 3614, Keroflux ES 6100 (manufacturer BASF), OFI 7650 (7660) (manufacturer Innospec) et al.

The principle of operation of traditional depressors is that n-alkanes contained in the fuel can easily crystallize when the temperature is lowered. The onset of crystallization is manifested in the turbidity of the fuel. Then the crystals grow and at a certain size and concentration, they form a spatial structure (lattice). As a result of this process, the fuel loses its mobility and is poorly pumped through the piping and filter systems. Depressor additives are sorbed on the surface of the emerging crystals and prevent their growth and association (Bormann & Stocker 1975, Ivanov et al. 1982, Khvostenko 1998, Mitusova et al. 2005).

2.3 Influence of depressor additives on low-temperature properties of biodiesel fuels

Since there are no paraffins in the composition of biofuels, the impact of depressant additives on them the expected result will not be not achieved (Mitusova et al. 2017). The pour point of esters depends on the radical of added alcohol. The pour point of alcohols having a normal structure is significantly higher than that esters of the same iso-structure.

To confirm the theoretical conclusions experiments were conducted. The influence of two additives was considered: one from the manufacturer Clariant and the other from a Russian manufacturer.

The additive from the manufacturer Clariant is a depressant-dispersant additive which improves the low-temperature properties of middle distillates. It can significantly reduce the filterability limit temperature and the freezing temperature of diesel and light heating fuels. The product also improves the dispersion stability of paraffin hydrocarbons in the diesel fuel during its storage below the cloud point. The characteristics of the additives are presented in Table 4.

Table 4.   The composition and properties of the depressant-dispersant additive from manufacturer Clariant.

Appearance	From transparent to turbid, viscous liquid from reddish to dark brown
Density at 40°C, g/sm^3	0.905
Viscosity at 40°C, mm^2/s	140
Pour point, °C	+9
Flash point in closed crucible, °C	56
Solubility	(at 30°C) Easily soluble in aromatic hydrocarbons (eg xylene, naphtha). Insoluble in water and methanol.

Table 5.   The value of the pour point and depression of blended diesel fuel with optimal additive concentrations.

	Russian	Clariant
The amount of additive, %	0.22	0.02
Pour point, °C	-36	-35
Magnitude of depression, °C	11	10

3   RESULT AND DISCUSSION

The presence of biodiesel in the mixed fuel influenced the pour point of the fuel, but only slightly. The values of pour point of epy mixed fuel, satisfying GOST 305-82, are achieved at the same concentrations of additives as for hydrotreated diesel fuel. The magnitude of depression in the mixed fuel at the optimum concentration at the time of introduction of additives is presented in Table 5.

Thus, the presence of biodiesel fuel in a mixture with petroleum hydrotreated diesel fuel in an amount of not more than 5% has a slight effect on the pour point of the mixture. This facilitates the introduction of depressant-dispersant additives in an optimal concentration of 0.02-0.22% into the mixed environmentally friendly diesel fuel improving its low temperature properties to the required standards.

4   CONCLUSION

To obtain biodiesel fuel, the optimal parameters of the process such as a process temperature of 195°C; synthesis time 240 minutes; agitator stirring speed 250 rpm; the ratio of raw oil: alcohol - 2: 1 were determined. The biodiesel yield in the process of transesterification was 65-66 %mass. The best component (alcohol) for biodiesel is butanol.

The use of depressant-dispersant additives in adequate concentrations in biodiesel has no effect on the pour point. With the introduction of 0.22 %mass of the Russian depressant-dispersant additive, the pour point of commercial diesel fuel decreases to -36°C, and with 0.2 %wt of the additive from manufacturer Clariant, the pour point of fuel decreases to -35°C.

REFERENCES

Bormann, K., Stocker, J. 1975 // *Chem. Techn. (DDR). Bd. 27. № 5.* P. 286.
Eremeeva, A.M. 2014. Ways to create environmentally friendly biofuels for diesel engines. *SEVER-GEOEKOTEKH-2014: Mater. XV Intern. youth science conf.*

Iovleva, E.L., Zakharova, S.S., Lebedev, M.P., & Popova, L.I. 2013. Prospects for improving the low-temperature characteristics of diesel fuel fractions. *Bulletin of the Saratov State Technical University 2. 2 (71)*. P. 116–120.

Ivanov, V.I., Torner, R.V., Fremel, T.V., & Shapkina L.N. 1982. *Works VNIINP. M. TSNIIT Eneftekhim. Issue. 41.* p. 100.

Kemalov, A.F., Kemalov, R.A., & Valiev, D.Z. 2010. Getting winter varieties of diesel fuel using depressant-dispersant additives based on petrochemical raw materials. *Bulletin of Kazan Technological University. 10.* P. 645–647.

Khvostenko, N.N. 1998. Development of low fouling diesel fuels with depressant additives. *Dis. Cand. tech. sciences. 05.17.07.* Yaroslavl, 174 p.

Kondrasheva, N.K., Eremeeva, A.M., & Nelkenbaum, K.S. 2016. Study of the influence of additives and additives on the operational and environmental characteristics of diesel fuel. *Izvestia SPbGTI (TU). № 38.* p. 86–89.

Kozin, V.G., & Bashkireva, N.Yu. 2009. Modern technologies for the production of motor fuel components. Kazan.

Mitusova, T.N., Kalinina, M.V., & Polina, E.V. 2005. Reducing the cloud point of diesel fuel due to a special additive. Oil refining and petrochemistry. *Scientific and technological achievements and best practices. № 2.* p.18.

Mitusova, T. N., Kondrasheva, N. K., Lobashova, M. M., Ershov, M. A., & Rudko, V. A. 2017. Influence of Dispersing Additives and Blend Composition on Stability of Marine High-Viscosity Fuels. *Journal of Mining Institute, 228(6)*, 722–725.

Sidracheva, I. I. 2009. Synthesis of anti-wear additive to diesel fuels based on rapeseed oil and n-butyl alcohol. *dis. PhD in Technical Sciences: 02.00.13* 117 p.

Smirnova, T. N., & Podgaetsky, V.M. 2007. Biodiesel - an alternative fuel for diesel engines. Receipt. Specifications. Application. Cost. *Engine: scientific and technical journal. Moscow. № 7 (49)*.

Tovbin, I.M., Melamud, N.L., & Sergeev A.G. 1981. Hydrogenation of fats. *Light and food industry* 296 p.

Vasiliev, I.P. 2005. Environmentally friendly areas of production and use of plant-derived fuels in internal combustion engines. *Ecotechnologies and resource saving. 1.*

Viter, V.N., & Zuboko, A.V. 2008. The principle of biodiesel production. *Chemistry and Chemists: electronic journal. № 3.*

Scientific and Practical Studies of Raw Material Issues – Litvinenko (Ed)
© 2020 Taylor & Francis Group, London, ISBN 978-0-367-86153-7

Environmental technologies in the production of metallurgical silicon

M. Glazev & V. Bazhin
Mining University, Saint-Petersburg, Russian Federation

ABSTRACT: To date, poorly resolved issues of disposal and the possibility of further use of waste silicon production. The use of waste allows not only to provide enterprises with additional resources, but also to reduce the environmental burden on the environment, as well as to stabilize the work of the main processes – the preparation of raw materials and blast furnace production.The article describes the main characteristics of microsilica (waste silicon production), as well as various applications for the creation of a wide line of products with high added value.

1   INTRODUCTION

Metallurgical silicon is one of the few industrial products that tends to increase production and consumption (Elkin 2014). To date, a significant amount of silicon dust (microsilica - waste of silicon production) of gas purification of electrothermal furnaces of silicon production is sent to the sludge storage, which worsens the environmental condition of the regions and requires material costs for transportation and storage of waste (Leonova 2015). Currently, both in Russia and in the world, there are no technologies for complete utilization and processing of microsilica in the production of metallurgical silicon, with its partial use of no more than 8-10% in some sectors of the construction industry and the refractory industry (Kondratyev 2010). The implementation of various proposed in the article technical directions with increasing production of metallurgical silicon will allow for further scaling of technological lines for processing of microsilica into commercial products with added value in industrial production (Kucherov 2012, Tyutrin 2011).

2   MAIN CHARACTERISTICS AND VARIOUS APPLICATIONS OF MICROSILICA

2.1   *Main characteristics microsilica (silicon production waste)*

Silicon dust in the form of waste, which is formed in the production of metallurgical silicon (in relation to 0.4-0.45 per 1 ton of silicon) during the melting of quartzite in ore-thermal furnaces in the presence of charcoal, is a conglomerate of solid particles ranging in size from nanoparticles to agglomerates of several tens of microns with a complex chemical and phase composition, as well as a highly developed surface shape (Nemchinova, Kletz 2008).

The object of research is microsilica JSC RUSAL in the city of Kamensk-Uralsky (Sverdlovsk region, Russia). The chemical composition of microsilica is presented in Table 1.

The table shows that the largest mass fraction is silicon dioxide (at least 98%), while the size of the dust particles in percentage terms allows us to say that the dust is more than half represented by nanoparticles (Table 2).

Chemical analysis of dispersed particles of carbon-containing silica, which are dust wastes of gas treatment plants of go, showed that they are more than 92-94 % composed of silica

Table 1. Chemical composition of microsilica.

The name of the substance	Mass fraction %	Measurement method
SiO_2	at least 98	EN 196-2
CaO	no more than 0,3	EN 451-1
SO_3	-	EN 196-2
K_2O	no more than 0,3	EN 196-2
Na_2O	no more than 0,1	EN 196-2
Fe_2O_3	no more than 0,1	GOST standard 2642.5-97
Al_2O_3	no more than 0,3	GOST standard 2642.4-86
MgO	no more than 0,2	GOST standard 2642.8-97
P_2O_5	-	GOST standard 2642.10-86
Cl	-	EN 196-2
H_2O	no more than 0,3	GOST standard 2642.1-86
SiC	-	GOST standard 26564.1-85
C	-	GOST standard 2642.15-97

Table 2. Granulometric composition of microsilica (RUSAL).

Granulometric composition of microsilica (after long-term storage and self-coagulation processes) mkm	Value of indicator %	Measurement method
Individual nanoparticles:		
- nanoscopic, less 1	at least 63,5	GOST standard 8.755-2011
Agglomerates:		
– small, more 1	no more than 30,0	
– middle, more 10	no more than 5,0	
– large, more 45	no more than 1,5	

fume SiO2, which confirms the feasibility of return (recycling) of this type of waste in the production process (Heggestad, Holm, Lonvik, Sandberg 1984).

The General property of waste silicon production (microsilica - carbon containing silica) is their extremely high dispersion of 4-10 nm, which is much higher than the values for artificially synthesized silica-carbon compositions. For example, the dimension of microsilica of technogenic origin is 100-200 nm, formed during the electric melting of ferroalloys, and natural shungite rocks - 1000-5000 nm.

Figure 1 shows electronic image of microsilica particles.

Microsilica silicon production can be considered an activated chemical additive, which due to its structural features, dimension and chemical composition can affect the structure and

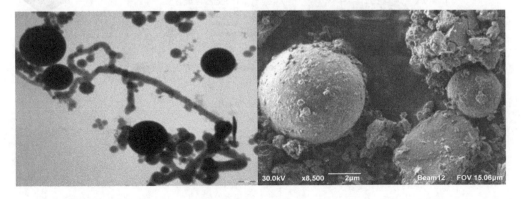

Figure 1. Electronic image of microsilica particles.

115

properties of materials for various applications with the creation of a wide line of products with high added value (Bernd 2006).

Figure 2 shows an approximate scheme of utilization of exhaust gases of silicon melting elements.

2.2 *Various applications of microsilica to create a line of products with high added value*

2.2.1 *The use in the construction industry*

Currently microsilica found wide application in the global construction industry, namely for obtaining of new generation concretes with special properties: obtaining high-strength concrete (the compressive strength of 80-100 MPa, and 240 MPa in the autoclave treatment); obtaining concrete with high durability (resistance to sulfate and chloride attacks, to the action of weak acids, sea water, high and low temperatures).

The addition of silica fume increases the water resistance of 25-50%, sulphate resistance of 90-100%. In addition, when using microsilica the opportunity to save up to 50% of the cement in the concrete without loss of technological properties. Thus, the use of microsilica allows to obtain concrete in real production with cement consumption of 200-450 kg/m3 and the following characteristics: brand strength — M300-M1000, water resistance — W12-W16, frost resistance — F200-F600 and up to F1000 with special additives, corrosion resistance is not lower than on sulfate-resistant cement.

2.2.2 *The use in additive technologies*

Due to its high dispersion, activity and overall dimension, this type of raw material after regeneration or separation into carbon and microsilica can be considered as a source of powders for use in 3D printers for the manufacture of complex building elements and blocks.

Small-scale (less than 1 %) in actual volume, but intensively developing and having a high cost estimate, are the markets for carbide-silicon powder materials: powder with a particle size

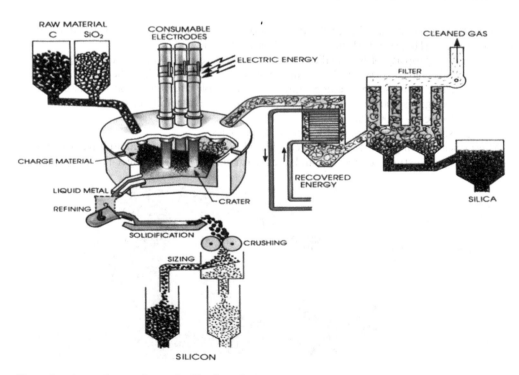

Figure 2.   Approximate scheme of utilization of exhaust gases of silicon melting elements.

of less than 1 μm (micronized carbide) for ceramics and nanopowder with a particle size of less than 100 nm (nanocarbide) for high-quality structural ceramics and electroplating.

There are two ways to develop technology for producing nanopowders of three types: pure refined silica powder, carbon powder and silicon carbide powder.

The introduction of silicon carbide in the form of monocarbide opens up new areas of application. It becomes possible to use nanopowders in 3D printers.

The use of new technologies is the main trend of recent years in any sphere of industrial production. Every company in Russia and the world strives to create cheaper, reliable and high-quality products using the most advanced methods and materials. The use of additive technologies is one of the clearest examples of how new developments and equipment can significantly improve traditional production.

2.2.3 *The use of microsilica in the manufacture of ceramic products*

The dispersed state of silica-containing waste in combination with physical and chemical characteristics determines the prospects of their use in the production of firing products. During the heat treatment of products, the amorphous highly active state of the microsilica leads to an early accumulation of the liquid phase and sintering of the material.

Amorphous silica-containing waste is applicable both as the main component of the raw mixture and as additives. In most cases, amorphous silica is introduced into the raw material charge to reduce the average density, thermal conductivity and improve the strength characteristics.

These wastes are effectively used in the production of thermal insulation materials. The refractory properties of silica-containing waste led to their use in the manufacture of refractory and heat-resistant materials with high strength, resistance to thermal shock, low shrinkage, the production of light fillers silica-containing waste is used as raw materials, as well as a powdering additive.

The structure and properties of silicon waste allow us to consider their application in the production of glass, technical glass and foam glass.

2.2.4 *The use of microsilica in the manufacture of foam glass*

The content of a significant number of particles and pores of nanosize leads to a decrease in thermal conductivity and an increase in reactivity, which makes it possible to use microsilica as an active additive in the production of foam glass, as well as as a filler of vacuum thermal insulation panels

The technical result of the use of microsilica is to increase the strength of foam glass, expand the raw material base and reduce energy costs in the implementation of the technological process.

The raw material mixture for the manufacture of foam glass is characterized in that as a silica-containing raw material is used mikrosilica with a silicon oxide content of at least 85% and a fraction size of less than 1 μm, which can be up to 68 m.%.

2.2.5 *Modifying additive for cement mixes*

The use of microsilica in grouting solutions can improve the condition of the cement ring behind the casing, the quality of cementing, there is an increase in the strength and tightness of the contact zones of the grouting stone, which avoids various complications. In addition to increasing the strength, microsilica does not have a negative effect on the properties of the solution, there is a slight increase in density and a decrease in the porosity of the cement stone, which, as a result, leads to an improvement in its water resistance and corrosion resistance (Monsen, Seltveit, Sandberg, Bentsen 1984).

It is also possible to use microsilica in other cement-containing mixtures, including for household needs. The use of microsilica as a mineral additive in sand concrete and other cement-containing mixtures provides for the compaction of cement mortar by filling the micro-cavities with particularly strong hydration products (Alrekabi, Cundy, Raymond, Whitby, Lampropoulos, Savina 2011). It has been proved experimentally that the compaction of the cement-containing mass through the use of microsilica is much more effective than in the case of other additives, such as blast furnace slag and zeolite tuff.

Technological application microsilica as mineral additives in sand concrete involves effective metering and flow directly to amantadinebuy mixture. Technological application in this case involves two methods, namely: the use of microsilicon as a dry additive or aqueous suspension. The dry version of the technological application involves the introduction of microsilicon in the composition of concrete in combination with superplasticizers. The liquid version provides for the use of microsilicon as water-based suspensions.

2.2.6 *Technological briquettes*

The technology of silica dust briquetting for its return to ferrosilicon production, which is carried out with the use of roller presses, has been developed. The obtained briquettes from ferrosilicon elimination can be used as a substitute for lump ferrosilicon. The composition of the briquetted mixture for silicon has a high reducing ability, and the resulting briquettes have the necessary strength and porosity.

Silicon briquettes obtained by the proposed technology, with a shortage of quartz raw materials can be used as a raw material additive to the main charge in the production of metallurgical silicon. The use of briquettes of the proposed composition allows to increase the technical and economic indicators of the process due to the increase in the involvement of waste (microsilica, shale and wood dust) in production.

3 CONCLUSION

Microsilica of silicon production can be considered an activated chemical additive, which, due to its structural features, dimension and chemical composition, can influence the structure and properties of materials and be used as additives, modifiers for various applications with the creation of a wide line of products with high added value. Currently, in the Mining University of Saint-Petersburg in some areas there is a scientific basis in the patent security (5 existing patents of the Russian Federation and 3 applications for invention in FIPS).

ACKNOWLEDGEMENTS

The authors express their gratitude and appreciation to the staff of the Department of metallurgy of St. Petersburg Mining University for their attention, assistance and support at various stages of writing the article.

REFERENCES

Alrekabi S., A. Cundy, Raymond L.D. Whitby, A. Lampropoulos, I. Savina. 2011. Effect of Undensified Silica Fume on the Dispersion of Carbon Nanotubes within a Cementitious Composite. IOP Conf. Series: Journal of Physics: Conf. Series 829 (2017).
Bernd F. 2006. Microsilica– characterization of an unique additive. IIBCC 2006 - Sao Paulo, Brazil. October 15–18, 2006. Universidade de Sao Paulo & University of Idaho: Sao Paulo, 2006.
Elkin K. S. 2014. Production of metallic silicon in Russia — state and prospects//"Non-ferrous metals and minerals 2014": materials of the sixth international.Congress (16–19 September 2014). Krasnoyarsk.
Heggestad, K., Holm, J.L., Lonvik, K., Sandberg, B. 1984. "Investigations of Elkem Microsilica by Thermosonimetry". Thermochimica Acta 72.
Kondratyev V. V. 2010. The ways of trapping and characterization of the particulate phase in the production of silicon: a monograph — Irkutsk: publishing house of ISTU.
Kucherov A.V. 2012. Comparative feasibility analysis of alternative energy sources of Russia/ A. V. Kucherov, A. V. Shipileva//news of higher educational institutions. Series: Economics, Finance and production management.
Leonova M. S. 2015. Return to the process of dust silicon production//Young scientist.

Monsen, B., Seltveit, A., Sandberg, B., Bentsen, S. 1984. "Effects of microsilica on physical properties and mineralogical composition of refractory concretes". Advances in Ceramics 13 (1984).

Nemchinova N. In. Kletz V. E. 2008. Silicon: properties, preparation, application: studies. benefit. Irkutsk: ISTU Publishing house.

Tyutrin A. A. 2011. Silicon — the basis for the production of solar cells [electronic resource]//youth Bulletin of ISTU. Irkutsk.

Scientific and Practical Studies of Raw Material Issues – Litvinenko (Ed)
© 2020 Taylor & Francis Group, London, ISBN 978-0-367-86153-7

Challenges in processing copper ores containing sulfosalts

A. Kobylyanski, V. Zhukova, G. Petrov & A. Boduen
Saint-Petersburg Mining University, Saint-Petersburg, Russian Federation

ABSTRACT: The current status of processing copper ores containing high concentration of arsenic and antimony for South Ural industry area was considered. Mining and industrial complexes with challenging operating conditions were identified by key figures. Based on current technological processes, problematic factors and its influence on finished product were determined. Leaching of concentrates' samples containing approximately 3.2% As and 19.2% Cu was studied in alkaline sulfide solutions containing sodium hydroxide and sodium sulfide.

1 INTRODUCTION

Copper plays a crucial role in modern global society – to our health and well-being, in the home as well as in the business and industry. Copper is a key component of some of the most important technological developments, including the CERN Large Hadron Collider, or technologies for renewable energy sources. In the short term, the consumption of this metal will grow (The European Copper Institute, ECI 2017 Annual Report) taking into account the development of industries and expansion of the number of copper products applications.

However, statistical database of International Copper Study Group and recent researches indicate that producers are turning to the mining of the off-grade mineral reserves in the deposits that comprise lower valuable components concentration in the ores. This situation is common for many sectors of the mining industry, including copper sector.

Quality characteristics of copper ores are divided into the following categories:

- High grade (copper content more than 3.0–5.0%);
- Rich (with the content of valuable component more than 2.0%, for copper-porphyry ore deposits – more than 1.0%);
- Medium quality (with copper content more than 1.0%, for ores of copper-porphyry deposits – more than 0.4%);
- Poor (with copper content from 0.7 to 1.0%).

Meanwhile, many producers have no alternative other than to develop ore deposits with valuable component content indicators in the ore ranging between 0.5 and 0.8%. Considering this fact, it is important to utilize and to process raw materials with various quality characteristics effectively in order to replenish the metallurgical companies' mineral resource base.

The ore raw materials changes in the characteristics, that are utilized in production process (especially steady decrease of the content of the valuable component in the ores) leads to the digression of the processing indicators of raw materials. The chemical and mineralogical composition of the processed concentrates get worse. All these factors will have influence on the final metallurgical products cost.

For complex utilization of raw materials with different quality characteristics in the production process - adjusting and upgrading the production schemes and processes is necessary.

An example of the situation in copper industry is the Uchalinskiy mining and processing plant (JSC "Uchalinsky GOK"), which is one of the largest industrial companies of Southern Ural, (production specialization - zinc and copper concentrate). Production and economic indicators of JSC "Uchalinsky GOK" are shown in Figure 1 and Figure 2.

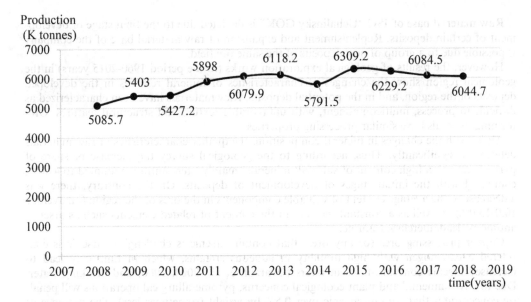

Figure 1.   Ore production levels of JSC "Uchalinsky GOK".

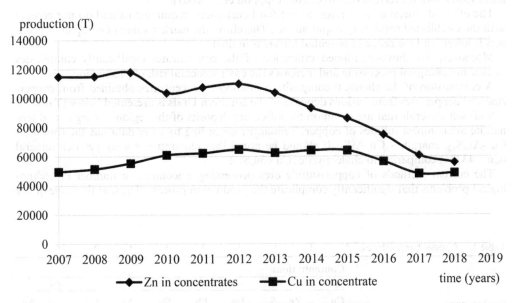

Figure 2.   Concentrations of metals in products of JSC "Uchalinsky GOK".

2   CHALLENGES IN ORES PROCESSING

In the period starting from 2007 -until 2017, according to the data of JSC "Uchalinsky GOK", there was an increase in ore production by 17.5% (Figure 1). However, the essential production statistics (the content of metals in concentrates) have been in steady decline - decrease in the proportion of zinc in concentrate was by 52.9%, for the copper concentrate by 3.0% (Figure 2).

Raw material base of JSC "Uchalinsky GOK" is depleted due to the final stage of development of certain deposits. Replenishment and expansion of raw material base of the company is possible due to a group of new deposits of the same ore field.

However, the results of geological exploration works (in the period 1960-2015 years) in the geological region show the changes of characteristics of mineral reserves in the developed deposits of the region, and in the potential deposits. Raw materials have been characterized as difficult to process, multicomponent, with unfavorable texture and structural features. Ore-forming minerals have similar processing properties.

Along with the changes in mineral composition, the quality characteristics of raw materials deteriorate significantly. Thus, according to the geological survey the increase of share of pyrite ores with a high content of sulfosalt minerals (mainly - tennantite) amounted to 41.0%, compared with the initial stages of development of deposits. On the contrary, there was a decrease in the average content of valuable components in the ores of the region (on average 10.0-15.0%), as well as a constant increase in the content of related elements such as arsenic, antimony, lead, mercury, fluorine.

Copper processing ores (or any ores) that contain arsenic is challenging; arsenic is considered a carcinogen with high mobility in aqueous streams, which in time may lead to increased arsenic concentrations in rivers (Oyarzun et al., 2004, 2006) and drinking water. Due to environmental and plant ecological concerns, pyrometallurgical operations will penalize concentrates that contain arsenic over 0.5% by weight (sometimes less). The presence of arsenic can also increase concentrates shipping costs, since concentrates are commonly imported/exported overseas (Castro, 2008; Filippou et al., 2007).

The off-grade metal concentrates do not fulfil customers' requirements and do not comply with the established regulatory requirements. Therefore, the market value of company's products is lower, and the range of potential buyers is limited.

Moreover, the above-mentioned utilization of the concentrates significantly complicates further metallurgical processing and increases the environmental risk of production.

A comparison of the chemical composition of copper concentrates obtained from copper-zinc and copper ores from various deposits of the southern Urals is presented below (Table 1).

Sulfosalt minerals that are common to sulfide ore deposits of the region - enargite and tennantite are complex sulfides of copper. Tennantite according to x-ray data has the formula - $Cu_{12}As_4S_{13}$, enargite - Cu_3AsS_4. It should be noted that tennantite is a copper-rich mineral (Cu – 43.0%) comparing to chalcopyrite (Cu - 34.5%).

The current methods of copper-sulfide ores processing, encompass a number of technological problems that significantly complicate the production process. Thus, at the stage of ore

Table 1.   Copper concentrates.

| Product name | Concentrations | | | | | | | | | |
| | % | | | | | | | | g/t | |
	Cu	Zn	S_{tot}	Fe_{tot}	Pb	Sb	As	Cd	Au	Ag
Chalcopyrite copper concentrate JSC "Uchalinsky GOK"	18.22	3.3	39.5	33.18	1.58	0.072	0.2	0.011	3.6	11.0
Tennatite copper concentrate JSC "Uchalinsky GOK"	19.19	5.3	40.7	38.5	2.5	-	3.2	-	5.4	17.6
Copper concentrate Sibaysky branch of JSC "Uchalinsky GOK"	14.10	3.0	41.9	33.10	0.074	0.07	0.075	-	3.8	50.7
Copper concentrate JSC "Gaysky GOK"	19.5	1.5	40.7	33.7	0.14	0.01	0.062	0.008	4.5	51.7
Copper concentrate Ltd "Bashkir Med (Copper)"	17.6	3.4	35.5	26.7	0.053	0.001	0.02	0.016	2.8	35.0

preparation sulfosalt minerals usually are over crushed due to the very thin interpenetration of the main ore-forming minerals: pyrite, chalcopyrite, and sphalerite and sulfosalt minerals.

Besides the similarity of the flotation properties of minerals and over crushing of sulfoarsenides, the process of selective flotation is complicated, because of the main copper sulfides combined with tennantite might be easily turned to sludge, and that leads to losses of copper with processing tails, challenges in obtaining a conditioned copper concentrate.

Currently, a serious amount (about 35.8%) of the total volume of ores utilized by the processing plant consists from pyrite ores with a high content of tennantite. The applied methods of processing of pyrite ores do not provide an opportunity to process effectively and to provide selective separation of the main sulfide minerals of copper and zinc in the presence of tennantite and enargite.

3 METHODOLOGY

The conventional method to reduce the content of arsenic and other impurities in copper concentrates is roasting, where the arsenic is removed by volatilization as either sulfide or oxide. Although roasting is very effective for arsenic elimination, roasting plants may also have problems to comply with the increasingly stringent norms that restrict the arsenic emissions in the atmosphere. Therefore, there is a growing interest in the copper industry for alternative processes to remove arsenic and antimony from copper concentrates (Anderson, C.G. & Twidell, L.G. 2008).

Amongst current technologies, an alkaline sulfide hydrometallurgical process has been demonstrated successful results in selective dissolution of arsenic and antimony from a copper concentrate containing enargite and tennantite (Prada et al., 2014; Tongamp et al., 2009; Baláž et al., 2006). The process undertakes an alkaline sulfide leaching with a concentrated solution of NaHS-NaOH to transform the arsenic sulfides into soluble compounds.

In the present paper, the behavior of a complex copper concentrate containing arsenic and antimony using NaHS-NaOH is discussed.

4 EXPERIMENTAL WORK

4.1 *Materials and methods*

The objects of the study were the concentrate and middling provided by the Uchalinsky GOK. The results of the chemical analyses (ICP) are presented in the Table 2.

The X-ray diffraction analyses on a Bruker D2 Phaser showed that the main minerals presented by tennantite, chalcopyrite and pyrite.

The alkaline sulfide leaching experiments were carried out with 50 g of the sample under atmospheric pressure in a 1000 ml glass jacketed cell. Temperature was controlled by thermostat connected with a hotplate. A freshly prepared solution containing the desired amounts of NaHS and NaOH (depending on liquid to solid ratio) was preheated over a hotplate in the reactor until it reached the temperature selected for the test. All chemicals were of analytical grade. The leaching was then started by adding the solid sample. During dissolution, the pulp was vigorously mixed using a paddled stir rod for the predetermined time. The leaching time was varied from 3 to 5 hours. After the leaching the solids were filtered, washed and finally

Table 2. Chemical analysis of the copper concentrate and middling (only basic elements).

Sample	Content (wt. %)										
	Cu	Pb	Zn	Fe	Co	SiO_2	Al_2O_3	Cd	As	Sb	Ag
Copper concentrate UGOK	16.0	23.8	0.8	5.3	0.1	1.94	0.82	0.03	1.36	0.21	0.009
Middling UGOK	6.22	0.74	7.3	24.4	0.1	4.02	1.32	0.03	1.36	0.19	0.009

dried at 80°C. Leaching results were evaluated by means of elemental determinations on the leach products using Inductively Coupled Plasma-Atomic Emission Spectrometry (ICP-AES).

The studied variables included temperature, concentration of sodium hydroxide and sodium sulfide and their effect on the dissolution of arsenic, antimony, copper and iron was investigated. Leaching results were evaluated by means of elemental determinations on the leach products using Inductively Coupled Plasma-Atomic Emission Spectrometry (ICP-AES).

The selected parameters: concentrations of 2.5 M, 3.5 M, 4 M for NaOH and 0.5 M, 1 M, 1.5 M for Na_2S, temperature (80 - 95°C).

4.2 *Results and discussion*

Results reported by other researches (Prada et al., 2014; Tongamp et al., 2009; Baláž et al., 2006) has shown arsenic extractions of up to 97% can be achieved from high-arsenic concentrates using the alkaline sulfide process. Achieving similar arsenic extraction results were one of the key objectives for proving it efficiency. The results are presented on Figures 3 and 4 and in Table 3 and 4.

A main reaction, which takes place during the process:

$$Cu_{12}As_4S_{13} + 6NaHS + 6NaOH \rightarrow 5Cu_2S + 2CuS + 4Na_3AsS_3 + 6H_2O \qquad (1)$$

Best results were obtained when both reagents are added in high amounts (4 M NaOH and 1.5 M Na_2S). This behaviour could suggest that both reagents are acting directly in the dissolution of arsenic. The reaction takes place quickly and reaching almost 100%, when the solution contains high concentration of NaOH and Na_2S (Figure 3).

As seen on Figure 4, the temperature has a significant effect on the extent of arsenic and antimony removal from the concentrate. This strong temperature dependence suggests that the controlling mechanism could be the chemical reaction taking place on the surface of the particles.

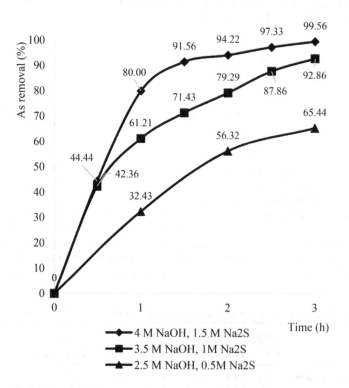

Figure 3.   Effect of NaOH and Na_2S on arsenic extraction at 95°C in 250 ml of solution.

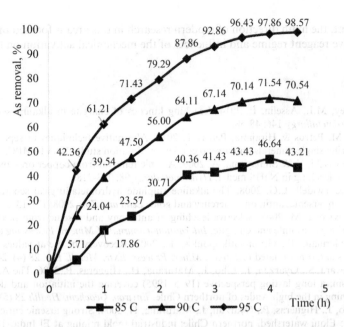

Figure 4. Effect of temperature on arsenic extraction in 250 ml of solution 3.5 M NaOH and 1 M Na$_2$S.

Table 3. Results of effect of NaOH and Na$_2$S after 3 hours.

Parameter	Concentration, M					
	Na$_2$S	NaOH	Na$_2$S	NaOH	Na$_2$S	NaOH
	1.5	4	1	3.5	0.5	2.5
As content in solution (mg/l)		1393		1300		890
As removal (%)		99.55		92.86		65.44

Table 4. Results of temperature effect after 4.5 hours.

Parameter	Temperature, °C		
	85	90	95
As content in solution (mg/l)	605	988	1380
As removal (%)	43.21	70.54	98.57

5 CONCLUSION

It is evident that the existing scientific and industrial experience in the study of regulatory of arsenic extraction from processing products, for further quality improvement of the final products is insufficient. In addition, in the application of certain technologies at processing plants do not comply with environmental standards. All this factors makes this area of research very relevant.

The method of flotation cannot separate copper and zinc sulfides from sulfosalt minerals completely, that increases the concentration of arsenic in the product.

The selective leaching of arsenic and antimony using sodium sulfide sulfides in alkaline solutions is an effective way of upgrading copper-gold-silver concentrates to make them suitable for smelting. This creates an opportunity for the mining of off-grade mineral reserves.

At the moment, the main direction of modern research in this area is focused on the identification of effective reagent regime and influences of the mechanical activation on the process.

REFERENCES

Parada, F., Jeffrey, M.I., Asselin, E. 2014. Leaching kinetics of enargite in alkaline sodium sulphide solutions. *Hydrometallurgy* 146: 48-58.

Himawan, T. B. M. Petrus & Hirajima, Petrus T. 2012. Alternative techniques to separate tennantite from chalcopyrite: single minerals and arseno copper ore flotation study. XXVI IMPC. New Deli.

Tongamp, W., Takasaki, Y. & Shibayama, A. 2009. Arsenic removal from copper ores and concentrates through alkaline leaching in NaHS media. *Hydrometallurgy* 98: 213-218.

Anderson, C.G. & Twidell, L.G. 2008. The alkaline sulphide hydrometallurgical separation, recovery and fixation of tin, arsenic, antimony, mercury and gold. *Lead and Zinc* 2008: 121-132.

Baláž, P., Achimovičová, M. 2006. Selective leaching of antimony and arsenic from mechanically activated tetrahedrite, jamesonite and enargite. *International Journal of Mineral Processing* 81: 44-50.

Filippou, D., St-Germain, P., Grammatikopoulos, T., 2007. Recovery of metal values from copper-arsenic minerals and other related resources. *Miner. Process. Extr. Metall. Rev.* 28 (4): 247-298.

Oyarzun, R., Guevara, S., Oyarzun, J., Lillo, J., Maturana, H., Higueras, P. 2006. The Ascontaminated Elqui river basin: a long lasting perspective (1975–1995) covering the initiation and development of Au–Cu–As mining in the high Andes of northern Chile. *Environ. Geochem. Health* 28 (5): 431-443.

Oyarzun, R., Lillo, J., Higueras, P., Oyarzun, J., Maturana, H. 2004. Strong arsenic enrichment in sediments from the Elqui watershed, northern Chile: industrial (gold mining at El Indio-Tambo district) vs. geologic processes. *J. Geochem. Explor.* 84 (2): 53-64.

Fornasiero, D. 2001. Separation of enargite and tennantite from non-arsenic copper sulfide minerals by selective oxidation or dissolution. *International Journal of Mineral Processing* 61 (2): 109-119.

Nadkarni, R.M & Kusik, C.L. 1988. Hydrometallurgical removal of arsenic from copper concentrates. *Arsenic Metallurgy Fundam. Appl. Proc. Symp:* 263-286.

Mining services

Scientific and Practical Studies of Raw Material Issues – Litvinenko (Ed)
© 2020 Taylor & Francis Group, London, ISBN 978-0-367-86153-7

Use of brown coal ash for the amelioration of dump soils

Carsten Drebenstedt
Technische Universität Bergakademie Freiberg, Freiberg, Germany

ABSTRACT: Since the mid-1950s brown coal filter ash has been systematically used to ameliorate the soil, in particular Tertiary dumped substrates otherwise bare of vegetation. In conjunction with a partial project the long-term effects of the ash to the soil were examined. From the results obtained it can be estimated that as regards the aspects of soil science, soil productivity and seepage waters from dump areas ameliorated with ash no detrimental effects are produced as compared with soils ameliorated with lime – in fact, quite the contrary. The substitution of natural lime has advantages but is however limited by transportation distance. In each individual case the basic amelioration of dumped soils with ash must be approved in accordance with the development plan licensing procedure stipulated by the mining law.

1 INTRODUCTION

180.4 mio. t of brown coal were produced in German Federal Republic in the year 2007. 92.6% were used for the electricity and district heating generation, 7.4% were used for the other raw coal outlet (Jahresbericht 2007, 2008).

Combustion impurity and ash remain after the energetic brown coal use. Ash yields in various chemical and physical combination and consequentially various qualities according to the natural lignitization and deposition conditions, to production, transportation and storage technology and to conditions and execution of the combustion process. These qualities are important for the ash use in the sense of the sustainable raw material economy and the waste and circuit economy law.

Lignite ash is widely used as e.g. raw material (concrete aggregate), for soil melioration/ stabilization, for water quality handling, in landscape construction (backfill material) or in disposal site construction in various forms. Principles and utilization experience of brown coal ash were documented comprehensively (Drebenstedt, 1994; Handbuch der Verwertung von Braunkohlenfilteraschen in Deutschland, 1995).

2 LAUSITZ BROWN COAL ASH

Brown coal is produced in Lausitz since more than 150 years. Nowadays the main extraction object is the second Miocene layer with the following characteristics:

Calorific value 8.500 – 9.300 kJ/kg
Ash contents 2.6 – 11.0%
Water contents 52 – 56%
Sulphur contents 0.3 – 1.1%

Reserves in the approved and developed brown coal fields account for ca. 2 bn. t.

At the Lausitz brown cowl area in 2007 were extracted ca. 60 mio. t raw brown coal and produced ca. 350.000 t briquette, 690.000 t powdered brown coal and 220.000 t fluidized brown coal.

2.1 Sorts, origin and combination of ashes

2.1.1 Fly and pit ash
Lausitz raw brown coal is mainly used for the electricity generation (95,7%) at three modern power stations Jänschwalde (3000 MW), Schwarze Pumpe (1600 MW) and Boxberg (at present 1900 MW), which are equipped with the flue gas desulfurization units (FDU). Brown coal fly ash and brown coal pit ash (fewer amount – 20%) are mainly used at the power stations. The main part of fly ash is used for the fastening of the FDU chloride containing water (FDU-ash). Vattenfall Europe Mining AG has an authorisation for the deposit of ash, FDU-ash and gypsum, which originates by desulfurizing and can not be used at once, in open-pit mining above the after-mining ground-water level. Redemption of ca. 4.35 mio. t of ash from the storage depot at the opencast mine Jänschwalde and at the opencast mine Nochten (ca. 2.45 mio. t FDU-ash) will take place yearly.

2.1.2 TAV- and WSV-ashes
Powdered brown coal is mainly used at the little heat and combined heat and power stations. Limestone powder or calcium hydroxide is added to powdered brown coal by combustion. Brown coal dry additive fly ash (TAV-ash) yields. Vattenfall Europe Mining AG committed itself to the redemption of this ash and has an approved establishment at the opencast mine Reichwalde. Brown coal fluidization ash (WSV – ash) yields when putting into operation of a combined heat and power station Cottbus with the stationary fluidization firing by addition of calcite for desulphurization. The annual ash redemption by small-scale consumers amounted to ca. 50.000 t in the year.

Ash from the briquette firing remains as a rule at a consumer as well as ash, which is used at the powdered brown coal input by cement works and asphalt blending systems. Large quantities of raw brown coal are currently delivered to the combined heat and power station Chemnitz. This ash is partly utilized locally.

2.2 Chemistry of ashes

Combination of charged coal ash and filter ash chemical composition from the intermediate purification is broadly identical at Lausitz, so the outlet conditions are comprehensible (without combustion with different materials!).

Ash quantity and ash compound in coal alternates depending on during moor development occurred marine substance input (in the northern part of the coal field) or terrestrial embankments (in the south-eastern part of the coal field). Ashes of the second Lausitz layer are limy-sulphatic, in the northern and south-eastern panels increasingly siliceous. Mixed ashes can be characterized as fine sand schluffs.

2.2.1 Allocation qualities according to LAGA
The chemical structure of fly ash is subjected to the deposit and specific mining fluctuations that is indicated per correlating belt width. Fly ashes correspond at least to the allocation quality Z2 according to German states working group for waste (LAGA). The surpassing of the allocation quality can emerge in pH-value, conductance and sulphate content (Drebenstedt & Rascher, 1998).

TAV and WSV ashes can be compared in the structure with fly ashes, but they display the tech-nically higher contents of sulphur and CaO. Because of chosen charge coal are fluctuations of the element contents as a rule obviously lower and comply to allocation qualities Z0 (Drebenstedt & Rascher, 1998).

2.2.2 Soil-physical and chemical identification of ash
The grain-size distribution characterizes the Lausitz ashes as sandy with obvious fine sand and schluff tops. They are slight carboniferous and have a meagre up to high sorption ability.

Ashes can be characterized as slightly carboniferous to carboniferous because of the high Ca- and Mg-contents. The content of the other macronutrients potassium and sulphur is slight to high. The micronutrients boron and manganese are present in high contents. Cu, Mo and Zn sufficiently exist in case of the equalized acid alkaline balance.

The chemical composition of ashes changes independently from deposit because of the increasing co-burning of the recycling materials in the brown coal power stations.

3 LAUSITZ' DUMPED SUBSTRATE

Naturally formed carbonate containing cohesive rock is widely absent in the barren ground of Lausitz. Sandy, pleistocene and tertiary deposits occur, that get to the dump closing embankment as the dump substrates and that should be after-used. The tertiary deposits contain sulphur and coal and are responsible for acid, heavy acid and even extremely acid soil milieu and for slight buffer ability of the substrates. The mobilization of sulphur happens because of the oxidation of sulphur combinations when oxygen accesses rock due to dewatering or overburden removal.

The dump substrate in Lausitz are generally characterized except of the disadvantageous soil reaction of the tertiary substrates with other negative qualities, such as:

– missing or lacking, accessible for plants nutrients
– slight sorption ability or slight cation exchange capacity
– lacking humus, lacking soil life, lacking soil structure
– soil heterogeneity on the surface and in the potentially rooted space
– high wetting resistance and erosion tendency

Besides the sandy substrates mostly have the slight water holding capacity, that proves itself as the further disadvantage for the cultivation under continental climate conditions with low precipitation.

Especially the sulphur containing tertiary soils with the strongly acid soil reaction have not had over decades any vegetation yet and are not appropriate for plants. If the encountered unusable or restricted usable dump substrates are cultivated, they should be ameliorated according to the usage goal; this process is defined as a basis melioration in the Lausitz area. It has as a main purpose to remedy the disadvantageous chemical soil characteristics in the potentially rooted horizon (ca. 100 cm), thereby substrate will be deeply loosened and homogenized at the same time to stimulate the circulation of nutrients and to create the start conditions for the plants and soil development.

4 PROCESS OF MELIORATION OF THE DUMPED SUBSTRATES
CHARACTERISTICS THROUGH THE ASH AMELIORATION

People know since decades that brown coal and its ashes can be used as fertiliser (e.g. Novalis in the year 1800).

Ashes were firstly used on the basis of empirically ascertained influences on the soil melioration; their qualities were scientifically substantiated later and then special methods, especially in the Lausitz brown coal area, were developed at the end of the 50-es years of the 20-th century.

The goals of the ash melioration are improvement of sorption ability, soil reaction and nutrients supply as well as homogenisation of substrates in the melioration horizon and, in case of higher quantities, increase of mesovoid ratio in sandy and cohesive substrates (structural ameliorative influence).

Brown coal fly ash was used as a basic carrier for the amelioration of qualities and for the capaci-ty increase of the dumped substrates in Schwarzkollmer (1959), Domsdorfer (1965), Koyne (1966) and Kleinleipischer (1974) processes. The data calculation on the basic carriers could be fulfilled with implementation of the acid alkaline balance as a method for the definition of lime demand of the carbonaceous, tertiary substrates in such a way that the durable influence, especially for the buffering of the acid additional delivery through the continually degradation could be stopped (Katzur, 1998).

Up to 1990 circa 10.000 ha of tertiary substrates and tertiary mix-substrates, which were pre-pared for the forestall or agricultural usage, were ameliorated with brown coal fly ash in Lausitz. Many million tons of ash were used because of the CaO need up to 1.000 dt/ha. Furthermore ca. 1.000 ha of the pure ash areas were created in the region Calau for the agricultural use. After 1990 ash was not regularly used for the substrate ameliorating.

5   RESULTS OF THE ESTIMATION OF THE AREAS AMELIORATED WITH ASH 20-30 YEARS AGO

Projects „Using of power station ashes for recultivation of the anthropogenically influenced soil" (Rascher et al., 1993) and „Human-toxicological analysis of brown coal ashes" (Dominok et al., 1994) were elaborated within the network research project "Utilization of ashes" with the objective to conduct the stocktaking at the Lausitz area, that had to estimate a long-term effect of the ash using during the basis melioration of the dump substrates on ecological compatibility, effectiveness and further procedural applicability. Ash producers and removers were also at the research group in charge of RWE AG (Handbuch der Verwertung von Braunkohlenfilteraschen in Deutschland, 1995).

Explorations in ten different forestally and two agriculturally used dump areas in five openworking areas, where fly ash from the neighbouring power stations was deployed, were conducted for the estimation of the ash amelioration. Ca. 50 prospects were laid out and seven main soil forms were found. Data were surveyed on the earth surface at the agricultural enterprises and seepage water analysis from 27 open land lysimeter was evaluated. Tests gave the following results.

5.1   *Pedologic analysis*

The original dump substrates show after 20-30 years of development a distinct, till 0,4 m in-depth ApC, Ah-C soil profile and can be classed as a development stage of regosol. Several millimetres deep humus layers were found.

The long-term influence of the neutralization mediums could be determined in the ameliorated sinking area, i.e. the permanent acid regeneration was softened, and there is the base excess yet. The differences of the areas ameliorated only with lime or ash are not identifiable.

Any heavy metal charges in the dumped regosol were not identified regardless the amelioration medium (ash or lime). Heavy metal contents in the dump raw soil lie under their grown locations. The partially higher heavy metal contents are registered by nature lime than by brown coal ash.

Increased boron contents could be proved neither in soil nor in plants.

5.2   *Efficient analysis*

Different influences of the ash or lime choice as amelioration medium on the return of forests or on the agricultural fruits cannot be proved. Gains achieve the level, which is equivalent to the natural ground of the same soil type.

The combination ash-lime had a yield-increasing effect at one location. A double wood growth rate was registered in this case comparing with the ash or lime amelioration.

The independent of amelioration mediums nutrient and light conditions are crucial for the natural resettlement of the dump locations. The encountered herb layer points to differentiated species settlement, where the protected species are also present (e.g. "Red list of Saxony").

5.3 *Analysis of seeping water*

In the big lysimeters analysed seeping water shows without melioration substrate induced the high substance freights under the sulphurous tertiary dump substrates, especially high elevated sulphate contents and heavy metal discharges.

Significant differences can be noticed between with ash or lime ameliorated tertiary, sulphurous dumped substrates after 2 experiment years. Lower substance concentrations appear in the ash ameliorated dump raw soil than in the lime ameliorated (the immobilisation characteristics of ash).

5.4 *Human-toxicological analysis*

Ash ingredients lie up to the 10th power under the relevant MAK (maximal permitted concentration) -figures. The elution reaction lies within drinking water regulation, excepting chrome, however that is in the encountered amount sanitary harmless. The organic aggressive substances (dioxins, furan, PAH and others) could not be proven or lie significantly under MAK-values. The radioactivity of ashes oscillates within the natural limit. A durable skin contact should be avoided because of the high pH-value.

5.5 *Summary of analysis results*

The concepts that were taken as a basis for the amelioration practice with ash decades ago could be approved in their long-term influence:

– The influences on the soil reaction are strongly positive
– The additional nutrients supply with macro- (K, Mg) and micronutrients (trace elements) takes place without the negative influence of the heavy metal contents in soil
– Production development is comparable with the natural locations
– Seeping water of the tertiary substrates is influenced significantly positive
– Soil hydrologic balance in improved because of the intensive moisture penetration
– Erosion disposition is diminished because of the additional ameliorative influence
– Contact with ash does not cause any threats for people
– Any disadvantages are not ecologically recognisable after the ash melioration

These results will be supported per further analysis, e.g. Advisory opinion on the estimation of free lime ashes from the combustion of Lausitz brown coal (Rascher, 1998), Investigative report waste material utilization BFA (Reststoffverwertung Braunkohlen-Filterasche (BFA), 1993) and incubation experiments with BFA and REA-gypsum (Inkubationsversuche mit BFA und REA-Gips, 1995). The utilization of brown coal ashes in the solum for recultivation was intensively discussed at the workshop with the same name (Verwertung von Braunkohlenaschen in Oberböden bei der Rekultivierung, 1998). Ash surfaces were analysed under the scientific attendance of the East German brown coal redevelopment also (Drebenstedt, 2001).

The with brown coal fly ash caused sulphate input is not relevant on the verge of the geogenic background of the massive substance transportation from the whole tertiary dump field. The developed during melioration dump-substrate-ash-composite partially enable the cultivation of these locations and catalyse the process of pedogenesis. The soils developed from dump substrate and brown coal ash are equivalent to the background values of the comparable natural soils. The basic melioration with soil auxiliary substance brown coal ash has proven itself as a process for the soil melioration of tertiary, sulphurous dumped substrates of Lausitz.

The experience gained on the basis of the fly ash use can be transferred to TAV- and WSV-ashes, because their compound is comparable, whereby TAV-ashes have a more favourable CaO-content.

The further synergy effects develop from the substitution of lime as a melioration medium:

– Conservation of the natural lime deposits as well as of the bound natural space
– Minimization of the environment pollution per superregional transport of lime-fertiliser
– Non-take-up of the landfill space for ash

The conducted analyses show utility and safety of the using of brown coal ash for rehabilitation. This using of ash complies with the imperative of the preference of utilisation towards deposition.

When we estimate the using of ash as the melioration medium, environmental protection, availability of ashes as well as procedural and economical boundary conditions should be considered nowadays.

6 LOOKOUT AND SUMMARY

Brown coal ash is currently used at Vattenfall Europe Mining AG mainly for backfilling of the after open-pit mining remained spaces, disposal site construction and slope stabilization. The covering of demand for lime for the soil melioration ensues through the customary lime-fertilizer. How-ever, the amount of the tertiary dump substrates on the surface at the open-pit mines of Vattenfall can be kept technologically relative slight, but it will increase contingent on deposits in the long term perspective (open-pit mine Welzow-Süd and Nochten). Tertiary dumped and mixed dumped surfaces should be cultivated in the Lausitz area of LMBV in the context of brown coal redevelopment in the coming years.

Brown coal ash is appropriate soil melioration medium, especially for tertiary locations ecologically as well as economically interesting.

The possibility of the using of brown coal fly ash as well as TAV- and WSV-ashes in the top-soil for recultivation is current for the Lausitz area and it should be kept open for the permission in the concrete case of application. Thereby attention should be paid to the availability of ashes because of the disposal site concepts and the ash quality due to the con-combustion of the external waste.

REFERENCES

Dominok B. et al. (1994): Humantoxikologische Untersuchungen von Braunkohlefilterasche, Abschlußbericht, Hygieneinstitut Cottbus.
Drebenstedt C. (1994): 30jährige Erfahrungen beim Einsatz von Braunkohleasche zur Melioration von Kipprohböden in der Lausitz, Braunkohle Heft 7, S. 40–45.
Drebenstedt C., Rascher J. (1998): Nutzung von Braunkohlenaschen zur Verbesserung von Kippsubstraten im Lausitzer Braunkohlenbergbau – Ergebnisse und Bestandsaufnahme, In: Verwertung von Braunkohlenaschen in Oberböden bei der Rekultivierung, DEBRIV, Köln.
Drebenstedt C. (2001): Boden. In: Wissenschaftliche Begleitung der ostdeutschen Braunkohlesanierung, Lausitzer und Mitteldeutsche Bergbau-Verwaltungsgesellschaft, Berlin.
Handbuch der Verwertung von Braunkohlenfilteraschen in Deutschland, RWE AG Essen, 1995.
Inkubationsversuche mit BFA und REA-Gips. Untersuchungsbericht, BGR, 1995.
Jahresbericht 2007, Deutscher Braunkohlen-Industrie-Verein, Köln, 2008.
Katzur J. (1998): Melioration schwefelhaltiger Kippböden. In: Braunkohlentagebau und Rekultivierung (Hrsg. W. Pflug). Springer Verlag Berlin Heidelberg New York, 1998, S. 559–572.
Rascher J. et al. (1993): Nutzung von Kraftwerksaschen bei der Rekultivierung anthropogen beeinflußter Böden - Bestandsaufnahme Lausitzer Revier, Abschlußbericht, G.E.O.S. Freiberg.
Rascher J. (1998): Gutachterliche Stellungnahme zur Bewertung freikalkhaltiger Aschen aus der Verbrennung Lausitzer Braunkohlen, LAUBAG.
Reststoffverwertung Braunkohlen-Filterasche (BFA). Untersuchungsbericht, Niedersächsisches Landesamt für Bodenforschung und Bundesanstalt für Geowissenschaften und Rohstoffe, 1993.
Verwertung von Braunkohlenaschen in Oberböden bei der Rekultivierung. Tagungsband zum Workshop, DEBRIV, Köln, 1998.

Scientific and Practical Studies of Raw Material Issues – Litvinenko (Ed)
© 2020 Taylor & Francis Group, London, ISBN 978-0-367-86153-7

Engineering and ecological survey of oil-contaminated soils in industrial areas and efficient way to reduce the negative impact

M.V. Bykova & M.A. Pashkevich
Saint Petersburg Mining University, Saint Petersburg, Russian Federation

ABSTRACT: The article is devoted to the problem of oil-contaminated soils formed on the territory of various production facilities. Describes the main sources of petroleum products in the geological environment and main negative consequences discusses the advantages and disadvantages of methods of disposal of oily waste. On the basis of the thermal method, a method of soil purification from oil products is developed. Experimental studies have been carried out, the results of which prove the effectiveness of this approach.

1 INTRODUCTION

The development of the industry is accompanied by a negative impact on the environment, creating a load on its components. Petroleum products are one of the main pollutants of the environment.

They are formed during transportation of crude oil and products of its processing, operation of various machines and mechanisms (Khalilova 2015). Any technical facilities in the fields are potential sources of anthropogenic flows, differing in composition, concentrations and type of substances (Shuvalov et al. 2008). Soil contamination in the areas of operation of hydrocarbon deposits can occur during drilling, fracturing, during transportation by road, as well as in case of violation of the casing of production wells, corrosion of pipelines and tanks (Gilmore et al. 2014). Contamination of soil with oil products in the areas of transportation is determined directly by the presence of trunk pipeline systems and their long length. In the areas of processing, there is a significant contamination of soils due to spills and leaks of condensate and lubricating oils, as well as various chemical reagents (Laden et al. 2006). In areas where petroleum products are stored, contamination is mainly due to systematic leakage from tanks or noncompliance with pumping or discharge processes. Petroleum products are used in all spheres of human activity. In addition to the inevitable losses, large losses of petroleum products occur due to poor labor organization and weak technological discipline, and sometimes simply because of the barbaric attitude to nature. An example is the discharge of waste oil from the car engine directly to the surface (Timonin 2003).

The work of motor vehicles, oil refining and petrochemical enterprises, gaseous emissions and waste water, numerous spills as a result of pipeline and oil tankers accidents, accidents and fires at oil storage facilities and refineries lead to pollution of the atmosphere, hydrosphere, the upper layer of the lithosphere and pose a serious threat to the environmental situation of the regions of the Russian Federation.

The peculiarity of oil pollution of the geological environment is that long-term local emergency and technological leaks and spills lead to the formation of lithochemical halos and affect the entire ecosystem as a whole. Contamination of soil with oil products leads to changes in their chemical composition, properties and structure. First of all, it affects the humus horizon. The amount of carbon in the humus increases significantly, which leads to a deterioration of the fertile properties of the soil (Kovaleva 2017).

Hydrophobic particles of petroleum products worsen the process of water flow to the roots of plants, and leads to their physiological changes. As a result of the transformation of petroleum products, the composition of soil humus changes dramatically. Due to the increase in the con-centration of carbon residue of petroleum products in the soil profile, there is a change in oxidation-reduction conditions and an increase in the mobility of humus components.

Soil contamination with oil leads to the disturbance of the soil microbiocenosis. Hydroca-bonoxidizing microorganisms react to the inflow of oil products by increasing the number and in-creasing their activity, which disrupts the balance of microorganisms in the soil in the natural state.

In addition to the impact on microorganisms, oil-contaminated soils inhibit photosynthetic activity and productivity of plants. This affects, first of all, the development of the most sensitive soil algae, which can manifest itself in the form of oppression or their complete destruction.

The severity of the consequences depends on many factors, including the concentration of petroleum products, the rate of decomposition and migration capacity, etc (Gennadiev et al. 2007).

According to this, it can be concluded that environmental impact assessment is an important part of sustained development. Environmental impact assessment and reclamation of disturbed industrial areas have been investigated by a number of scholars (Pashkevich & Petrova 2014, Mezhibor 2016, Pashkevich & Petrova 2017, Alekseenko et al 2017, Sobolev 2018).

2 MATERIALS AND METHODS

When carrying out pollution control, it is necessary to take into account the nature and location directly depending on the specification of the industrial facility, its technological processes and the probability of leaks and spills. Conventionally, there are two main stages in the integrated assessment: field and laboratory studies of soil contamination.

Field research consisted of visual and organoleptic assessment of the territory and sampling. Thus, designated areas for sampling were installed in one or more characteristics, namely:

- the territory is devoid of vegetation either in part or in whole;
- the soils have a characteristic smell of oil products;
- the proximity of production facilities with increased probability of leakage.

In the summer of 2015, an engineering and environmental survey of the territory of the object for the extraction of hydrocarbons (exhausted well 9 Kumzhinskoe gas condensate field) was carried out. The location of the sample areas is shown in Figure 1.

Based on the results of engineering-ecological survey in the form of a visual estimate, in place was chosen a narrow area for sampling, seeking to line (soil-geomorphological profile of the long-term approximately 100 meters), crossing a drilling site of the well. 6 samples were taken at intervals of about 20 meters. One sample (2) is benthal deposits. The mass of each sample was not less than 200 g. All samples were placed in the containers of dark glass and isolated from each other to prevent the transformation of petroleum products during transportation and storage.

In the summer of 2017, an engineering and environmental survey of the territory of the object for transportation and storage of petroleum products (Leningrad region) was carried out.

Considering the fact that the major part of the production facility territory has been concreted, or asphalted, the areas of potentially contaminated soils are located along the perimeter of the loading/unloading rack and main pipelines. The location of the sample areas is shown in Figure 2.

For a more detailed study of the change in oil products concentration with depth, the sampling was carried out on three horizons: 0-5, 5-15 and 15-20 cm. Consequently, on each of the seven sample areas, 3 combined samples were picked for each of the horizons A0, A1 and A2: A0 = 0-5 cm, A1 = 5-15 cm and A2 = 15-20 cm, respectively. A total number of soil samples was twenty-one (21). The mass of each sample was not less than 200 g. All samples were placed in the containers of dark glass and isolated from each other to prevent the transformation of petroleum products during transportation and storage.

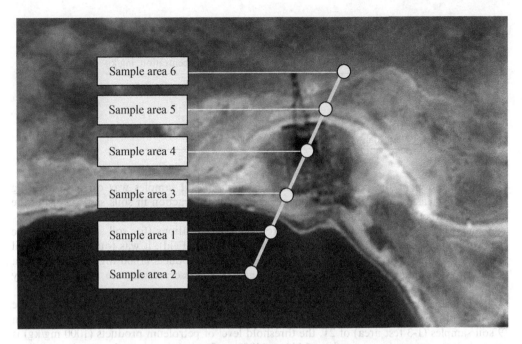

Figure 1. Layout of the test areas (exhausted well 9 Kumzhinskoe gas condensate field).

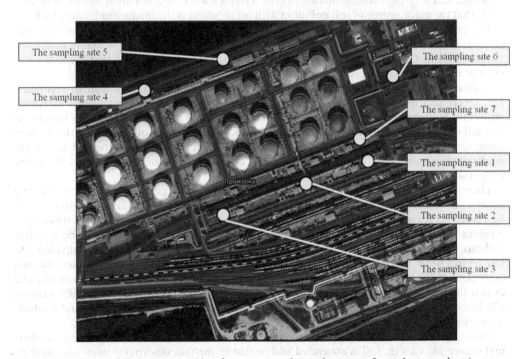

Figure 2. Layout of the test areas (object for transportation and storage of petroleum products).

The purpose of laboratory studies of selected soil samples was to establish the gross amount of petroleum products. Table 1 presents the threshold levels of the oil products concentration to characterize the soils technogenic contamination degree currently used in the Russian Federation.

Table 1. Threshold levels of the oil products concentration in soils (in Russian Federation).

Contamination level	Oil products concentration, mg/kg
Acceptable	<1000
Low	1000-2000
Medium	2001-3000
High	3001-5000
Very high	>5000

Determination of the petroleum products content in soil samples was carried out using a standard measurement of the petroleum products mass fraction in soils samples using the fluorimetric method with Fluorat-02 liquid analyzer.

Comparing the obtained results with threshold concentrations (for soil samples from the territory of exhausted well 9 Kumzhinskoe gas condensate field), it was found that in 5 soil samples (1-3,5,6) of 6, the threshold level of petroleum products (1000 mg/kg) is exceeded. Concentration range from 4800 to 372500 mg/kg and these samples are characterized by a high and very high contamination degree.

Comparing the obtained results with threshold concentrations (for soil samples from the territory of object for transportation and storage of petroleum products), it was found that in 9 soil samples (1-3 test area) of 21, the threshold level of petroleum products (1000 mg/kg) is exceeded. Concentration range from 3540 to 24850 mg/kg and these samples are characterized by a high and very high contamination degree. The location of 1-3 test areas allows us to conclude that the main source of soil pollution with oil products is the loading/unloading rack.

The existing methods of oil-containing wastes disposal are numerous and diverse, but each of them has its advantages and disadvantages, which determines their applicability under certain conditions. Comprehensive analysis of all existing methods suggested that thermal treatment of oil-contaminated soils may serve as a widely-accepted purification technic.

Unlike chemical and physical-chemical methods, thermal ones do not lead to secondary lithosphere and hydrosphere contamination due to the lack of reagents, which is extremely important in areas with an ecosystem vulnerable to chemical impacts. The main advantage of thermal methods over physical ones is the lower residual content of oil products, which determines the decontamination efficiency. The disadvantage of the biological method is that the used microorganisms are extremely sensitive to environmental conditions, especially to the temperature regime (Bykova et al. 2019).

Thermal desorption processes were used to clean the soil from petroleum products. Thermal desorption technologies are important in the recovery of soils from various hydrocarbons, however, in Russia these technologies are not widespread enough. Thermal desorption involves the application of heat to contaminated soils with the intention of desorbing hydrocarbons, which are then carried away by a sweep gas. Thermal desorption can be divided into low-temperature thermal desorption (LTTD, 100-300°C) and high-temperature thermal desorption (HTTD, 300-550°C) (Khan et al. 2004, Baker & Kuhlman 2002). Both in situ and ex situ thermal desorption are highly effective. LTTD and HTTD can result in >99% removal efficiency, although treatment time varies significantly by process configuration and contaminant (Stegemeier & Vinegar 2001).

Distinguish the thermal desorption ex situ and thermal desorption in situ. During the thermal desorption ex situ, soil is excavated and heated in thermal desorption units such as thermal screws or rotary drums. Desorbed hydrocarbons are carried away from the main reactor chamber by a sweep gas and incinerated or adsorbed onto activated carbon for final disposal and air pollution control (Troxler et al. 1993). Thermal desorption is achieved in situ through the application of dual heater/vacuum wells to desorb and remove contaminants via vapor extraction (Baker & Kuhlman 2002). In addition, the implementation of these technologies requires the use of large size equipment and energy that increases costs. For example, reported costs for thermal desorption ex situ range from $46 to $99 per metric ton (adjusted for 2019

USD), while thermal desorption in situ costs between $70 and $462 per metric ton (adjusted for 2019 USD) (Pellerin 1994, Barnes et al. 2002, Vermeulen & McGee 2000).

Thermal desorption cleaning applied to a wide range of volatile and semi-volatile hydrocarbons, including refined fuels, tars, creosote, rubber wastes, and TPH (Yeung 2010). Thermal desorption in situ can take weeks or years, while thermal desorption ex situ can take several minutes to complete treatment. Known research reports that 45% of benzo(a)pyrene was removed in two years in situ, and it has been suggested that high molecular weight PAHs cannot significantly desorb in less than one year of in situ treatment. Light hydrocarbons can be desorbed much faster. It is known that thermal desorption in situ was shown to remove >99% of coal tar in several days (Hansen et al. 1998). The results of laboratory studies on the cleaning of oil contamination soils using thermal desorption processes are known. Studies have been conducted on model soil of the following composition: sand-oil; loam (5% humus) - oil; soil (20% humus) - oil with different oil content. As a result of research, it was established that the degree of extraction of petroleum products from soils is 80-98%, and the residual content of petroleum products does not exceed 0.5-1% of the mass of soil at any initial content of petroleum products (Trushlyakov 1999).

When choosing a thermal method as a universal one for oil-contaminated soils disposal, not only its competitive advantages were taken into account. Given that the concentration of petroleum products in soils is significantly lower than in oil sludge, it can be assumed that the desired result can be achieved even with a slight temperature effect.

As an object of experimental studies, soil from the territory of the object for transportation and storage of petroleum products with a pollutant content above the permissible level was used. The aim of the experiment was to establish the optimal temperature regime taking into account the thermal properties of petroleum products.

3 RESULTS

The temperature treatment of contaminated soils was carried out using the LECO TGA701 thermogravimetric analyzer (Figure 3). The basis of the method is the measurement of weight loss as a function of temperature in the environment under control. The thermogravimetric analyzer was used only to create the necessary parameters for the thermal action.

To determine the optimal temperature for oil-contaminated soils purification, it is necessary to specify the thermophysical properties of possible oil components. The study of soil samples for the content of oil products has shown that exceeding of the permissible contamination level is observed directly near the loading/unloading rack due to the spills and leaks taking

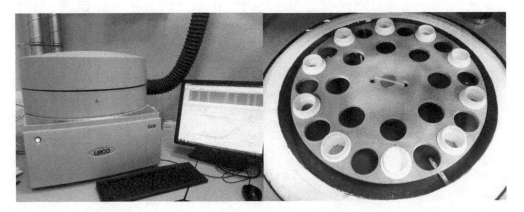

Figure 3. Thermogravimetric analyzer LECO TGA701, from left to right: general view; sample carousel.

place during operation processes. Gasoline boiling begins at a relatively low temperature (31-42°C) and proceeds very intensively.

The boiling point of diesel fuel begins at higher temperatures: on average 10% of the volume is distilled at 250°C.

It should be noted that a mixture of gasoline and diesel fuel is not a pure oil-product solution. The theory of solutions states that the boiling point of such mixtures always lies be-tween the boiling points of the mixture components (Akhmetov 2002). Consequently, it becomes possible to reduce the distillation temperature of 10% of diesel fuel from 250 to 150-170°C.

After a series of experiments, the optimum temperature was established (Figure 4). The decrease in the concentration of pollutants averaged 4-6 thousand times for all samples, and the residual oil content was significantly below the threshold level. Maximum temperature of treatment of the soil below that the end point of humate annealing (450°C).

Table 2 presents the results of residual oil content in soils. With a visual evaluation of ther-mally treated soils, no signs of particle sintering or structural damage were observed.

Experimental studies have shown that thermal impact on soils is an effective method not requiring sample preparation.

The implementation of given soil purification method in large industrial enterprises includes the possibility of equipment (for example, incinerator, multi-core, and drum-type furnaces) upgrade. The obtained results can be used to design a mobile apparatus that al-lows soil puri-fication at hard-to-reach or relatively small objects of the petroleum products transportation and operation. This is a significant difference from the known methods of cleaning soils with thermal desorption using large-size equipment.

Consequently, despite the advantages of this method, further research should suggest the ways of the process impact minimization, including operation regimes optimization and emis-sion gas control system integration.

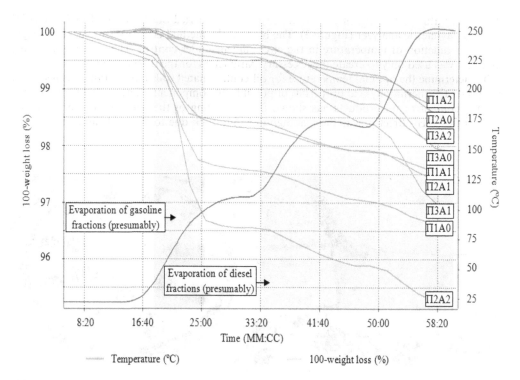

Figure 4. Results of thermogravimetric analysis with a final temperature 250°C, uniform heating rate of 15°C/min and a 1-minute holding at 170°C and 250°C.

Table 2. Residual content of oil products in soil samples.

Test area		Horizon	Petroleum products content, mg/kg	Petroleum products concentration level
1	1	A0	0	Acceptable
	2	A1	0	Acceptable
	3	A2	0	Acceptable
2	4	A0	36	Acceptable
	5	A1	410	Acceptable
	6	A2	130	Acceptable
3	7	A0	20	Acceptable
	8	A1	1	Acceptable
	9	A2	1	Acceptable

4 CONCLUSION

The problem of soil contamination with hydrocarbons is acute in all sectors of the economy. Spills and leaks of petroleum products, as practice shows, are inevitable in all life cycles of hydrocarbons, so it is necessary to organize monitoring of soil quality and sanitation of oil-contaminated areas.

In the Russian Federation and around the world there is a problem of rationing the content of oil and petroleum products in soils, which complicates the task of assessing the impact of production facilities on the geological environment. However, with the content of oil products more than 1000 mg/kg in soils, the processes of self-purification slowdown, which can lead to partial or complete destruction of vegetation, which was confirmed by the results of engineering and environmental survey of the territories of various production facilities.

As a result of the activity of the investigated production facilities, a significant contamination of soil with oil products occurs. The results of laboratory studies have shown that in the territory of exhausted well 9 Kumzhinskoe gas condensate field the excess of the permissible content of oil products in the samples is about 16-310 times; in the territory of object for transportation and storage of petroleum products the concentrations exceed the permissible content by 3-25 times;

Most enterprises in the Russian Federation transfer the treatment of this type of waste to third parties and pay them money. With the implementation of this method of cleaning in enterprises, a positive economic effect can be achieved.

REFERENCES

Akhmetov S.A. (ed.) 2002. *Technology of oil and gas processing: a textbook for high schools*. Ufa: publishing house «Guillem».

Alekseenko V.A. et al. 2017. Geochemical Barriers for Soil Protection in Mining Areas. In J. Bech, C. Bini, M.A. Pashkevich (eds), *Assessment, Restoration and Reclamation of Mining Influenced Soils: 255-274*. London: Academic Press.

Baker R.S & Kuhlman M. A. 2002. Description of the mechanisms of in-situ thermal destruction (ISTD) reactions. In Al-Ekabi H (ed.), *Proceedings of the 2nd International Conference on Oxidation and Reduction Technologies for Soil and Groundwater*; 17-21 November 2002. Toronto: Canada.

Barnes D.L. et. al. (ed.) 2002. *Treatment of petroleum-contaminated soil in cold, wet, remote regions*. Missoula: USDA Forest Service.

Bykova M.V. et al. 2019. Assessment and abatement of the soil oil-contamination level in industrial areas, *Topical Issues of Rational Use of Natural Resources*: 347-359.

Gennadiev A.N. & Pikovskii Yu. I. 2007. The Maps of Soil Tolerance toward Pollution with Oil Products and Polycyclic Aromatic Hydrocarbons: Methodological Aspects. *Eurasian Soil Science 40(1)*: 70–81.

Gilmore K. R. et al. 2014. Transport of hydraulic fracturing water and wastes in the Susquehanna river basin, Pennsylvania. *Journal of Environmental Engineering 140(5)*: 10–21.

Hansen K.S. et. al. 1998. In situ thermal desorption of coal tar. In Proceedings of the IGT/GRI (ed.). *International Symposium on Environmental Biotechnologies and Site Remediation Technologies*, 7-9 December 1998. Washington, DC: US Environmental Protection Agency.

Kovaleva E.I. 2017. Ecological Evaluation of Oil-Contaminated Soils (Sakhalin) Using Enchytraeidae. *Eurasian Soil Science 50(3)*: 350-358.

Khalilova H. Kh. 2015. The Impact of Oil Contamination on Soil Ecosystem. *Biological and Chemical Research*: 133-139.

Khan F.I et al. 2004. An overview and analysis of site remediation technologies. *Journal of Environmental Management 71(2)*: 95-122.

Laden F. et al. 2006. Reduction in Fine Particulate Air Pollution and Mortality. *American Journal of Respiratory and Critical Care Medicine*: 667-672.

Mezhibor A.M. et al. 2016. Geochemical features of sphagnum mosses and epiphytic lichens in oil and gas exploitation areas (the case of Western Siberia, Russia). *Environmental Earth Sciences 75(18)*: 1260.

Pashkevich M.A. & Petrova T.A. 2014. An integrated approach to the creation of air monitoring system on the objects' oil and gas industry [Kompleksnyj podhod k sozdaniyu sistemy monitoringa atmosfernogo vozduha na ob"ektah neftegazovoj promyshlennosti]. *Oil Industry (4)*: 110-111.

Pashkevich M.A. & Petrova T.A. 2017. Reclamation by Containment: Polyethylene-Based Solidification. In J. Bech, C. Bini, M.A. Pashkevich (eds), *Assessment, Restoration and Reclamation of Mining Influenced Soils: 235-253*. London: Academic Press.

Pellerin C. 1994. Alternatives to incineration: there's more than one way to remediate. *Environmental Health Perspectives 102(10)*: 840-845.

Shuvalov Yu.V. et. al. 2008. *Rational methods of contaminated site treatment by hydrocarbon compounds* [Racional'nye sposoby sanirovaniya ochagov tekhnogennogo zagryazneniya uglevodorodnymi soedineniyami]. St. Petersburg: X-print.

Sobolev I.S. et al. 2018. Chemical diagenesis in near-surface zone above oil fields in geochemical exploration. *Applied Geochemistry. 95*: 33-44.

Stegemeier G.L. & Vinegar H.J. (ed.) 2001. *Thermal conduction heating for in-situ thermal desorption of soils*. Boca Raton: CRC Press.

Timonin A.S. 2003. *Engineering and environmental handbook* [Inzhenerno-ekologicheskij spravochnik]. Kaluga: Publishing house N. Bochkarevoi.

Troxler W.L. et. al. 1993. Treatment of nonhazardous petroleum-contaminated soils by thermal desorption technologies. *Journal of the Air & Waste Management 43(11)*:1512-1525.

Trushlyakov V.I. et. al. 1999. Development of technology and equipment for detoxication of soil, soil and oil scraps from petroleum products [Razrabotka tekhnologii i oborudovaniya dlya detoksikacii pochv, gruntov i nefteshlamov ot nefteproduktov]. *Omsk scientific Bulletin*: 22-24.

Vermeulen F. & McGee B. 2000. In-situ electromagnetic heating for hydrocarbon recovery and environmental remediation. *Journal of Canadian Petroleum Technology 39(8)*: 24-28.

Yeung A.T. 2010. Remediation technologies for contaminated sites. In Chen Y., Zhan L., Tang X. (eds), *Advances in environmental geotechnics: 328-369*. Hangzhou: Zhejiang University Press.

Scientific and Practical Studies of Raw Material Issues – Litvinenko (Ed)
© 2020 Taylor & Francis Group, London, ISBN 978-0-367-86153-7

Landscape geochemical consequences of ore mining

V.A. Alekseenko
Institute for Water and Environmental Problems of Siberian Branch of the Russian Academy of Sciences,
Barnaul, Russia;
Admiral Ushakov Maritime State University, Novorossiysk, Russia
Southern Federal University, Rostov-on-Don, Russia

N.V. Shvydkaya
Kuban State Agrarian University, Krasnodar, Russia

A.V. Puzanov
Institute for Water and Environmental Problems of Siberian Branch of the Russian Academy of Sciences,
Barnaul, Russia

ABSTRACT: The distinctive features of technogenic geochemical landscapes arising from mining of mineral deposits are considered. Mine dumps cause pollution of the soil and plants of the neighboring landscapes with heavy metals. This is mainly due to water migration (often as ionic solutions) and insignificantly owing to atmospheric transport. Particular attention is paid to the changing prevalence of major ore-forming and associated chemical elements in various parts of industrial landscapes and pollution during the development of natural biogenic landscapes. Associations of elements found jointly in elevated concentrations in soils and plants often do not coincide, which indicates the role of biogeochemical processes. The possibilities of self-restoration and self-cleaning of geochemical landscapes after the closure of mines are assessed.

1 INTRODUCTION

The study of man-made landscapes of mines allows prediction of their further development. This is of particular importance for developing programmes of the most balanced reclamation, as well as for the subsequent changes in mining landscapes (primarily of the soil-vegetation layer). In different regions and countries, these problems are solved very differently, which largely depends on socio-economic conditions. In states with a high level of economic development, a high population density, and a shortage of agricultural areas, high costs for soil remediation and restoration of disturbed land are acceptable.

In countries and regions where there is no such possibility or need, engineered reclamation is usually not performed. At the same time, it is believed that due to natural processes of self-restoration and self-cleaning of landscapes and, first of all, their soils are possible. It is this approach to soil rehabilitation in the overwhelming majority of cases that is characteristic of the regions under study, where metallic ore mining was performed for years. The most recent findings reflect, however, certain similar traits in environmental impact caused by the mining of non-metallic deposits (Buzmakov et al. 2019; Boente et al. 2019; Hasanuzzaman et al. 2018; Ondar et al. 2018; Zhan et al. 2019; Zhang et al. 2018).

2 MATERIALS AND METHODS

We studied the landscape-geochemical features of the areas of the developed deposits using the classification scheme of terrestrial geochemical landscapes based on the use of classification levels. According to the taxonomic hierarchy, at the first two levels of landscape identification, the prevalence of the leading type of migration of substances and its features are taken into account.

We emphasise that with the start of the development of the deposit, industrial landscapes new for the considered regions appear on the site of the biogenic. Depending on the mining methods, among them, the landscapes of mines with underground mining and the landscapes of quarries (open-cast mines) are distinguished. They differ from each other and from the surrounding biogenic landscapes by the characteristics of the technogenic (leading in this case) type of migration. With the closure of mines and the cessation of anthropogenic activity, the industrial landscapes formed during their work begin to turn back into biogenic ones.

For studying the industrial landscapes, we conducted areal testing of soils and plants at the mines of the Northwestern Caucasus. The sampling grid cell size at background sites, dumps of quarries and sedimentation tanks varied from 2-5 m to 250 m. Control testing was carried out in the amount of 3-5 %. All the 350 samples of soil and plant ash were subjected to emission spectral analysis for 14-25 chemical elements.

For a more complete and logical study of the most important landscape-geochemical changes caused by a specific anthropogenic activity – development of ore deposits – we deliberate these changes in accordance with the most important classification features of the landscapes and the features of the assessment of environmental-geochemical changes in the biosphere.

All of the above, as well as the number of samples, control of their collection, internal and external control of analyses, allow us to consider the data obtained as objective and reliable.

3 RESULTS

The landscapes of quarries and mines with underground mining of mineral raw materials are very significantly different from each other by geomorphological, geochemical and botanical patterns. Let's consider it in more detail.

Under the open-pit mining of deposits, the most important differences are those caused by land topography, geochemical, and vegetal changes, including changes in plant communities. The development of these changes is associated with the creation of quarries and dumps of rocks and the accompanying destruction of the soil-plant horizon of geochemical landscapes and the disruption of the natural processes of migration-concentration of elements. Environmental and geochemical changes are often significantly affected by rocks that have been exposed during quarrying and have not previously been exposed to an oxidizing environment. Features of emerging landscapes are taken into account at almost all taxonomic levels of the said classification.

Under the conditions of the surrounding mountain forest landscapes, in deep ($n \cdot 10$ m) open pits 50-80 years after the closure of the mines, the thickness of the loose formations at their bottom did not exceed the first tens of cm on average. This is despite the fact that in the early years, the primary forest soils crawled quite intensively from the landscapes surrounding the quarry. The soils formed at the bottom of the quarries can be somewhat conditionally divided into sliding from neighbouring landscapes and formed from temporary water flows (Aleksandrova et al. 2019; Bolshunova et al. 2017; Pashkevich & Petrova 2019).

When forming biogenic landscapes of dumps, plants largely determine the rate of soil formation and geochemical characteristics of soils, especially Inceptisols, i.e. the soils that, according to the USDA soil taxonomy, show an average degree of development and lack significant clay accumulation in the subsoil. Plants also affect greatly the rate of formation of natural landscapes from man-made ones and their geochemical features (Bezel' et al. 2015; Zhuikova et al. 2017).

Figure 1. An edge of low-slope dumps at the Sakhalinskoe mine. Conventional signs: 1 – edge height; 2 –edge length; 3 – plants on dumps, common for this landscape (a – *Pinus pallasiana*, b – *Swida australis*, c – *Carpinus orientalis*, d – *Rubus caucasicus*).

Ca. 90 % of the accumulated deposits are due to the flora of the adjacent zonal ecosystems, the remaining species are weedy. Plant communities formed at the developed deposits are not uniform in composition and structure. *Phragmites australis* (Cav.) Trin ex Steudel dominates on mercury deposits in the waterways of temporary flows and depressions of deposits, forming full-term highly productive cenopopulations.

The increased wetting of the substrate of transaccumulative landscapes (the Perevalnoye deposit) caused the territory to be populated with local moisture-loving plants, *Alnus glutinosa* (L.) Gaerth.

At the non-reclaimed dumps (the Sakhalinskoe deposit) among undisturbed forest landscapes, small-species tree-shrub communities with sparse forest stands of *Pinus kochiana, Pyrus caucasica* Fed. have formed and a minor undergrowth level of *Swida australis* (C.A. Mey.) Pojark. ex Grossh., etc. have emerged as well (Figure 1).

4 DISCUSSION

The thin soils formed under the influence of temporary water flows differ at mercury deposits in a slightly higher average content of Mo, Ni, and Ti. The accumulation of the first two elements can be explained by deposition at the sorption barrier created by clay material, the content of which in these areas is increased by an average of 5 times. The deposition of Ti, usually migrating in the form of oxide, occurred at the mechanical barrier when the velocity of the bearing water flow along the almost vertical walls of the quarry significantly decreased at its bottom.

The content of Ag, Ba, Co, Mn, Pb, Sn, Sr, and V in soils crawling to the bottom of the quarries is higher than in soils formed under the influence of temporary water flows and in soils of the surrounding forest landscapes. The formation of this association of elements cannot be explained by any single general geochemical process (Kabata-Pendias & Pendias

145

Table 1. Several crystal-chemical indicators of elements that are in high concentrations in soils that slid to the bottom of the quarries.

Element	Atomic radius	Ionic radius	Valence	Ionization potential of the first electron	Cartledge ionic potential	EC[1] (EC[2])
Ag	1.44	1.13	+1	7.54	0.89	0.60
Ba	2.25	1.43	+2	5.19	1.40	1.35
Co	1.26	0.82	+2; +3	7.81	2.44	2.15
Mn	1.30	0.91/0.70/0.52	+2; +3; +4	7.40	2.20/4.28/7.70	1.95-9.10
Pb	1.74	1.32/0.84	+2/+4	7.39	1.51/4.76	1.65/7.95
Sn	1.58	(1.04)0.74	+2/+4	7.36	- 5.40	- 7.90
Sr	2.16	1.27	+2	5.67	1.58	1.53
V	1.36	0.65/0.61/0.40	+3/+4/+5	6.76	4.61/6.56/12.50	5.32-5.12

2001; Kasimov et al. 2016; Perel'man 1986; Syso et al. 2016). Not all of the elements under consideration were precipitated from ionic solutions.

For instance, Sn and V are characterised by the formation of complex ions, poorly soluble hydrolyzed compounds, as shown by the values of the Cartledge ionic potential from 3 to 12. Judging by the values of the radii of the ions and the values of their energy coefficients, Ba, Co, and V could not be concentrated together from the true solutions. For Ag, Pb, and Sn, biogenic accumulation is insignificant (Table 1). Pb is characterised by the concentration at a hydrogen sulphide barrier, Mn accumulates at an oxygen one, etc. For the elements under consideration, the most probable concentration scenario is as follows (Aleksander-Kwaterczak & Helios-Rybicka 2009; Alekseenko et al 2017; Kremcheev et al 2018): accumulation starts in sliding soils at the bottom of the quarry as a result of deposition on mechanical and sorption geochemical barriers, while for Co, Mn, Sr, and V a certain role of biogenic accumulation was important.

The composition of the solutions, from which the precipitation occurred, has been influenced by the soils located above the quarry and exposed as a result of technogenesis, indigenous rocks of the sides and bottom of the quarries. Based on the foregoing, it is possible to predict a continuation of the concentration of these chemical elements in the soils under consideration.

5 CONCLUSIONS

The development of ore deposits leads to the formation of new technogenic landscapes with significant changes in the topography, composition and development of plant associations, and the concentration of a number of metals in soils and plants.

The geochemical patterns of the arising technogenic landscapes and the degree of their influence on neighbouring ones are determined by the combined influence of natural and technogenic factors. Of the first, the largest role belongs to the mineralogical features of the ores and the rocks enclosing them, moisture, terrain, plant associations, and geochemical barriers. Of the technogenic factors, we note the type of deposit mining; the implementation of at least the most primitive reclamation; area and occupied by dumps and their volume; creation of a hydrological regime localising the migration of contaminated solutions.

ACKNOWLEDGEMENTS

The analyses of soil and plant samples were carried out in the certified and accredited Central Testing Laboratory at Kavkazgeolsyomka using emission spectral analysis.

REFERENCES

Aleksander-Kwaterczak, U. & Helios-Rybicka, E. 2009. Contaminated sediments as a potential source of Zn, Pb, and Cd for a river system in the historical metalliferous ore mining and smelting industry area of South Poland. *J. Soils Sediments* 9 (1), 13–22.

Aleksandrova, T.N., Talovina, I.V. & Duryagina, A.M. 2019. Gold–sulphide deposits of the Russian Arctic zone: Mineralogical features and prospects of ore benefication. *Chemie der Erde* DOI: 10.1016/j.chemer.2019.04.006

Alekseenko, V.A., Maximovich, N.G. & Alekseenko, A.V. 2017. Geochemical Barriers for Soil Protection in Mining Areas. In: Bech, J., Bini, C. & Pashkevich, M.A. (eds), *Assessment, Restoration and Reclamation of Mining Influenced Soils*: 255-274. Elsevier Inc.

Bezel', V.S., Zhuikova, T.V. & Gordeeva, V.A. 2015. Geochemistry of grass biocenoses: Biogenic cycles of chemical elements at contamination of the environment with heavy metals. *Geochemistry International* 53(3),241-252.

Boente, C., Gerassis, S., Albuquerque, M.T.D., Taboada, J. & Gallego, J.R. 2019. Local versus Regional Soil Screening Levels to Identify Potentially Polluted Areas *Mathematical Geosciences* DOI: 10.1007/s11004-019-09792-x

Bolshunova, T.S., Rikhvanov, L.P., Mezhibor, A.M., Baranovskaya, N.V. & Yusupov, D.V. 2017. Biogeochemical features of epiphytyc lichens from the area of the tailing of a gold-polymetallic deposit (Kemerovo region, Russia) comparative to a reference area. *International Multidisciplinary Scientific GeoConference Surveying Geology and Mining Ecology Management, SGEM* 17(51),165-172.

Buzmakov, S., Egorova, D. & Gatina, E. 2019. Effects of crude oil contamination on soils of the Ural region. *Journal of Soils and Sediments* 19(1),38-48.

Hasanuzzaman, Bhar, C. & Srivastava, V. 2018. Environmental capability: a Bradley–Terry model-based approach to examine the driving factors for sustainable coal-mining environment. *Clean Technologies and Environmental Policy* 20(5) 995-1016.

Kabata-Pendias, A. & Pendias, H. 2001. *Trace elements in soils and plants*. CRC Press LLC, Boca Raton.

Kasimov, N.S., Kosheleva, N.E. & Timofeev, I.V. 2016. Ecological and Geochemical Assessment of Woody Vegetation in Tungsten-Molybdenum Mining Area (Buryat Republic, Russia). *IOP Conference Series: Earth and Environmental Science* 41(1),012026.

Kremcheev, E.A., Gromyka, D.S. & Nagornov D.O. 2018. Techniques to determine spontaneous ignition of brown coal. *Journal of Physics: Conference Series* 1118(1) 012021.

Li, F., Li, X., Hou, L. & Shao, A. 2018. Impact of the Coal Mining on the Spatial Distribution of Potentially Toxic Metals in Farmland Tillage Soil. *Scientific Reports* 8(1),14925.

Ondar, S.O., Khovalyg, A.O., Ondar, U.V. & Sodnam, N.I. 2018. Monitoring of the state of the left-bank confluents of the Upper Yenisei basin in the zone of impact of the coal industry enterprise. *International Journal of Engineering and Technology (UAE)* 7(3) 206-214.

Pashkevich, M.A. & Petrova, T.A. 2019. Recyclability of ore beneficiation wastes at the Lomonosov Deposit. *Journal of Ecological Engineering* 20(2),27-33.

Perel'man, A.I. 1986. Geochemical barriers: Theory and practical applications. *Applied Geochemistry* 1 (6),669–680.

Syso, A.I., Syromlya, T.I., Myadelets, M.A. & Cherevko, A.S. 2016. Ecological and biogeochemical assessment of elemental and biochemical composition of the vegetation of anthropogenically disturbed ecosystems (based on the example of Achillea millefolium L.). *Contemporary Problems of Ecology* 9 (5),643-651.

Zhan, F., Li, B., Jiang, M., Li, Y. & Wang, Y. 2019. Effects of arbuscular mycorrhizal fungi on the growth and heavy metal accumulation of bermudagrass [Cynodon dactylon (L.) Pers.] grown in a lead–zinc mine wasteland. *International Journal of Phytoremediation* DOI: 10.1080/15226514.2019.1577353

Zhang, Z., Sui, W., Wang, K., Tang, G. & Li, X. 2018. Changes in particle size composition under seepage conditions of reclaimed soil in Xinjiang, China. *Processes* 6(10), 201.

Zhuikova, T.V., Gordeeva, V.A., Bezel', V.S., Kostina, L.V. & Ivshina, I.B. 2017. The Structural and Functional State of Soil Microbiota in a Chemically Polluted Environment. *Biology Bulletin* 44 (10),1228-1236.

Scientific and Practical Studies of Raw Material Issues – Litvinenko (Ed)
© 2020 Taylor & Francis Group, London, ISBN 978-0-367-86153-7

Reduction of nitrous gases from blast fumes in underground mines using water based absorbency solutions

A. Hutwalker*, T. Plett & O. Langefeld
Clausthal University of Technology, Germany

ABSTRACT: State of the art in dealing with blast fumes in underground mining operations is ventilation, following the slogan: dilution is the solution to pollution. Today, blasting is one of the key procedures in modern underground mines; it is used for drift development as well as for extraction of raw materials. The resulting blast fumes are one of the major sources for nitrous gases (NO and NO_2) in the mines. The reduction of Occupational Expose Limits (OEL) for these gases in Germany by about 90% in 2016, with a transition period until 2021 for the mining sector and 2023 for the rest of Europe, leads to a big challenge for the mining industry. To keep re-entry times after blasting at least at the present level and thereby ensure an economic operation of the mine, using ventilation solely is not sufficient. For this reason, a new approach to reduce nitrous gases from blast fumes in underground mining operations using spraying systems with aqueous absorbency solutions is described. Laboratory experiments using water, sodium hydroxide (NaOH) and sodium sulfite (Na_2SO_3) were performed. A reduction of NO and NO_2 was observed.

1 INTRODUCTION AND MOTIVATION

The reduction of occupational exposure limits (OEL) for the gases nitrogen oxide (NO) and nitrogen dioxide (NO_2) in Europe, and especially in Germany – no other topic was discussed more often than this in the field of mine ventilation within the last years (Hutwalker 2016; Kübler 2016; Kuhn 2017; Neumann 2016; Steinhage 2014; Triebel 2016; Triebel 2017; Weyer 2019).

Blasting is one of the major processes in modern underground mining, used for drift development as well as for production. State of the art in dealing with blast fumes is ventilation; the fumes are diluted with fresh air and ventilated out of the mine. This treatment of blast fumes is more than sufficient with the former OEL.

The adaption of the OEL in Germany to the recommendations of the Committee on Occupational Exposure Limits (SCOEL), a committee of the European Commission, leads to big challenges for the mining industry in reducing the concentrations of these gases at the workplaces underground. On its meeting on May 2nd and 3rd 2016, the Committee on Hazardous Substances (Ausschuss für Gefahrstoffe, AGS), an advisory body of the German Federal Ministry of Labour and Social Affairs (BMAS), advised the BMAS to copy the OELs, as recommended by the SCOEL, to the Technical Rules for Hazardous Substances (TRGS 900). BMAS followed this advice, consequently the new OEL where published on November 4th 2016 and became part of the TRGS 900 (BAuA 2016). Table 1 shows the former and new OELs.

These new OEL became law after publishing, but two exceptions were made: For tunneling operations they became law on October 31st 2017, for mining operations they will become law on October 31st 2021 (TRGS 2017).

To meet these new requirements while at the same time allow the workforce to enter a workplace after blasting in re-entry times similar to today, and thereby ensuring a profitable

---

* Corresponding author: alexander.hutwalker@tu-clausthal.de

Table 1. Occupational Exposure Limits in Germany.

	Nitrogen oxide (NO)	Nitrogen dioxide (NO$_2$)
Former OEL	25 ppm	5 ppm
New OEL	2 ppm	0.5 ppm
Reduction	92%	90%

operation, research is done to identify measures to reduce the nitrous gases in the blast fumes to match the new OELs in a reasonable time.

One idea is to use spraying systems to absorb the nitrous gases from the blast fumes. Therefore, the absorption of nitrous gases from blast fumes by water based spraying systems was tested in a lab plant using different absorbencies.

In the next chapter, this paper will give a brief outline of the situation in the mine and the resulting boundary conditions, followed by the description of the laboratory equipment as well as the tests performed. At the end, results will be shown and discussed; also, an outlook on the following research will be given.

2 ENVIRONMENT AND BOUNDARY CONDITIONS

After blasting, the blast fumes create concentrations of CO, NO and NO2 exceeding the respective OEL's many times over. These gaseous substances mix up with the air in the mine, following the airways to the exhaust shaft. The mining companies have to ensure, that on the way from the place of blasting to the shaft, no workforce is faced with concentrations of these gases exceeding the OEL's. However, not only these gases, also OEL's of other substances need to be considered. Following this, it needs to be ensured that all substances used as absorbencies are not harmful or the respecting OEL's will not be exceeded in areas where workforce is present. To be on the safe side, it was decided that droplets created by the spraying system needs to be sedimented before reaching areas where people might be present.

To protect the spraying system from the shockwave of the blast as well as from flyrock, the system should be installed at least 50 m from the face. The blast shelter, where the workforce rests during and after blasting, is located 300 m from the face. With an average airspeed of 0.8 m/s the blast fumes take about 5 min (312 s) from the spraying device to the blast shelter. To ensure no absorbency droplets reach the area of the shelter, droplet sizes have to be calculated to a diameter that they will settle down after 4 minutes, accordingly 192 m. This means, the absorbency needs to reduce the concentration of nitrous gases within 4 minutes of contact significantly.

3 MATERIALS AND METHODS FOR LAB SCALE TESTS

To test different absorbencies a lab scale absorption plant was installed (Figure 1). This test plant is used to measure absorption rates of nitrous gases by different absorbencies and under various conditions regarding temperature and pump pressure. In addition, different nozzles are used to test different flow rates of absorbency solution and droplet sizes (Hutwalker 2016).

The gas used for the tests, the "artificial blast fumes", matches the concentration ratio of CO/NO/NO2 as measured after underground blasts for drift development in German hard coal mines with a Class I permissible explosive. This gas was produced by Westfalen AG.

Sodium sulfite (Na2SO3) used is purity ≥ 98%, p.a., ACS reagent, anhydrous. Sodium hydroxide (NaOH) used is purity ≥ 99%. Na2SO3 and NaOH were obtained from Carl Roth GmbH.

For measurement of CO, NO and NO2 a Dräger X-am® 7000 is used, equipped with following Dräger sensors: XS EC CO, XS EC NO and XS EC NO2. The X-am® 7000 is the

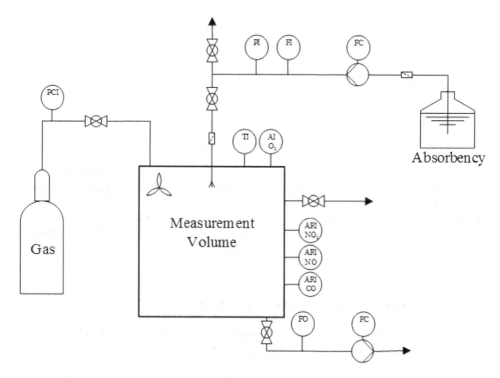

Figure 1. Concept of the laboratory absorption test plant.

standard measurement device in German hard coal mines. In a later stage of this research, the absorbencies should be tested in underground hard coal mining blasts. For this reason, the Dräger device is chosen for laboratory test also. Oxygen concentration is monitored using a GOX 100t, manufactured by GHM-GREISINGER. Temperature is monitored by a TM25 digital thermometer (EXTECH Instruments), the flow rate of absorbency with a ROTA YOKOGAWA rotameter, type RAGK41.

For the measurements the test gas (Figure 1, left) is led into the 216 l measurement volume (Figure 1, middle), until a concentration of 40 ppm NO and 4 ppm NO2 is reached. The pump for the absorbency (Figure 1, right) is switched on and runs for 12 minutes. During this time, concentration of nitrous gases in the volume is recorded and later analyzed using Dräger CC-Vision and spreadsheet software. To avoid cross-contamination, a separate pump is used for every absorbency solution. Absorbency solution is removed from the measurement volume using an electric pump. Between measurements the cell is thoroughly cleaned and rinsed with clear water.

To vary the flow rate and droplet size of the absorbency solution two different nozzles, 0.3 mm and 0.5 mm full cone nozzles (obtained from M.R.S.; Germany), were used. Furthermore, the pump pressure was varied from 0.7 to 1.5 MPa.

4 THEORY

4.1 *Blast fumes composition*

Generally, explosives consist of the elements nitrogen (N), hydrogen (H), carbon (C) and oxygen (O). In an ideal thermodynamic reaction, detonation would result in water (H_2O), nitrogen (N_2) and carbon dioxide (CO_2). However, data from underground measurements show that blast fumes also contain numerous other gases than those mentioned. Additional gases occurring are carbon monoxide (CO), nitrogen oxide (NO) and nitrogen dioxide (NO_2)

as well as traces of ammonia (NH₃), sulfur dioxide (SO₂) and volatile organic compounds (VOC) (Abata 1982; Garcia 1986; Garcia 1989; Heinze 1993; Mainiero 1997; Maurer 1963).

Furthermore, there are changes in the composition of the blast fumes after blasting due to chemical reactions with ambient mine air. The nitrogen oxide is oxidized to nitrogen dioxide (Berner 1979; Graefe 1975; Koplin 1979):

$$2\,NO + O_2 \leftrightarrow 2\,NO_2$$

The above given equation is an equilibrium reaction. At atmospheric pressure and temperatures below 423.15 K, the entire NO reacts to NO₂. Figure 2 shows the progress of the nitrous gas concentrations, as measured in the dry lab plant without absorbency spraying.

These data measured match the above given chemical equation, so the NO and NO2 measuring device is applicable.

4.2 Reducing nitrous gases in blast fumes by absorption

To reduce the nitrous gases in the blast fumes they must dissolve in the absorbency solution. This mass transfer is based on the physical process of diffusion, which is described by Fick's first law of diffusion (Fick 1855). The diffusion flux J is calculated by:

$$J = -D \cdot \frac{\partial c}{\partial x} \left[mol \cdot m^{-2} \cdot s^{-1} \right]$$

with

$$D = Diffusion\ coefficient\ \left[m^2 \cdot s^{-1} \right]$$
$$\partial c / \partial x = concentration\ gradient\ \left[mol \cdot m^{-4} \right]$$

As the concentration of nitrous gases in the fumes is very low (ppm), and the diffusion coefficient is constant for given substance systems, a low diffusion flux J is expected. From the unit of the diffusion flux can be derived that the amount of gas molecules absorbed in the liquid phase depends on time and the area of common phase boundary. To achieve a high absorbency rate, the common phase boundary should be as large as possible. Therefore, a spraying system is used to create small droplets with a high surface/volume ratio.

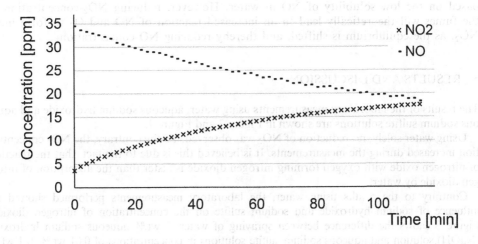

Figure 2.   Time-dependent concentrations of nitrous gases in the laboratory test plant.

The experiments were performed using aqueous solutions of sodium sulfite and sodium hydroxide (NaOH) as well as and water without added absorbency. NO2 dissolves in the solution and reacts with the absorbency in a second step. In case of using water solely, multiple reactions with the water take place (Gann 1998; Logan 2003; Mortimer 2008; Topchiev 1959; Warneck 1999):

$$NO + NO_2 + H_2O \rightarrow 2\,HNO_2$$

$$2\,NO_2 + H_2O \leftrightarrow HNO_2 + HNO_3$$

$$3\,NO_2 + H_2O \rightarrow 2\,HNO_3 + NO$$

$$3\,NO_2 + O_2 + 2\,H_2O \rightarrow 4\,HNO_3$$

$$N_2O_4 + H_2O \rightarrow HNO_2 + HNO_3$$

The chemical reactions of all above given equations are rather slow, taking place in hours. Therefore, these reactions are not relevant for this study, with planned reaction times of 4 minutes.

When using sodium hydroxide as absorbency, sodium nitrite and sodium nitrate are produced (Chambers 1937; Sun B 2017):

$$2\,NO_2 + 2\,NaOH \rightarrow NaNO_2 + NaNO_3 + H_2O$$

With sodium sulfite as absorbency, NO2 forms sodium sulfate (Chang 2004):

$$2\,NO_2 + 4\,Na_2SO_3 \rightarrow N_2 + 4\,Na_2SO_4$$

Using NaOH or Na_2SO_3 as absorbency, the products will remain as a white solid substance when the water is evaporated, along with that amount of sodium hydroxide or sodium sulfite which did not react.

Considering Sun et al. (2017) and Chang et al. (2004), it is not expected that there is any chemical reaction between sodium hydroxide or sodium sulfite and nitrogen oxide, based on the low solubility of NO in water. However, reducing NO_2-concentration in the fumes will theoretically lead to an increased reaction of NO and O_2 to form more NO_2, as the equilibrium is shifted, and thereby reducing NO-concentration.

5 RESULTS AND DISCUSSION

The results of the laboratory measurements using water, aqueous sodium hydroxide and aqueous sodium sulfite solutions are shown in Figure 3 and Figure 4.

Using water solely, no reduction of NO_2 was observed. To the contrary, the NO_2 concentration increased during the measurements. It is believed this is due to the fact, that the reaction of nitrogen oxide with oxygen forming nitrogen dioxide is faster than the absorption of nitrogen dioxide by water.

Contrary to the results using water, the laboratory measurements performed showed an influence of sodium hydroxide and sodium sulfite on the concentration of nitrogen dioxide. Figure 3 shows the difference between spraying of water, 1 wt.% aqueous sodium hydroxide (NaOH) solution and aqueous sodium sulfite solutions in concentrations of 0.01 wt.%, 0.1 wt.%

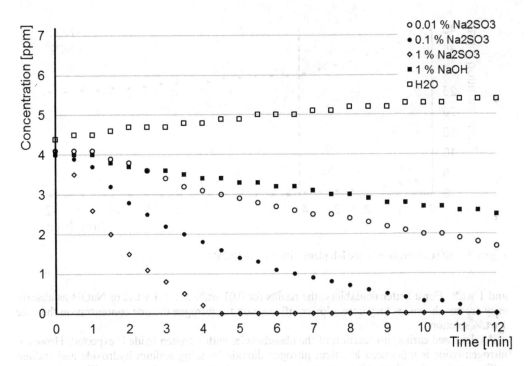

Figure 3. Time-dependent concentration of NO₂ with spraying of various absorbencies.

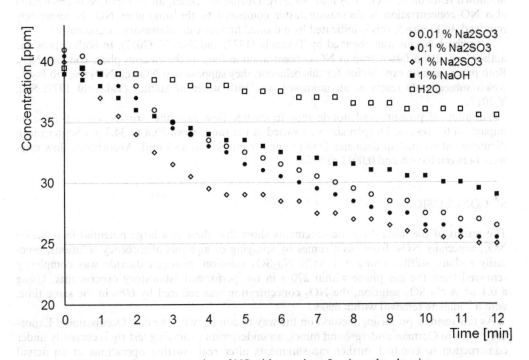

Figure 4. Time-dependent concentration of NO with spraying of various absorbencies.

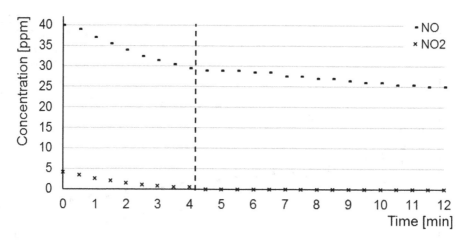

Figure 5.   NOx absorption in the lab plant with 1 wt.% Na2SO3 aqueous solution.

and 1 wt.%. For a better readability, the results for 0.01 wt.% and 0.1 wt.% of NaOH as absorbency are not shown, as they have a lower influence on the nitrogen dioxide concentration than the 1 wt.% solution.

As discussed earlier, no reaction of the absorbencies with nitrogen oxide is expected. However, nitrogen oxide is influenced less than nitrogen dioxide by using sodium hydroxide and sodium sulfite solutions, but there still is a reduction compared to the usage of water (see Figure 4).

Chang et al. (2004) state that there is no significant change in NO concentration by using Na_2SO_3 in a wet scrubber. But our laboratory measurements indicate that there is a yet unknown reaction of NO, NO_2 and Na_2SO_3. During the timespan of rapid NO_2 absorption also NO concentration is decreasing faster compared to the time after NO_2 is completely removed. In Figure 5, this is indicated by a dashed line for our laboratory measurements.

A similar behavior was observed by Takeuchi (1977) and Sun, Y. (2017). In both papers, the authors measure a faster drop in NO-concentration as long as the gaseous phase contains NO_2. Both papers give no explanation for this behavior, they suppose that NO and NO_2 react to N_2O_3, which subsequently reacts in an unknown way with water or sulfites (Takeuchi 1977; Sun, Y 2017).

Variation of pressure and nozzle type to change flow rate and droplet size did have no impact on the results. Droplet size was varied in the range from 25.8 to 34.3 μm Sauter mean diameter (Mass median diameter D50 ranging from 32.9 to 43.4 μm). Absorbency flow rates were between 0.0008 and 0.0012 m³/min.

6   CONCLUSION

The results of the laboratory measurements show that there is a large potential in reducing NO_x, especially NO_2, from blast fumes by spraying of aqueous absorbency solutions, especially sodium sulfite. Using a 1 wt.% Na_2SO_3 solution, nitrogen dioxide was completely removed from the gas phase within 270 s in the performed laboratory experiments. Using a 0.1 wt.% Na_2SO_3 solution, the NO_2 concentration was reduced by 60% in the same time, with a complete removal within 690 s.

As this seems a promising method on the way to comply with the new Occupational Exposure Limits in German underground mines, an underground spraying test rig is currently under construction to conduct further measurements after real blasting operations at an actual underground workplace. The aim of these measurements is to compare the times needed to reduce NO_x to the new OELs; on the one hand using ventilation solely, on the other hand using ventilation (with the same airflow) and additional absorbency-spraying.

When the proposed spraying rig is used in an underground operation, the products as well as the excess absorbency will remain on the floor as fine solid crystals after the water is evaporated. Therefore, contact of the miners with the solid absorbencies and products cannot be precluded. Taking into account the potential hazards of sodium hydroxide and sodium sulfite, the sodium sulfite is even more favorable. There are no H- and P-phrases for Na_2SO_3, but several for NaOH. Furthermore, the products of the chemical reaction of NO_2 with NaOH are sodium nitrite and sodium nitrate, with the sodium nitrite being a severe health hazard to humans (Chow 2002; NIH 2001). The product of the reaction of NO_2 with sodium sulfite is sodium sulfate (Na_2SO_4), a harmless substance also known as Glauber's salt.

Furthermore, there should be research conducted on the effect of sodium sulfite on the reduction of nitrogen oxide in presence of nitrogen dioxide.

REFERENCES

Abata, D.I.; Bunting, B.G.; Johnson, J.H.; Robb, J. 1982. Monitoring of Gaseous Pollutants from Six Explosives Tested in an Underground Mine. Transactions of Society of Mining Engineers of AIME (272), pp 1936-1944.

BAuA – Federal Institute for Occupational Safety and Health 2016. Bek. v. 19.9.16,Bekanntmachung von Technischen Regeln; TRGS 900 „Arbeitsplatzgrenzwerte". Gemeinsames Ministerialblatt (Joint Ministerial Gazette), 45, pp 886-889.

Berner, J. 1979. Anleitung für das Messen und die Bewertung von Sprengschwaden untertage. Nobel-Hefte (45) 4, pp 157-167.

Chambers, F. S.; Sherwood, T. K. 1937. Absorption of Nitrogen Dioxide by Aqueous Solutions. Industrial and Engineering Chemistry, Vol. 29 (12), pp 1415–1422.

Chang, M.B.; Lee, H.M.; Wu, F.; Lai, C.R. 2004. Simultaneous Removal of Nitrogen Oxide/Nitrogen Dioxide/Sulfur Dioxide from Gas Streams by Combined Plasma Scrubbing Technology. Journal of the Air & Waste Management Association, 54 (8), pp 941–949.

Chow, C.K.; Hong, C.B. 2002. Dietary vitamin E and selenium and toxicity of nitrite and nitrate. Toxicology, Vol. 180 (2), pp 195–207.

Fick, A. 1855. Über Diffusion. Poggendorff's Annalen der Physik. 94, pp 59–86.

Gann, R.; Friedman, R. 1998. Principles of Fire Protection Chemistry and Physics - 3rd Edition. National Fire Protection Association, ISBN 978-0877654407.

Garcia, M.M.; Harpalani, S. 1986. Gases from Explosives Detonated in Underground Mines. Proceedings of the 3rd Mine Ventilation Symposium, Pennsylvania State University.

Garcia, M.M.; Harpalani, S. 1989. Distribution and Characterization of gases Produced by Detonation of Explosives in an Underground Mine. Mining Science and Technology, 8, pp 49-58.

Graefe, G. 1975. Sprengschwaden im Steinkohlenbergbau. Glückauf (111) 8, pp 368-374.

Heinze, H. 1993. Sprengtechnik. 2. Überarbeitete Aufl., Dt. Verlag für Grundstoffindustrie, Leipzig, ISBN 3-342-00653-6.

Hutwalker, A.; Clausen, E.; Langefeld, O.; Pötsch, J. 2016. Niederschlagung von Nitrosegasen aus Sprengschwaden im Steinkohlenbergbau durch Absorption – Konzeption einer Messzelle. In Mischo, H.; Müller, T.; Herold, J.: Beiträge zum 2. Internationalen Freiberger Fachkolloquium, Grubenbewetterung und Gase, ISBN 978-3-86012-537-3.

Koplin, H. 1979. Die Gehalte an Kohlenmonoxid und an nitrosen Gasen in den Schwaden gewerblicher Sprengstoffe bei Sprengversuchen in einer Gesteinskammer. Nobel-Hefte (45) 4, pp 129-156.

[NIH 2001] TOXICOLOGY AND CARCINOGENESIS STUDIES OF SODIUM NITRITE IN F344/ N RATS AND B6C3F1 MICE. NTP Technical Report 495, U.S. Department of Health and Human Services, National Institutes of Health, 2001.

Kübler, G.; Triebel, R.; Knappe, M. 2016. Neue Anforderungen aus der Grenzwertdiskussion zu Stickoxiden NOx. Kali & Steinsalz, 2/2016, S. 7-17, Verband der Kali- und Salzindustrie e.V., ISSN 1614-1210.

Kuhn, M. 2017. Aktuelle Diskussion von Grenzwerten für den Bergbau unter Tage am Beispiel der Stickoxide. Sprenginfo, 1/2017, Deutscher Sprengverband e.V., ISSN 0941-4584.

Logan, S.R. 2003. Physical Chemistry for the Biomedical Sciences. CRC Press, ISBN 978-0748407101.

Mainiero, R.J. 1997. A Technique for measuring toxic gases produced by blasting agents. Proceedings of the Twenty-third Conference on Explosives and Blasting Technique, Las Vegas, International Society of Explosive Engineers.

Maurer, W.C. 1963. Physical and chemical properties of AN. Quarterly of the Colorado School of Mines (58), 2, pp 2-6.

Mortimer, R.G. 2008. Physical Chemistry and the Periodic Table. Benjamin-Cummings Publishing Company, 3rd Edition, ISBN 978-0805345599.

Neumann 2016. Grenzwerte und Grenzwertfestlegung am Beispiel der Stickoxide. In Mischo, H.; Müller, T.; Herold, J.: Beiträge zum 2. Internationalen Freiberger Fachkolloquium, Grubenbewetterung und Gase, ISBN 978-3-86012-537-3.

Steinhage, M.; Triebel, R. 2014. Arbeitsplatzgrenzwerte: Gesundheitsschutz der Arbeitnehmer verbessern und die Wettbewerbsfähigkeit der Industrie wahren. Kali und Steinsalz, 3/2014, pp 6-13, Verband der Kali- und Salzindustrie e.V., ISSN 1614-1210.

Sun, B.; Sheng, M.; Gao, W.; Zhang, L.; Arowo, M.; Liang, Y.; Shao, L.; Chu, G.-W.; Zou, H.; Chen, J.-F. 2017. Absorption of Nitrogen Oxides into Sodium Hydroxide Solution in a Rotating Packed Bed with Preoxidation by Ozone. Energy & Fuels, Vol. 31 (10), pp 11019-11025.

Sun, Y., Hong, X., Zhu, T., Guo, X., & Xie, D. 2017. The Chemical Behaviors of Nitrogen Dioxide Absorption in Sulfite Solution. Applied Sciences 7 (4), Article 377.

Takeuchi, H., Ando, M., & Kizawa, N. 1977. Absorption of Nitrogen Oxides in Aqueous Sodium Sulfite and Bisulfite Solutions. Industrial & Engineering Chemistry Process Design and Development 16 (3), pp 303–308.

Topchiev, A.V. 1959. Nitration of Hydrocarbons and Other Organic Compounds. Pergamon Press, ISBN 978-0080091549.

[TRGS 2017] Technical Rules for Hazardous Substances, TRGS 900, Version 2006/2017, Section 3, List of OEL, „Nitrogen dioxide"and „Nitrogen oxide", Column „Remarks", BAuA.

Triebel, R.; Kübler, G.; Knappe, M. 2016. Expositionsminderung im Kali- und Steinsalzbergbau – Anforderungen und Maßnahmenpaket. In Mischo, H.; Müller, T.; Herold, J.: Beiträge zum 2. Internationalen Freiberger Fachkolloquium, Grubenbewetterung und Gase, ISBN 978-3-86012-537-3.

Triebel, R. 2017. Aktuelle bohr- und sprengtechnische Entwicklungen in den Bergwerken der K+S Gruppe. In Langefeld, O.; Tudeshki, H.: 20. Kolloquium Bohr- und Sprengtechnik – Tagungsband. ISBN 978-3-86948-547-8.

Warneck, P. 1999. Chemistry of the Natural Atmosphere. International Geophysics (Book 71), 2nd Edition, Academic Press, ISBN 978-0127356327.

Weyer, J. 2019. New Threshold Limit Values for Gases in Europe and Resulting Problems in Mines. Proceedings of the 17th North American Mine Ventilation Symposium, Montreal, Kanada, pp 754-759.

Scientific and Practical Studies of Raw Material Issues – Litvinenko (Ed)

Geoecological justification of reclamation of outer coal waste piles

A. Mukhina
Saint-Petersburg Mining University, Saint-Petersburg, Russian Federation

ABSTRACT: The forming and functioning of outer waste piles is connected with different geomechanical processes that are dangerous to the environment and infrastructure. This circumstance requires special studies, including geoecological assessment of the conditions of forming and transformation of waste piles. Considering the big amount of coal in the waste piles, the forming of technogenic eluvium should be reviewed during the studies, because technogenic eluvium is the basis of biological reclamation. The article presents the analysis of conditions and factors that are react on forming of outer coal waste piles and the assessment of applicability of technogenic eluvim as the basis of reclamation.

1 INTRODUCTION

Russian Federation is the fifth producer of coal in the world after China, USA, India and Australia. The surface coal mining comes with the retirement of agricultural and forest grounds and disturbance of them while forming of the waste piles.

In relation to increasing volumes of surface stripping, it is necessary to study the forming of waste piles to justify their reclamation (Figure 1). The study is important because of frequent losses of stability of high wall slope, lack of viable lands and absence of stable ecosystem of mining facilities. With this, the waste piles effect negatively on components of environment and adjoining lands.

The research of the parameters of waste piles is conducted on example of Kuznetsk coal basin. The object of research is the lands, disturbed by the outer coal waste piles and technogenic rock from surface stripping.

According to the data from Ministry of Energy, the stable growth of coal mining, especially on the Kuznetsk coal basin (Ministry of Energy of the Russian Federation 2019). At the present moment, annual coal mining on the Kuznetsk coal basin is 221 million tons. According to estimates, to 2030 the annual mining will be at 275-330 million tons. With this, approximately 2/3 of total volume of coal comes from the surface coal mining. Surface coal mining is characterized by the high volumes of stripped soils, that are stocked in the waste piles. Mining facilities is 80% of disturbed lands and cover the area of thousands hectares. Surface mining is the most productive and cheap way to mine, but this way does not consider severe disturbance of environment.

To solve the problem of natural environment protection and far-sighted use of land resources it is important to discuss the questions of shrinkage of land allotment and reclamation of outer waste piles.

Usually, the reclamation of lands made in two stages: technical (reclamation leveling, covering with rich soil layer, construction of roads, hydrotechnical and meliorative facilities etc.) and biological (agrotechnical and phytomeliorative measures for recovery of flora and fauna on disturbed lands). A great variety of outer waste piles requires a different approach to soil acquisition and remediation options.

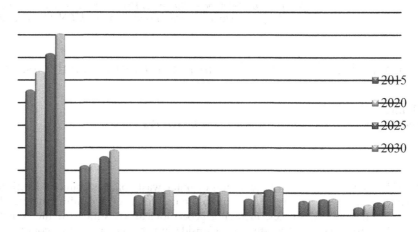

Figure 1. Estimates of coal mining on a country-by-country basis to 2030, million tons *(data of 2018)*.

2 LITERATURE REVIEW

All works devoted to the problems of reclamation of disturbed areas can be divided into three parts. Some are dedicated to researching technologies of recultivation without surface planning and without soil reclamation, others are exploring technologies with the restoration of vegetation only. And few are dedicated to technologies with the use of various soil improvers or special reclamation techniques, such works are usually narrowly focused (Paulo 2005; Semina & Androkhanov 2014; Shipilova 2017).

The studies of directions of reclamation had positive effect and reviewed in monograph (Motorina & Ovchinnikov 1975).

Geologic and engineering studies related to waste piles, particularly, related to their parameters, were conducted by many scientists (Galperin & Kutepov 2012; Kutepov 1999; Kutepov & Kutepova 2003; Kryachko & Norvatov 1980; Ovchinnikov 1975 and others).

There are many studies, related to geomechanic justification of the parameters of waste piles. Among the studies, the most interesting is related to justification of methods of calculation of stability of highwall slopes (Fisenko 1965; Dyomin 2009; Pevzner & Popov 2012; Streltsov, Ilyin & Galperin 1985 and others).

The researches, related to different collectors of factory wastes that are effect on the environment is also important (Pashkevich & Petrova 2019; Kutepov & Kutepova; Mironenko, Zharikov 2005).

A large number of publications are devoted to the formation and evolution of technogenic soils on the dumps of mining enterprises (Androkhanov & Kurachev 2004; Gadzhiev 2001; Kupriyanov, Manakov & Barannik 2010; Brom et al. 2012; Frouz & Novakova 2005; Uzarowicz & Skiba 2011and others). Particularly, the studies related to forming of technogenic soils and technogenic landscapes in Kuzbass made by Ragim-Zade, Bragina and others (Popov, Ragim-Zade& Trophimov 1970; Bragina & Zibert et al. 2014).

The thickness of the humus and organic horizons are indicators of the development and age of the soils on dumps. By now, numerous data on the soil formation on dumps has been gathered. The composition of the organic matter, the mineralogy of these soils, and their hydrological regime have been studied (Androkhanov et al. 2004; Zharikov 2005; Brom et al. 2012; Misz-Kennan & Fabianska 2012; Uzarowicz & Skiba 2011).

The impact of dumps is manifested in the formation of multielemental anomaly in dumps and the accumulation of pollutants with their subsequent migration into the components of the environment. Geochemical monitoring of disturbed lands is necessary to quantify the changes in landscapes and to give practical recommendations for improving the environmental situation (Alekseenko, Maximovich & Alekseenko 2017; Pashkevich & Petrova 2017;

Alekseenko, Pashkevich & Alekseenko 2017; Androkhanov & Kurachev 2009; Motuzova & Karpova 2013; Swartjes 2011).

Reviewed studies are sufficient for engineering and geological conditions of forming of waste piles. But the experience of assessment of reclamation of waste piles (from geomechanic and biologic point of view) nearly does not exist.

3 METHODOLOGY

During the geoecologic studies the complex method is used. The complex method includes the analysis of studies that are made earlier; field, outdoor and laboratory methods of determination of consistence, condition and characteristic of rock and eluvium; parameters of technogenic masses; methods of analytic processing of results; modeling.

The field methods should be oriented on the massive assaying of waste pile body. First of all, it is needed for the assessment of character of heterogeneity and transformation of composing rock.

The laboratory methods should be aimed on the establishment of the way of change of conditions and characteristics of rock under the stress, because with the increasing height of waste pile the stress is also increasing.

The result of study related to geoecologic conditions is the development of model of the object. The model allows to perform the assessment of parameters of waste piles, their stability considering different natural and technogenic factors. Also, the model gives the opportunity to study the applicability of technogenic eluvium for the purpose of biologic reclamation.

4 ANALYSIS

4.1 *Analysis of reformation of overburden rock to technogenic outer waste piles.*

Generally, the overburden rock of Kuznetsk basin is Permian coal sediments and Neogene-Quaternary rock (Mining Encyclopedia 2015). With this, coal bearing layers consist sandstones and siltstones. The difference between the sandstones and siltstones is in the cement of grains and strength. These characteristics depend on the lithification and metamorphosis of coal. The structure of cement depends on the stage of catagenesis and metamorphosis of coal. The structure of waste piles represented not only with the fragmental material of coal bearing layers, but with the Neogene-Quaternary rock too (Table 1). Moreover, a small amount of coal present in the waste piles.

Upper and lower layers of overburden rock contain loams and eluvium clays (crust of weathering). The similar results were obtained by other researchers (Zharikov 2005; Manakov 2008).

After the transportation and stocking up the material into the waste pile, the sequential transformation of overburden into the rock occurs. The transformation caused by the gravitational, physical and chemical solidification (the stages of diagenesis and catagenesis). The rocks mentioned before are mixing up in the waste pile and forming into the new technogenic rocks. Their condition and characteristics depends on lithological structure of rocks, cement in grains, age of initial material, strength of material, height of waste pile, technological process of forming of waste pile (Kutepov 1999). The most intensive transformations occur in waste piles that are contain fragmental material of coal bearing rock with clay cement. The

Table 1. Composition of overburden rock of surface coal mine.

Name	Sandstone	Siltstone	Argillite	Clay	Coal and other
Quantity, %	31-80	14-57	to 17	to 22	2-6

forming of technogenic water-bearing layer is the most common event for the high (~ 100 meters) and super high (more than 100 meters) waste piles. This factor defines the geoecological conditions of the waste pile.

4.2 Study of parameters of technogenic rock mass

The outer waste piles of Kuznetsk region are technogenic rock mass. The shape of waste piles is nearly truncate cone. The waste piles are situated on the area of hundreds hectares, the height ranges from 30 to 200 meters.

The engineering and geological conditions are changing during the forming and functioning of the waste pile. These conditions are: structure of waste mass, condition, characteristics of technogenic and natural soils. Such conditions determine the development of geomechanic and hydrodynamic processes that are dangerous for the environment and infrastructure (Figure 2).

There is a tendency to increment the parameters of waste piles last years. At the present, the creation of waste piles with area from 600 to 2000 hectares and height of 300 m is planned on surface coal mines Bachatskiy and Taldinskiy. Moreover, with increased depth of Bachatskiy surface coal mine the increase of height of the waste pile is possible.

The exploitation of high and super high waste piles is connected with transformation of rock in waste pile body. The granulometric composition, porosity and openness have the biggest impact. The transformation of rock may cause the complication of water removal due to the infiltration recharge. As the result, the water-bearing layer is forming and the hydrodynamic regime of this layer will define the condition of highwall slope stability of the waste pile (Kutepov Yu.I. & Kutepova N.A. 2003; Korobanova 2015).

In addition, wide use of hydro mechanization on some surface mines of Kuznetsk basin is predetermined the forming and functioning of hydraulic waste pile. The mass of "weak" water-bearing rock and water-collecting pond may cause the hydrodynamic accident. The accident may be followed with spreading-out of water-bearing material out of the borders of the hydraulic waste pile. It is common for some companies to combine the waste pile with hydraulic waste pile (Kutepov 1999; Zharikov 2005; Galperin & Kutepov 2012).

4.3 Analysis of forming of technogenic eluvium

The assessment of soil fertility is important geoecological topic of study of technogenic rock mass. The rocks that are moved from the depth to surface have quite low potential of fertility. Low fertility caused by the negligible content of minerals and especially nitrogen. The rocks

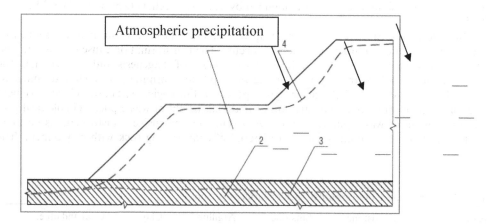

Figure 2. Forming of water-bearing layer: 1- body of waste pile, 2- foundation of waste pile, 3- level of water without stress, 4- completed water-bearing layer.

are transforming to technogenic eluvium with different speed and the reason of transformation is the erosive influence of water (Raghim-zade 1977; Zharikov 2005). The technogenic eluvium of outer waste piles is not homogeneous and contains sands, clay sands, clay loams, clays that are formed during the weathering process.

The presence of coal in the rock and oxidation of coal affect positively in the fertility of eluvium. That is why technogenic eluvium is applicable for the biological reclamation and for development of plants. The meliorative measures are necessary.

It is important to notice that keeping the formed eluvium and soil depends on laudable conditions in waste pile rock mass.

With the optimal hydrothermal regime and ribbed terrain the laudable conditions for plant formation are establishing. The steepness of slope of the waste pile, absence of plant formation, shaded surface leads to severe warming and dewatering of fertile layer. High summer temperatures and low amount of water lead to intensified xeromoprhism. Due to the water flow the edaphic conditions are more pleasant in lower parts of terrain (Johannes & Schmitt. 2006; Steinhuber 2005; Jochimsen, Wiggering & Kerth 1991).

The main indicators of soil condition include: the content of organic matter, the depth of the humus horizon, filtration capacity of the soil, aggregation, pH, electrical conductivity, concentration of main pollutants (Semina & Androkhanov 2014; Sourkova et al. 2005; Maiti 2013).

Granulometric composition of soil-forming rock have a big impact on the speed of forming process of soil on the surface of technogenic rock mass. With this, granulometric composition affects on intensity of soil-forming processes, defines the physical and mechanical characteristics of soils. Under the weathering influence, the rock is transforming to the soils, but the difference between these soils and zonal soils will remain for hundred years.

The selection of key indicators and their thresholds to be maintained for the normal functioning of the soil is necessary to monitor changes and identify trends of improvement or deterioration of soil quality for different agroecological zones.

5 DISCUSSION

Mining-engineering stages of reclamation of the waste piles, situated on a big areas in Kuznetsk basin are connected with geomecanical side of topic. Geomecanical side includes the stability of highwall slope of waste piles and solidification of technogenic rock mass. The study requires the special engineering and geological researches, geomechanical calculations, numeric modeling. Mining-engineering stage of reclamation comes with the solidification of rock mass. Solidification leads to the low water permeability, low moisture capacity, aging of airing. The wide vertical and horizontal penetration of roots increases the stability of the substrate, necessary for the successful completion of reclamation (Ros et al. 2004; Busse et al. 2006). The causes of such events are in engineering-geological and hydro geological conditions and in technological character of waste pile forming.

The biological stage of reclamation is the planting. Planting depends on characteristic of rock of waste pile and on the weathering. The recommendations to biological reclamation are based on the studies, related to transformation of Neogene-Quaternary rock, to forming of technogenic eluvium etc. The morphological requirements to surface that allows to create laudable ecosystems will have an impact on the parameters of waste piles. The maintenance of soil quality is critical to environmental sustainability.

The tendency of increased height of waste piles is positive factor in case of natural environment protection and far-sighted use of resources. However, it is important to justify the stability of highwall slope of the waste pile, because the stability provides the safety of waste pile forming. The stability of waste piles defined by many factors and the main are: engineering-geological and hydro geological conditions of the waste pile forming, landscape of the waste pile placing. Considering the facts mentioned above it is very important to provide special researches in case of the study.

The maintenance of ecological safety is not only in geomecanical assessment of the stability. Ecological damage is also in the change of quality of land resources used. In this manner, further existence of the waste pile is determined by the reclamation as the stage liquidation of mining operations.

Thus, the reclamation should be performed on the basis of complex study of geoecological conditions. These conditions are: geometric parameters of waste piles; content, condition and characteristics of the technogenic rock mass; hydro geological structure of technogenic rock mass; fertility of technogenic eluvium and others.

Due to the presence of different technologies of waste piles forming it is important to justify the recommendations for the reclamation operations.

5 CONCLUSION

Due to the deficit of land resources and detrimental effect on the environment the reclamation is the topic of especial importance. Dumps are well-studied objects from the point of view of dumps formation and geomechanical justification of optimal parameters. However, the geoecological aspect of such objects remains unplumbed.

The miningtechnical stage of recultivation is difficult and expensive, therefore it is often excluded or minimized. Artificial plantings on the unprepared surface of the dumps are unstable and so quickly degrade, returning the soil to its original state, i.e. to the technogenic desert. To improve the ecological situation, a new approach is needed in the conduct of reclamation works, especially in the scientific support and design of the developed reclamation technologies.

In the course of work done the conditions that are have an impact on the forming of outer waste piles from the surface coal mining are studied.

In addition, the basic characteristics of waste piles such as lithological composition of rock, geometric parameters are reviewed. Also, the possibility of transformation from the technogenic rock to technogenic eluvium as the basis for the biological reclamation is reviewed.

In this manner, provided analysis allows to determine the regularity of forming of geocological and geomechanical conditions of waste piles and gives the opportunity to create the recommendations for elaboration of the reclamation stages.

REFERENCES

Alekseenko, V.A., Maximovich, N.G. & Alekseenko, A.V. 2017. Geochemical Barriers for Soil Protection in Mining Areas. In: Bech J., Bini C., Pashkevich M.A. (Eds.): *Assessment, Restoration and Reclamation of Mining Influenced Soils. Academic Press*: 255–274.

Alekseenko, V.A., Pashkevich, M.A. & Alekseenko, A.V. 2017. Metallisation and environmental management of mining site soils. *Journal of Geochemical Exploration*: 121–127.

Androkhanov, V.A., Kulyapina, E.D. & Kurachev, V.M. 2004. Soils of technogenic landscapes: genesis and evolution. [Pochvy tekhnogennykh landshaftov: genezis i evolyutsiya]. Novosbirsk: *Publishing House of the Siberian Branch of the Russian Academy of Sciences*: 181.

Androkhanov, V.A. & Kurachev, V.M. 2009. Principles of assessment of soil-ecological state of technogenic landscapes. [Printsipy otsenki pochvenno-ekologicheskogo sostoyaniya tekhnogennykh landshaftov]. *Siberian Journal of Ecology 16(2)*:165-169.

Bragina, P. S., Zibert, A. S. & Zavadsky, M. P. 2014. Degradation, restoration and protection of soils. Soils on the overburden dump in the forest-steppe and mountain-taiga zones of Kuznetsk basin. [Degradatsiya, vosstanovleniye i okhrana pochv. Pochvy na otvalakh vskryshnykh porod v lesostepnoy i gorno-tayezhnoy zonakh Kuzbassa]. *Soil Science 7*: 878-889.

Brom, J., Nedbal, V. & Procházka, J. 2012. Changes in vegetation cover, moisture properties and surface temperature of a brown coal dump from 1984 to 2009 using satellite data analysis. *Ecological Engineering 42*: 45–52.

Busse, M.D., Beattie, S.E., & Powers, R.F. 2006. Microbial community responses in forest mineral siol to compaction, organic matter removal, and vegetation control. *Can. J. Forest Res 36(3)*: 577-588.

Dyomin, A. M. 1973. *Stability of open mine workings and waste dumps*. [Ustoychivost' otkrytykh gornykh vyrabotok i otvalov]. Moscow: Nedra.

Dyomin, A. M. 2009. *Landslides in quarries: analysis and forecast.* [Opolzni v kar'yerakh: analiz i prognoz]. Moscow: GEOS.

Dr. Johannes, A. & Schmitt. 2006. Mountain and industrial dumps as secondary biotopes in the Saarland with special consideration of the coal-mining heaps of Grube Reden. [Berge- und Industrie-Halden als Sekundärbiotope im Saarland unter besonderer Berücksichtigung der Steinkohlen-Bergehalden von Grube Reden]. Germany.

Fisenko, G.L. 1965. *Stability of sides of quarries and dumps* .[Ustoychivost' bortov kar'yerov i otvalov]. Moscow: Nedra.

Frouz, J. & Novakova, A. 2005. Development of soil microbial properties in topsoil layer during spontaneous succession in heaps after brown coal mining in relation to humus microstructure development. *Geoderma 129*: 54-64.

Galperin, A.M. & Kutepov, Yu.I. 2012. Development of technogenic massifs at mining enterprises. [Osvoyeniye tekhnogennykh massivov na gornykh predpriyatiyakh]. Moscow: *Mountain book*: 336.

Hajiyev, I. M., Kurachev, V. M. & Androkhanov, V. A. 2001. *Strategy and prospects of solving problems of reclamation of disturbed lands.* [Strategiya i perspektivy resheniya problem melioratsii narushennykh zemel']. Novosibirsk: CARIS.

Jochimsen, M., Wiggering, H. & Kerth, M. 1991. *Ecological aspects of vegetation development on mountain dumps.* [Ökologische Gesichtspunkte zur Megetationsentwicklung auf Bergehalden]. Germany: Piles of coal mining.

Korobanova, T.N. 2015. *Dangerous geodynamic processes accompanying dump's formation.* Applied and Fundamental Studies: Proceedings of the 8th International Academic Conference. St. Louis: Science and Innovation Center Publishing House. 84-90.

Kryachko, O. Yu. 1980. *Management of open pit mining.* [Upravleniye otvalami otkrytykh gornykh rabot]. Moscow: Nedra.

Kupriyanov, A. N., Manakov, A. Yu. & Barannik, L. P. 2010. *Restoration of ecosystems on dumps of the mining industry in Kuzbass.* [Vosstanovleniye ekosistem na otvalakh gornodobyvayushchey promyshlennosti Kuzbassa]. Novosibirsk: Geo.

Kutepov, Yu.I. 1999. *Scientific and methodological foundations of engineering and geological support of dumping in the development of coal deposits.* [Nauchno-metodicheskiye osnovy inzhenerno-geologicheskogo obespecheniya otvaloobrazovaniya pri razrabotke ugol'nykh mestorozhdeniy]. Moscow: MGGU. Diss. Dr. tech. sciences.

Kutepov, Yu.I. & Kutepova, N.A. 2003. Technogenesis of alluvial deposits. [Tekhnogenez namyvnykh otlozheniy]. Moscow: *Geoecology (5):* 405-413.

Maiti, S.K. 2013. *Ecorestoration of the Coalmine Degraded Lands.* New York: Springer.

Manakov, Yu. A., 2008. Disturbed land of Kuzbass. Rehabilitation problems. *Promyshl. Ekol.* 4: 29-34.

Ministry of Energy of the Russian Federation, 2019. [Ministerstvo Energetiki Rossiyskoy Federatsii]. Statistics, viewed 21.05.2019, https://minenergo.gov.ru/activity/statistic

Misz-Kennan, M. & Fabianska, M. 2010. Thermal transformation of organic matter in coal waste from Rymer cones (Upper Silesian Coal Basin, Poland). *Int. J. Coal Geol. 81:* 343–358.

Motorina, L. V. & Ovchinnikov, V. A. 1975. *Industry and land reclamation.* [Promyshlennost' i rekul'tivatsiya zemel']. Moscow: Thought: 240.

Motuzova, G. V. & Karpova, E. A. 2013. *Chemical pollution of biosphere and its ecological consequences.* Moscow: MGU Publishing House: 304.

Mountain Encyclopedia. Mineral deposit. Kuznetsk coal basin, 2015. [Mestorozhdeniya. Kuznetskiy ugol'nyy basseyn]. 2015, viewed 14.05.2019, http://www.miningenc.ru/k/kuzneckij-ugolnyj-bassejn

Pashkevich, M.A. & Petrova, T.A. 2017. Technogenic Impact of Sulphide-Containing Wastes Produced by Ore Mining and Processing at the Ozernoe Deposit: Investigation and Forecast. *Journal of Ecological Engineering* 18: 127-133.

Pashkevich, M.A. & Petrova, T.A. 2019. Recyclability of Ore Beneficiation Wastes at the Lomonosov Deposit. *Journal of Ecological Engineering 20*: 27-33.

Paulo, A. 2004. Economical and natural conditions applicable to the development of postmining areas. Conference "Valorisation of the Environment in the Areas Exposed to Long Term Industrial and Mining Activities". *Pol. Geol. Inst* 2005(17): 49-69.

Pevzner, M.E., Popov, V.N. & Makarov, A. B. 2012. *Geomechanics.* [Geomekhanika]. Moscow: MSMU: 345.

Popov, V. M., Rahim-zadeh, F. K. & Trofimov, S. S. 1970. Classification of overburden rocks of the Kuznetskbasin for suitability for biological recultivation. [Klassifikatsiya vskryshnykh porod Kuzbassa po prigodnosti dlya biologicheskoy rekul'tivatsii]. *Recultivation of Siberia and the Urals.* Novosibirsk: *Science*: 25-41.

Ragim-Zade, F.K. 1977. *Technogenic eluvium of overburden of coal deposits of Siberia, assessment of their potential fertility and suitability for restoring their soil cover.* [Tekhnogennyye elyuvii vskryshnykh porod ugol'nykh mestorozhdeniy Sibiri, otsenka ikh potentsial'nogo plodorodiya i prigodnosti dlya vosstanovleniya ikh pochvennogo pokrova]. Novosibirsk: diss. Cand. biol. Sciences.

Ros, M., Garcia, C., Hernandez, T., & Andres M. 2004. Short-term effects of human trampling on vegetation and soil microbial activity. *Commun. Soil Sci. and Plant Analit* 35(11-12): 1591-1603.

Semina, I. S. & Androkhanov,V. A. 2014. About recultivation of disturbed lands in Kuzbass mines. [O rekul'tivatsii narushennykh zemel' na razrezakh Kuzbassa]. *Mining information and analytical Bulletin (scientific and technical journal) 12*: 307-314.

Shipilova, A. M. & Semina, I. S. 2017. Assessment of the soil-ecological state of technogenic landscapes the Kuzbass depending on the reclamation technology of recultivation lands. [Otsenka pochvenno-ekologicheskogo sostoyaniya tekhnogennykh landshaftov Kuzbassa v zavisimosti ot tekhnologiy rekul'tivatsii narushennykh zemel']. *Izvestiya UGGU 3(47)*: 53-56.

Sourkova, M., Frouz, J. & Santruckova, H. 2005. Accumulation of Carbon, Nitrogen, and Phosphorous during Soil Formation on Alder Spoil Heaps after Brown-Coal Mining, Near Sokolov (Czech Republic). *Geoderma 124*: 203–214.

Steinhuber, U. 2005. *One hundred years of mining recultivation in Lusatia.* [Einhundert Jahre bergbauliche Rekultivierung in der Lausitz]. Diss. Berlin.

Streltsov, V. I., Ilyin A. I. & Galperin, A. M. 1985. *Management of long-term stability of slopes in quarries.* [Upravleniye dolgosrochnoy ustoychivost'yu sklonov v kar'yerakh]. Moscow: Nedra: 245.

Swartjes, F. A. (2011). Dealing with contaminated sites. Dordrecht: Springer.

Uzarowicz, L. & Skiba, S. 2011. Technogenic soils developed on mine spoils containing iron sulphides: Mineral transformations as an indicator of pedogenesis. Geoderma 163: 95–108.

Scientific and Practical Studies of Raw Material Issues – Litvinenko (Ed)
© 2020 Taylor & Francis Group, London, ISBN 978-0-367-86153-7

Disinfection of waste waters of industrial enterprises by vibroacoustic method

Gennady Fedorov & Yuri Agafonov
Mining Institute, National University of Science and Technology "MISIS", Moscow, Russia

Carsten Drebenstedt
Technische Universität Bergakademie Freiberg, Freiberg, Germany

ABSTRACT: The authors studied a physical mechanism of the disinfection process under the influence of elastic vibrations through analyzing hydrodynamic situation when the treated liquid flows through the acoustic line, as well as the disinfection process dependence on the modes and parameters of elastic vibrations. We established that vibroacoustic oscillations do improve disinfection process variables. The process effectiveness and productivity depend on the oscillation speed amplitude, frequency and the acoustic line cross-section. A method how to calculate an optimum mode for vibroacoustic disinfection has been developed. We developed a vibroacoustic apparatus model which has been successfully tested in laboratory conditions. The results of research demonstrated its application potential in water-sludge facilities of industrial enterprises.

1 INTRODUCTION

The issue of disinfection of waste waters of mining enterprises is quite important. Ineffectiveness of the existing technologies is well known and there are constantly appear new technologies and apparatus for their improvement. A new vibroacoustic method is proposed to solve this problem. The novelty consists in passing of the to-be-processed liquid through the acoustic line with a periodically varying cross-section area with simultaneous impact of vibroacoustic oscillations. As a result, well-developed cavitation originates in the whole liquid volume killing bacteria and microorganisms. This is the solution to the problem of waste waters disinfection.

2 STATEMENT OF BASIC MATERIALS

The industrial enterprises produce huge volumes of waste waters whose drain brings about contamination of water resources of the planet with finely-divided suspensions and microorganisms. Sanitary measures provide for disinfection of industrial waste waters prior to their secondary use in processes or prior to draining them into natural water bodies, with the aim of preventing various diseases. Widely used for the disinfection of waste waters are chemical methods based on the use bactericidal properties of chemical substances. Out of the multitude of existing methods and means of chemical disinfection most widely used is trivial disinfection with chlorine and chlorine-containing substances. Ozone is also used due to its great bactericidal activity as compared with chlorine. Nevertheless, these methods have a number of essential drawbacks: they have harmful effect on the flora and fauna of natural reservoirs, and high expenses of reagents delivery, storage and dosing.

The alternatives to chemical treatment of waste water are reagents-free methods. These methods are based on the use of various physical effects emerging in liquid under the influence of physical fields. These may be thermal, electrical, ultraviolet, hydraulic, ultrasonic, etc.

Most perspective are the ultrasonic and ultraviolet methods. Under the influence of ultrasonic oscillations in liquid, there appear cavitation, degassing, vibroturbulent streams, acoustic flows thanks to which disinfection of waste waters takes place (Yamshchikov, 1987; Webster, 1993). However, the bactericidal effect of ultrasound propagates only within a minor volume of water and therefore ultrasonic installations have low productivity (Kulsky, 1978). The ultraviolet irradiation has found practical use and is the best-investigated method out of other physical methods, but high economic expenses restrict its wider application.

Proposed here is a vibroacoustic device for disinfection of waste waters (Feydorov et al., 2016). Please see Figure 1.

The device contains the acoustic line 1 with a wave-like side wall. In the end portions of the wave guide, there are pistons 2 and 3 rigidly interconnected by rod 4. The pistons are secured to the wave guide with flexible membranes 5. The pistons 2 and 3 are driven to oscillate with the help of vibrator 6 with varying frequency and amplitude of oscillations. The vibrator frequency varies from 5 to 300÷350 Hz. The piston oscillations mode is measured with accelerometer (acceleration meter) 7. Tensiometers 8 (liquid pressure meters) allow measuring pressure in the area of narrowing and in the area of maximum expansion. The signals from the accelerometer and tensiometers are furnished to the electronic measuring unit (EMU) 9. Pipe connections 10 and 11 deliver and drain the liquid to/from the acoustic line; the flow rate is regulated with shutters 12 and 13. Flow meter 14 determines the rate of water passing through the apparatus.

The principle of operation of the vibroacoustic apparatus is as follows. The pistons 3 and 4 are driven with the help of vibrators 7 into in-phase oscillatory movement. Along the whole of the acoustic line 1, there happens oscillation of liquid with different speeds. This results in the emergence of zones with different pressures. In cross-sections with minimum area, rarefaction zones develop in which gas is emitted from the liquid and gas bubbles start originating. Further getting of liquid into the higher-pressure zone (maximum area of cross-section) results in the collapse of gas bubbles accompanied with

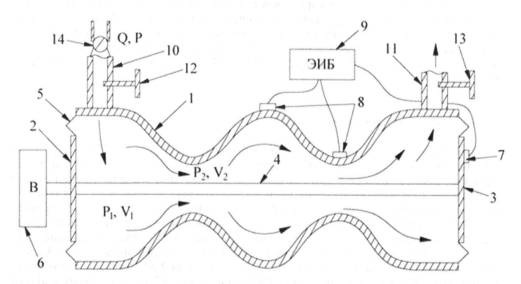

Figure 1.   Schematic diagram of a vibroacoustic device for disinfection of waste waters.

1 – acoustic line with wave-like side walls; 2, 3 – pistons; 4 – rod; 5 – flexible membranes; 6 – vibrator; 7- accelerometer (acceleration meter); 8 – tensiometers (strain gage indicators); 9 – electronic measuring unit; 10, 11- supply and drain connecting pipes; 12, 13 – shutters; 14 – flow meter.

the formation of shock waves and cumulative micro jets (Presset, 1989; Pirsol, 1975). The liquid oscillation mode and creation of conditions for cavitation depend on the oscillation speed of the piston dipole system.

A pipe is needed to organize steady flow of liquid through acoustic line. Upon that the liquid finds its way from the rarefaction zone into compression zone. The liquid passes from one section of acoustic line into another one. The pump productivity should be considerably less (by an order of magnitude) than the piston surface oscillating speed, that is $V\sim S1 \gg Q$ (where $V\sim$ - piston oscillating speed; $S1$ – piston surface area). The optimum pump productivity is selected taking into account the time of liquid stay in acoustic line necessary for waste waters disinfection.

So, the vibroacoustic device for waste waters disinfection operates on the principle of a "speed transformer". Depending on the change in the pipeline cross-section, the liquid movement speed varies.

Figure 1 shows our experimental installation used to determine conditions for emergence of cavitation depending on the amplitude of the pistons oscillation speed. Ttensiometers 8 were used to measure pressure in the narrow and wide portions of the acoustic line. Accelerometer 7 was used to determine the frequency and amplitude of oscillation speed of pistons 2 and 3.

It is known that the process of cavitation is accompanied with shock waves and cumulative micro streams which create acoustic noise in liquid. Emergence of this noise indicates the emergence of cavitation in liquid.

Figure 2 shows pressure diagrams in the narrow and wide portions of the acoustic line. The vibroacoustic oscillations were excited on frequency of f=10 Hz. These oscillograms show that the vibroacoustic noise in liquid and, correspondingly, cavitation depend on the piston

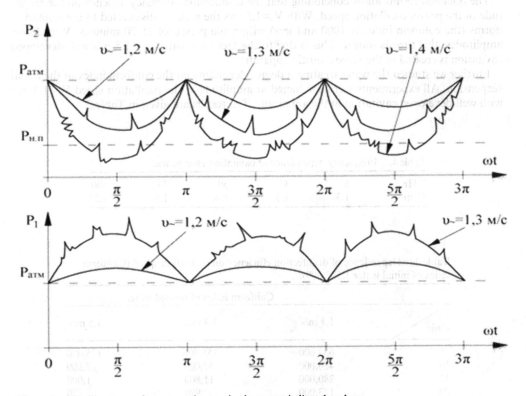

Figure 2.   Oscillograms of pressure changes in the acoustic line chambers.

Рн.п. – saturated vapor pressure; V~ - piston oscillatory speed

167

oscillation speed. With amplitude of the oscillation speed less than 1.3 m/s, there is no cavitation. At a speed of V~ = 13 m/s separate spikes appear on the oscilloscope pattern – this is the beginning of cavitation. With high values of the oscillation speed in liquid well-developed cavitation starts. Many shock waves and cumulative micro streams appear.

Thus, for creation of powerful cavitation fields within the whole volume of the vibroacoustic apparatus, it is necessary to have oscillation speed of pistons higher than 1.3 m/s.

The results of the frequency dependence study for cavitation in the range from 5 to 100 Hz are given in Table 1. With the increase of oscillation frequency, higher values of oscillation speeds are required for cavitation fields. Evidently, it is associated with the fact that at higher frequencies the inertia of liquid in the apparatus manifests itself, and it does not fully follow the piston oscillations.

The Table demonstrates that with the increase of frequency the amplitude of the oscillation speed increases and energy consumption goes up. We suggest conducting vibroacoustic disinfection in the frequency range from 5 to 35 Hertz.

Further experimental research involved determination of efficiency of water disinfection in the vibroacoustic apparatus depending on the mode of vibroacoustic oscillations (amplitude of the oscillation speed V~) and on the initial content of bacteria in water.

At the first stage, the water with a coliform index of 1525 was processed for 30 minutes at various amplitudes of piston oscillation speed (V1=1.1 m/s, V2=1.3 m/s and V3=1.5 m/s). In the first mode of impact, there was no cavitation, in the third mode we observed developed cavitation.

The results of disinfection with the vibroacoustic method and chlorine were compared. The coliform index (number of Escherichia coli bacteria in one liter of water) after contact with chlorine for 30 minutes should not exceed 1000. The obtained results are presented in Table 2.

The obtained results allow concluding that the disinfection efficiency depends on the amplitude of the piston oscillation speed. With V~=1.5 m/s the water is disinfected to the sanitation norms (the coliform index is 1000 and less) within the period of 15-20 minutes. With lower amplitudes, this time is longer. This is due to the fact that with V~ = 1.5 m/s well-developed cavitation is created in the vibroacoustic apparatus.

Further we studied the water treatment degree depending on the coliform index of the initial suspension. All experiments were conducted at amplitude of the oscillation speed V~=1.5 m/s with well-developed cavitation in the apparatus. The results are given in Table 3.

Table 1.   Frequency dependence of cavitation emergence.

f, Hz	5	15	50	75	100
V, m/s	1.3	1.3	1.4	1.5	1.65

Table 2.   Dependence of disinfection efficiency on oscillation speed (Coliform index of initial water 1.525.000).

	Coliform index of treated water		
V~ t min	1.1 m/s	1.3 m/s	1.5 m/s
5	633,000	155,000	125,000
10	420,000	57,000	7,800
15	280,000	11,800	1,000
20	123,000	980	770
25	61,000	810	360
30	970	630	120

Table 3. Dependence of the water treatment degree on the initial coliform index.

Coli index of initial water t min	Coliform index of treated water		
	4,350,000	2,800,800	1,525,000
5	575,000	207,000	125,000
10	240,000	85,000	7,800
15	87,500	10,500	1,000
20	11,500	4,000	770
25	6,000	1,000	360
30	1,000	870	120

As we see from the Table, the time required for water disinfection to sanitation norms (the coliform index is 1000) goes up with the rise of the coliform index in the initial water.

An important parameter of disinfection is its efficiency. It is determined as per the formula:

$$E(t) = \frac{N_0 - N_1(t)}{N_0} \cdot 100\% \tag{1}$$

where: N_0 – number of bacteria in waste water at the initial period;

$N_1(t)$ – number of bacteria at time moment t;

t – time of impact of the bactericidal factor.

On the basis of the obtained results, we have a dependence curve of efficiency of waste water disinfection on the duration of impact and the coliform index of the initial water. Please see Figure 3.

Curve 1 – the initial coliform index is 4,350,000

Curve 2 – the initial coliform index is 2,800,000

Curve 3 – the initial coliform index is 1,525,000

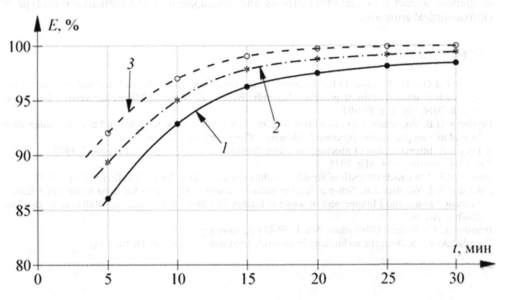

Figure 3. Dependence of disinfection efficiency on the time of impact and the coliform index of the initial suspension:

These dependencies have similar pattern and starting from the time of 15-25 min they begin to flatten out. Dependence of disinfection effectiveness on the coliform index of the initial suspension is minor.

The investigations have demonstrated that the process of the bacterial die-away in the vibroacoustic field is governed by the following law (Webster, 1993):

$$N_1(t) = N_0 \cdot 10^{-Kt} \tag{2}$$

where K – coefficient of bacterial die-away in the vibroacoustic field.
Taking the logarithm of this expression, we obtain:

$$t = \frac{1}{K} lg \frac{N_0}{N_1(t)} \tag{3}$$

From the formula (3) it follows that the time required for bringing the waste liquid to the necessary condition depends on the ratio between the number of bacteria in the initial suspension N_0 and the required number of bacteria $N_1(t)$ in the disinfected liquid.

Our investigations permit to conclude that the proposed vibroacoustical apparatus is capable of disinfecting waste waters and bringing them to sanitation norms. Moreover, disinfection depends on the amplitude of the piston oscillation speed. There is a threshold value of this speed (1.3÷1.5 m/s) starting from which one can observe strong effect of vibroacoustic oscillations. This is due to cavitation phenomena originating in the acoustic line under such conditions.

It has been found that the speed of waste water treatment increases. To bring the water down to sanitation norms, the treatment time decreased 2-3 times and makes up 15÷25 minutes. The results of testing have demonstrated potential of the vibroacoustic apparatus in the treatment facilities of industrial enterprises. The proposed design of the vibroacoustic apparatus is highly productive and can treat considerable volumes of waste waters. Optimum modes for conducting vibroacoustic water disinfection are also identified. Our technical solution seems to have no equivalents.

Further investigations should be aimed at the study of the mechanism of vibroacoustic oscillations impact on disinfection process and development of the industrial prototype of vibroacoustical apparatus.

REFERENCES

Agafonov J.G., Dudchenko O.L., Fedorov G.B. Infrasonic methods and technology: innovative approach to intensify mining practices. Scientific Bulletin of National Mining University, Dnepropetrovsk, 2014, No. 2, p. 99-104.

Feydorov G.B., Agafonov Y.G., Artemyev V.B. et al., Vibroacoustical methods and means of intensification of mining production processes/- Moscow: "Gornoye Delo", 256 p., 2016.

Kulsky L.A. Intensification of processes of water disinfection. Kiev, "Naukova Dumka", 1978.

Pirsol I. Cavitation. - M: Mir, 1975.

Presset M.S. The tensile strength of liquids. Cavitation state of knowledge, ASME.15-25 pp., 1989.

Shkundin S.Z. Woolsey J.R. Self-purification acoustic screen filter systems for mine waste water//Environmental Issues and Management of waste in Energy and Mineral Production, - Balkems, Rotterdam, 2000. p. 361-367.

Webster E., Cavitation. Ultrasonics, No. 1, 39-48 pp., January, 1993.

Yamshchikov V.S. Acoustic technology in minerals treatment, – Moscow: Nedra, 1987.

Scientific and Practical Studies of Raw Material Issues – Litvinenko (Ed)
© 2020 Taylor & Francis Group, London, ISBN 978-0-367-86153-7

Mining 4.0 in developing countries

Jan C. Bongaerts

Professor Umwelt- und Ressourcenmanagement, Freiberg, Germany

ABSTRACT: This paper presents an overview of some findings related to mining 4.0 which resulted from a DAAD International Alumni Seminar on the subject which was held at TU Bergakademie Freiberg from 30 April to May 5, 2019. The seminar was attended by twenty five Alumni from developing countries who had studied or/and were researchers at German universities earlier. All of them currently hold positions in Research and Development Institutions, the Mining Sector, Consultancy and Government. The seminar was entitled:

GERMANY'S CONTRIBUTION TO MINING 4.0: ALGORITHMS, SOFTWARE, SENSORS, AUTOMATED MINING, BUSINESS CONCEPTS FOR DEVELOPING COUNTRIES

This paper is divided in three parts: (i) an overview of some data related to the significance of developing countries to the generation of natural resources, (ii) some elements on the current state on MINING 4.0 and (iii) a case study about a fully automated mine in Mali.

1 DEVELOPING COUNTRIES AND MINING AND MINERALS

Developing Countries generate around 60 percent of all mineral resources globally. So-called Transition Countries provide 12.7 percent of minerals and Developed Countries generate the remainder of around 27 percent. These shares have remained constant since 2010 within the last upwards resource cycle. (See World Mining Data 2018). Figure 1 shows these relationships in a diagram.

Since many developing countries export minerals in substantial volumes, it is of interest to inspect their terms of trade. They can be defined as the Export value/Import value, in other words: the purchasing power of one unit of export value in terms of import value.

Obviously, the higher the Terms of Trade compared to some the base year, the higher the purchasing power. Table 1 shows the Terms of Trade for four continents from 2000 to 2015.

From Table 1, it can be seen that especially Africa, Latin America and Oceania had favourable Terms of trade. This corresponds to the so-called Resource Boom of the time period.

A third aspect of mining is reflected by the spending of mining companies in the countries in which they operate. Table 2 contains the results of a survey among the members, held by ICMM (International Council of Mining and Metals), an international industrial NGO.

It can be seen that most expenditures are on CAPEX and OPEX and, to a much lesser extent, on employment. In terms of CAPEX and OPEX, it should also be noted that smaller or larger shares of these amounts are for capital goods and equipment why may be imported into the mining countries and, hence, do not contribute to the national economies. The data of Table 2 reflect typically the capital intensive structure of the mining sector. As such, within OPEX, energy expenditures are contained.

When it comes to the TOP TWENTY of the mining countries in terms of production, we see, as shown in Table 3, that eleven of them are developing countries.

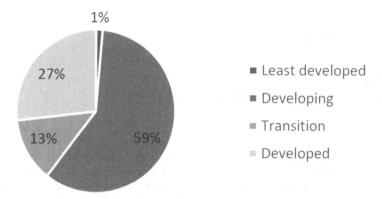

Figure 1.   Natural resources contributions (Source: Author based on World Mining data 2018).

Table 1.   Terms of Trade for four continents.

	2000	2002	2004	2006	2008	2010	2011	2012	2013	2014	2015
Africa	100	99.2	113.4	148.8	173.0	167.2	185.6	191.7	188.0	180.4	153.7
Asia	100	94.9	95.9	96.4	93.8	93.6	92.9	94.0	94.7	95.5	98.2
Oceania	100	104.9	115.0	139.0	164.7	169.1	187,8	171.3	168.9	160.8	145.9
Latin America	100	98.0	108.0	124.0	131.6	136.4	147.1	142.7	139.7	133.1	122.4

Source = Bundeszentrale für politische Bildung

Table 2.   Expenditures of mining companies in percent (ICMM, 2018).

CAPex and OPEX	50 - 65
Salaries and wages	15 - 20
Taxes and royalties	15 – 20
Local shareholders and financial costs	15 - 20
Community development	0.1 – 1.0

Table 3.   Eleven developing countries in the TOP TWENTY Mining Countries (Source, ICMM 2018).

rank 2016	country	US$ (mill)	previous rank (2013)
5	India	77.0	7
6	South Africa	48.9	5
7	Indonesia	47.5	10
9	Brazil	36.6	9
11	Mexico	28.9	12
12	Peru	27.1	11
13	Kazakhstan	18.6	13
14	Turkey	17.2	22
17	Colombia	10.1	19
18	Ukraine	9.9	14
19	Congo, Dem. Rep.	7.9	17

As a fifth element, we notice that many Developing Countries are also Resource Dependent. ICCM (International Council on Mining and Metals - 2018) classifies a country as Resource Dependent whenever
- Its natural resources contribute more than 20 % to total export earnings
or
- Its net income from natural resources (revenue minus extraction costs) is more than 10 % of GDP

Leaving out hydrocarbon resource dependent countries (75% of the resource dependency is from oil and gas), such as Algeria, the Middle East, Norway, Russia and concentrating on minerals and metals, a group of Mining Dependent Countries (75 % of resource dependency is from minerals and metals) can be defined. Most of these countries are in the Developing World, with Gross National Income per capita levels below 1 500 US$ (ICMM 2018). For these countries, mining is a significant contributor to economic and social development.

This can be seen in the following Figures 2, 3 and 4.

It is obvious that Africa has to be considered as resource dependent according to the export contribution criteria and that all other continents do not qualify as such. For Africa and only for Africa, this resource dependency has been increasing during the observed time period. In addition, the well-known cyclicality of mineral rents revenues for governments is also clearly visible.

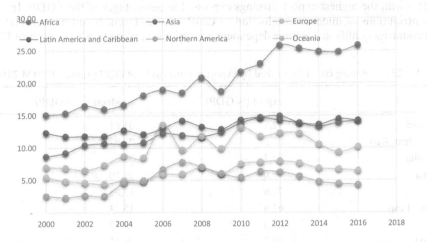

Figure 2.   Production volume of minerals as % of GDP (Source: adapted from ICMM 2018).

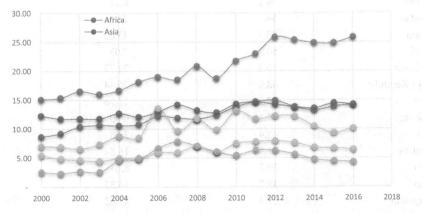

Figure 3.   Export contributions of minerals by continent as % of GDP (Source: adapted from ICMM 2018).

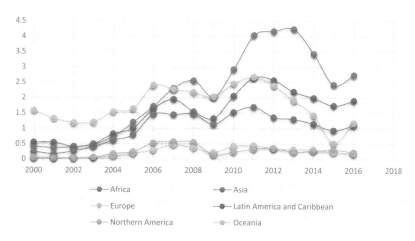

Figure 4. Mineral rent contributions of minerals as % of GDP (Source: adapted from ICMM 2018).

Lastly, resource dependency by country and not by continent delivers a much more detailed and diverse overview as shown in Table 4. It lists 23 developing countries – out of the World's top 25 – with the highest export earnings expressed in percentages of their GDP. In addition, the contributions of minerals production to GDP are also given. It can be seen that most of these countries qualify as resource dependent countries according to both criteria of ICMM.

Table 4. 23 Developing countries ranked by export earnings as % of GDP (Source: ICMM 2018).

	export (% GDP)	production (% GDP)
Botswana	92.7	31.81
Congo Dem. Rep	87.3	22.62
Mongolia	82.7	53.61
Zambia	75.0	21.10
Mali	74.6	13.85
Burkina Faso	63.6	15.24
Peru	61.0	14.12
Guyana	59.7	26.11
Sierra Leone	54.5	34.37
Namibia	54.2	8.76
Tajikistan	54.2	24.88
Surinam	53.3	33.02
Armenia	50.7	7.03
Mauritania	50.5	24.73
Kyrgyz Republic	48.0	36.34
Ghana	47.2	12.76
Uzbekistan	47.1	6.60
Sudan	42.4	2.58
Zimbabwe	41.5	16.84
Bolivia	39.7	8.53
Liberia	33.7	14.33
Madagascar	27.3	41.62
Senegal	20.6	16.90
Dominican Republic	20.2	2.37

In essence, MINING 4.0 stands for a set of technologies in combination with management concepts leading to fully automated mining in terms of production. Table 5 lists some topics related to MINING 4.0 which are currently in the focus.

There are many reasons and motivations for MINING 4.0, which is seen by many as a disruptive innovation, leading to a fundamental restructuring of the mining sector. In this paper, one key reason is identified and illustrated. According to two studies by McKinsey of 2016 and 2018, productivity in the mining sector has drastically decreased since and, with some improvement since 2016, is still very low compared to prior historical levels. This is shown in Figure 5.

While inspecting Figure 5, it should be noticed that the productivity of the mining sector, as measured by McKinsey's MineLens Productivity Index, is not accounting for geological factors, such as grades of deposits or depths of ore bodies. The Index only considers Capital input, Employment and Operating Expenditures other than for Labour, because these factors can be controlled by the mine operators. For that reason, the output of the Index is defined as total material moved and not saleable mine product. With productivity, as measured in this way, performing so low, it becomes understandable that a structural innovation in the mining sector is necessary. Further data from the studies also show that the moderate increase in productivity for the years 2014 and 2016 result from a decrease of the workforce of 3.1 % per year and a small increase of material moved of 1.8 % per year.

A more detailed explanation of the behaviour of the productivity of a mine operation is shown in Figure 6.

Table 5. Topics related to MINING 4.0 currently in the focus (Source: Unpublished).

Topics	
1 Robotics in surveying and virtualization of mineral deposits	8 Enhanced Maintenance of Equipment in Mining Operations
2 Visualization of Block Modelling in Mining Operations	9 Enhanced Safety at Work
3 Algorithms for Localization and Mapping	10 Management of Big Data for Automated Mining
4 Visualization of underground and open-pit mines	11 Automated Mining and Compliance
5 Real time analysis, operation and decision-making	12 Optimal Use of Planned Time
6 Remote Operation and Monitoring of Mining Equipment	13 Automated Mining and Vocational and Educational Training
7 Sensor Materials	14 Community Development

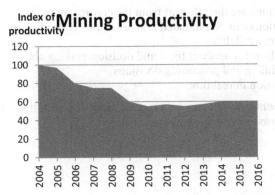

Index of productivity — Mining Productivity

Figure 5. Productivity in the mining sector (Source: adapted from McKinsey, May 2015 and McKinsey, June 2018).

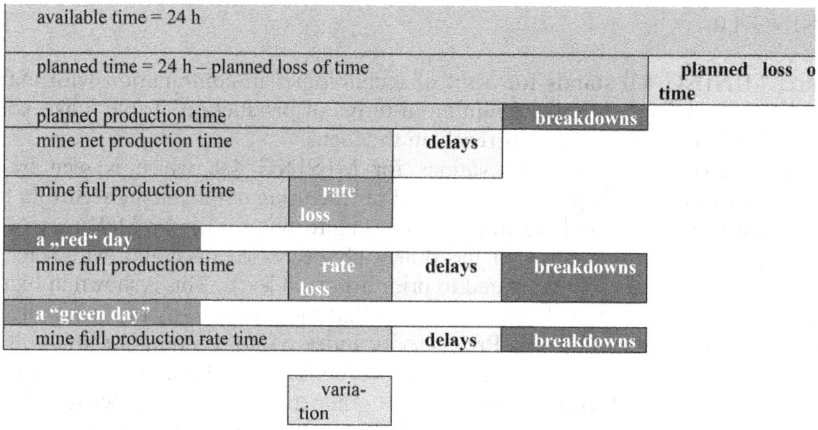

available time = 24 h					
planned time = 24 h – planned loss of time					planned loss o time
planned production time			breakdowns		
mine net production time		delays			
mine full production time	rate loss				
a „red" day					
mine full production time	rate loss	delays	breakdowns		
a "green day"					
mine full production rate time		delays	breakdowns		
	varia-tion				

Figure 6. Productivity of a mine operation with "red day" and "green day" (Source: adapted from Resolute).

As can be seen, use of mine time planning starts with a "planned" loss, to which more losses of time are added, as various real-world conditions of operation become apparent or not. In the end, so-called "red day" and "green days" represent real-world productivity. One disadvantage of this traditional mode of operation consists in the fact that a diagnosis of break downs, delays and failures is only possible in the aftermath and not on a real-time basis.

MINING 4.0 constitutes this structural innovation. It is characterized by technologies and management systems which enable a new way of mine operation which significantly differs from current and past practices, as follows:

- (Underground and open pit) machines communicate directly with the control room of a mine
- The state and location of all equipment are known and documented at any time
- Smart sensors in machines allow for
  →prediction of failure of components
  →communication between machines

- All data and information are directly transmitted
- Mine operators are remote from the locations of the machines
- Mine operators monitor machines in the control room
- Mine operators configurate cutting cycles according to the varying conditions of the (long-wall) face
- Machine instructions are directly sent from the control room
- Machine instructions can be simulated and tested before sending to the machines
- Real-time data are available
- Real-time data allow for analysis tools and decision-making about:
  • improved scheduling and processing decisions
  • the whole operation in real time

Results and outcomes which can be expected from the adoption of MINING 4.0 can be summarized as follows:

Real-Time dispatch
Real-Time production reporting
Production optimization
Production efficiency
Reduced personnel in mine

Proximity awareness
Traffic management
Electronic tagging
Automatic ventilation control

Reduced personnel
Production over shift change

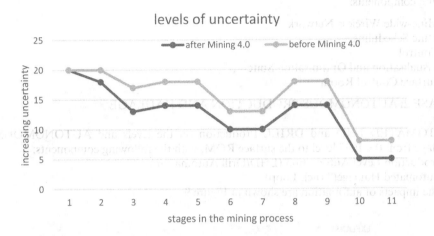

Figure 7.   Reduced uncertainty from automated mining (Source: adapted from McKinsey 2015).
1 = Mine modelling, 2 = Surveying, 3 = Rock recognition, 4 = Blasting, 5 = Core logging,
6 = Face inspection, 7 + 8 = Truck loading and hauling, 9 = Assaying, 10 = Stockpiling

Advantages of this structural innovation are supposed to be numerous, but essentially lefted on economic benefits through better use of available time, costs savings from reduced employment, energy savings and less equipment maintenance efforts – all of which have to be balanced against the higher capital outlays – and on reduced uncertainty, as illustrated in Figure 7

3   (IIII) CASE STUDY: SYAMA GOLD MINE IN MALI

The Syama gold mine in Mali is owned by Société des Mines de Syama S.A. (SOMISY) with two Shareholders:

- Resolute Mining 80%
- Government of Mali: 20 %

The mine is operated by Resolute Mining and it also has two processing plants in operation:
2.4 Mtpa sulphide processing circuit: thee-stage crushing, milling, flotation, roasting, calcine leaching, elution
1.5 Mtpa oxide processing circuit: conventional crushing, SAG milling, leaching
Resolute Mining is an Australian mining company with more than 30 years of experience as an explorer, developer, and operator of gold mines in Australia and Africa which have produced more than 8 Moz of gold. Currently, the company owns three gold mines, the Syama Gold Mine in Mali, the Ravenswood Gold Mine in Australia and the Bibiani Gold Mine in

Ghana. Resolute's Global Mineral Resource base is now 16.8 Moz of gold. Syama is a world class, long-life asset capable of producing more than 300,000 oz of gold per annum from existing processing infrastructure. Resolute has recently concluded the commissioning of the world's first fully automated underground gold mine at Syama with a mine life beyond 2032. Figure 8 shows a map of the mine location.

In a presentation given to Austmine 2019, the structure of the project for a fully automated mine operation was explained as follows:

PHASE 1: MINE DIGITALISATION

A complete control room with connection to the underground wireless network and the ability to schedule, control and monitor MANUAL underground activities in real-time, with the following components:

- Mine-wide Wireless Network
- Mine Scheduling
- Control
- Visualisation and Optimisation Suite
- Surface Control Room

PHASE 2: AUTONOMOUS PRODUCTION AND HAULAGE

AUTOMATED LHD and DRILL Production on the levels and AUTONOMOUS truck haulage from the 1055 level to the surface ROM, with the following components:
Production Level Automation (LHD/Drill Automation)
Automated Haulage(Truck Loop)
The impacts of automation are shown in Figure 9.

Figure 8. Syama mine map location (Source: adapted from USGS and Resolute).

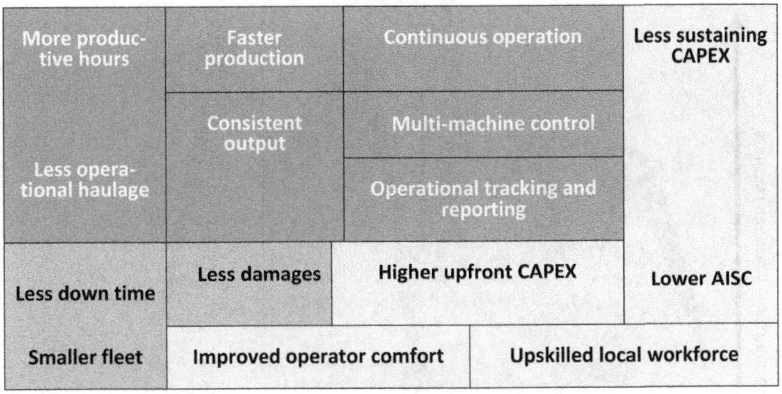

More productive hours	Faster production	Continuous operation	Less sustaining CAPEX
Less operational haulage	Consistent output	Multi-machine control	
		Operational tracking and reporting	
Less down time	Less damages	Higher upfront CAPEX	Lower AISC
Smaller fleet	Improved operator comfort	Upskilled local workforce	

Figure 9. Impacts of automation (Source: adapted from Resolute 2019).

The extension of the Life of Mine is illustrated in Figure 10 showing the original mine extension (in blue), as represented in the Definite Feasibility Study of 2016 and the Review (in yellow) of that Study in 2018. The impacts on improved use of production time are shown in Figure 11.

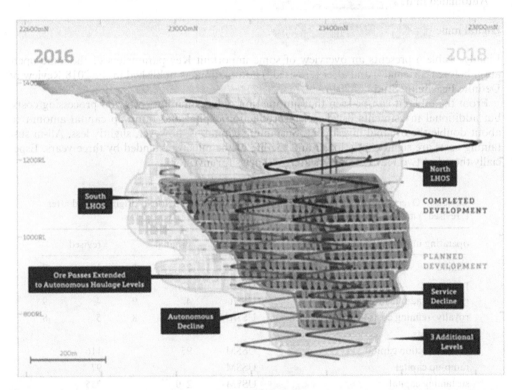

Figure 10. Original mine extension (in blue) – 2016 – and Review (in yellow) in 2018 (Source: Resolute 2019).

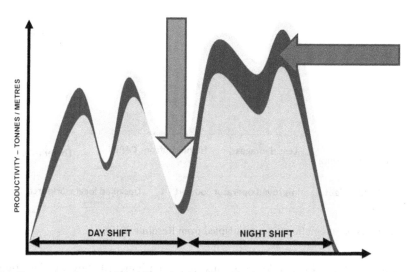

Figure 11. Impacts on improved use of production time (Author's design based on Resolute 2019).

Finally, Table 6 presents an overview of some important Key parameters of the mine operation before ("original") and after ("revised") automation, as published in the 2018 Review of Definite Feasibility Study of 2016.

From the table, it can be seen that automation reduces mining costs and processing costs, but additional investments listed as pre-production capital and ramp-up capital amount to about double the original investment. Sustaining capital, is, however, slightly less. All-in sustaining costs are significantly lower and the life of the mine is extended by three years. Especially these last two Key Parameters are good news for investors.

Table 6. Overview of Key Parameters of the mine operation before ("original") and after ("revised") automation (Source: Resolute 2019).

operating unit costs	unit	original		revised	
mining	US$/t	25.	2	19.	9
processing	US$/t	25.	0	19.	4
general & administrative	US$/t	4.	9	4.	9
royalty, refining costs and silver credits	US$/t	5.	8	5.	8
capital costs					
pre-production capital	US$M	95		116	
ramp-up capital	US$M			97	
sustaining capital	US$M	270		255	
all-in sustaining costs (AISC)	US$/oz	881		784	
Life of mine	years	13		16	

180

4 SUMMARY

This paper presents an overview of some findings related to mining 4.0 which resulted from a DAAD International Alumni Seminar on the subject held at TU Bergakademie Freiberg from 30 April to May 5, 2019. The paper has shown the relevance of the generation of natural resources in developing countries and their contribution at global level, the low productivity in the mining sector as an important motivation for MINING 4.0 as a structural innovation in the sector and some important elements of MINING 4.0 currently in the focus. A case study on a fully automated mine in Mali has been used for illustrative purposes.

REFERENCES

Bundeszentrale für politische Bildung: https://www.bpb.de/nachschlagen/zahlen-und-fakten/globalisierung/52667/terms-of-trade

ICMM: https://www.icmm.com/en-gb/society-and-the-economy/role-of-mining-in-national-economies

McKinsey 2015: https://www.mckinsey.com/industries/metals-and-mining/our-insights/productivity-in-mining-operations-reversing-the-downward-trend

McKinsey 2018:https://www.mckinsey.com/industries/metals-and-mining/our-insights/productivity-across-the-global-mining-sector-is-starting-to-improve

Resolute:https://clients3.weblink.com.au/pdf/RSG/02108094.pdf

USGS: Chirico, P.G., Barthélémy, Francis, and Koné, Fatiaga - Alluvial Diamond Resource Potential and Production Capacity Assessment of Mali: U.S. Geological Survey Scientific Investigations Report 2010–5044, https://pubs.usgs.gov/sir/2010/5044/pdf/sir2010-5044.pdf

Unpublished DAAD Alumni Seminar documents by TU Bergakademie Freiberg World Mining Data http://www.world-mining-data.info/wmd/downloads/PDF/WMD2018.pdf

Scientific and Practical Studies of Raw Material Issues – Litvinenko (Ed)
© 2020 Taylor & Francis Group, London, ISBN 978-0-367-86153-7

Risk management and its contribution to sustainable development of mining enterprises

D. Ivanova
Saint-Petersburg Mining University, Saint-Petersburg, Russian Federation

ABSTRACT: Modern mining companies operate in a volatile external and internal environment. This causes a threat of losses and stability in the production of quality products, which in its turn has a negative impact on the development of the company and its future. In this article we conduct an analysis of risk management process of mining companies and its particular features. We characterize the methods of qualitative and quantitative assessment and their applicability in modern working conditions of mining. This is done through the systematization of processes necessary for the risk assessment. Finally, we argue that the advantages of well-designed risk management system not only can be expressed in material terms but they also play a significant role in the sustainable development system of a mining company.

1 INTRODUCTION

At the present time, it is particularly important for mining companies to pay attention to the principles of sustainable development. This ensures their functioning in conditions of uncertainty of the external environment and allows them to ensure the inflow of investments and, as a result, to increase their profitability.

Investments in mining companies involve a risk of an income deficiency within the established deadlines. In modern conditions, for the purposes of sustainable development and maximization of profits, enterprises need to increase the production of competitive products, introduce modern equipment and technology, and develop sales channels. It all increases the likelihood of risks. The probability of each type of risk and the losses caused by them are different. In this regard, there is a need for a more detailed risk assessment, so that the person who makes the decision to invest in a particular enterprise had a clear picture of the real prospects for the return of funds and receiving of profit.

In Western practice, risk management has long been recognized as one of the most effective tools of modern management. Actually, largest European and American enterprises have a well-organized system of corporate risk management (Aven 2016).

It should be noted that recently, in the economic activities of Russian enterprises, there has been a tendency to organize an integrated risk management system as well. From this position, risk management is one of the most dynamically developing areas of today's management (Vasilieva 2015).

With the development of market relations in Russia, the role of competition is increasing, and therefore, companies in order to succeed need creative decisions and actions, mobility, readiness to introduce new technical tools, which inevitably gives rise to certain risks.

Therefore, risk management is one of the most important components of ensuring the economic independence of the enterprise, creating conditions for achieving business goals. Understanding of this on the part of owners, long-term investors and the management of companies, as well as the lack of clearly regulated management tools, diversification and optimization of operational risks in the enterprises' activities is becoming an increasingly serious problem.

In this paper, we attempt to conduct a comprehensive analysis of the risk management system and to present its structure and organization at a mining enterprise in such a way as to take into consideration all the possible risks, especially, the special ones, typical only for this mining industry. We don't limit the research to any certain mineral resource or field development system in order to present an aggregate picture and complete idea of the process. In the final part of the paper, we evaluate the impact of risk management on the sustainable development of resource companies and highlight the bullet points typical for the mining industry.

2 LITERATURE REVIEW

Despite the wide range of existing papers on different aspects of risk management (Aven 2016; Greiving 2006; Vasilieva 2015, Jonkman et al. 2003, Neil 2000), there are no papers discussing the impact of risk management system on sustainable development of mining enterprises, especially taking into consideration modern complicated working conditions.

The existing literature focuses on general approaches to the risk management at an enterprise, without specifying the particular characteristics of business environment (Kappes et al. 2012; Makarova 2014; Badalova 2006; Meulbroek 2001).

Other researchers discuss the risks of modern companies considering the current economical situation but usually focusing more on market and financial conditions of enterprises and not on the industry's features and production processes' special aspects (Romanova & Krutova 2019; Kashinova 2014; Arslanova 2010, Ivascu 2011).

As for the concept of sustainable development of a mining company, it's a topic of a great current interest but it's usually analyzed without any reference to a risk management system as it can be seen in many papers (Kleine A.& Hauff M. 2009; Korchagina 2011; Dyllick & Hockerts 2002; Pryanichnikov 2012; Biryukova 2014; Belousov 2013, Tonelli et al. 2013).

This paper sets out to offer a complete analysis of the well-designed risk management system at a mining enterprise which include not only the description of a general system but contains the characteristics of mining industry and their impact on the process as well as the whole influence of risk management on the sustainable development of a mining enterprise.

3 RISK MANAGEMENT SYSTEM

For the first time, risk management was mentioned in 1956 (Pruschak 2014), when the idea was put forth that firms, in order to minimize economic losses, should bring to bear appropriate expertise, i.e. risk managers. Reflections on risk have become especially intense since the second half of the 20th century. In predicting how the company will behave in the markets under risk, methods of mathematical statistics were used. However, only the in 1970s, risk management began to gain wide popularity in the business environment. It was the time of the emergence of consulting services in the field of risk assessment, and the main focus was on how the factors of economic instability in a particular country could affect the business of a foreign company functioning there.

However, in most non-financial companies, the practice of risk management was limited to certain industries, such as large-scale industry, energy, and transport, or to certain investment projects. Actually, this was caused by the complexity of production technologies, large possible economic losses and increased security requirements.

It should be noted that the institutionalization of risk management in non-financial companies began only in the 90s, when the creation of risk management structures was stemmed from the motives for reducing insurance costs.

These structures were endowed with risk control functions, focusing on increasing the share of self-insurance (creating a reserve risk fund). Thus, the most thoughtful loss prevention strategy allowed companies in some cases to reduce insurance costs significantly. At the same time, the majority of non-financial risks of organizations continued to use the system of insurance of economic risks.

Significant changes in the system of management of economic risks occurred in the mid-90s, when the top management of companies realized the need for an integrated or systematic approach to developing solutions for optimizing the risks of enterprises (Neil 2000).

Currently, much attention is paid to the development and adaptation in the Russian context of modern concepts of risk management of financial and economic activities of enterprises and organizations in various industries, including the mining sector.

The ability of a corporation to evaluate, control, and effectively manage its risks is a strong competitive advantage and contributes to an increase in value for shareholders and stakeholders, as well as the formation of a financially sustainable company structure.

Increasing the risk management status at the organizational level; the need for a full-scale risk management strategy with regard to the technical, economic, political, legal and environmental risk component; the possibility of attracting Western partners to the investment process - all this aspects force modern companies to restructure their corporate governance system in order to create a risk-oriented model.

The modern concept of corporate governance in companies in the real sector of the economy, as well as in any commercial organizations, combines the efforts of auditors and directors and is aimed not only at reducing the risk level, but also at building the potential for company value growth in light of risks. Therefore, the modern corporate governance of a company should be organized in such a way as to capture the benefits of all possible risks in its core business processes.

To ensure this, the risk management program should include: risk assessment, selection of specific measures to prevent losses, putting these measures into practice and monitoring their implementation.

3.1 Risk identification

Risk identification is an activity aimed at identifying the possibility of the realization of adverse events, changes in the operating conditions of an object, making wrong decisions, etc., that can lead to loss or damage. The purpose of risk identification is to compile a complete list of risks that may affect the company's performance. This list should be exhaustive, as unidentified risks can pose a significant hazard to the operation of an enterprise. Risk identification involves the collection of information on the composition and nature of possible hazards, their sources, causes and factors contributing to the occurrence of adverse events.

In the mining industry, the problem of identifying risks is particularly acute because today enterprises have to work in difficult mining and geological conditions, which cause additional risks (Melnikov 2013). First of all, we are talking about a specific mining risk, which means the likelihood of rock collapse, as well as water invasion into the mine workings when conducting work underground. Unlike other risks accompanying the activities of the enterprise, the consequences of the mining risk are not only financial losses, but also losses of life (Zakharov 2013); therefore, the companies of the mineral resource complex must intelligently approach this problem and pay special attention to its solution.

3.2 Risk assessment

Risk assessment includes two groups of methods, and the choice between them depends on current conditions and information.

3.2.1 Qualitative methods

For qualitative risk assessment in the activities of companies, various methods can be applied, for instance statistical, regulatory, analytical, expert, methods of analogies. Statistical methods are used to ascertain (or negate) the existence of a risk (its frequency, level of losses). Among the specific methods of such identification correlation analysis is widely used.

Methods of analogies are complementary to statistical methods. They are based on the use of statistical data on adverse events (frequency, strength) and the damages caused by them at sites which are analogues with the same characteristics as the object under study.

Regulatory methods are based on a comparison of the real values of risk factors with the standards adopted for them.

Methods of analytical risk assessment are usually applied in situations where it is possible to establish the existence of a risk by analyzing the causality of an adverse event that could harm the company. In risk studies, so-called "decision trees" are widely used.

Expert methods are used in cases when the necessary statistical database on the frequency of negative events or on damage caused by them is not yet collected or when the systematic presentation of risk formation processes is difficult. These methods are of importance to mining enterprises because of the uniqueness of their working conditions.

In some cases, complex methods can be used, as they are the combination of methods from different groups and allows the companies to get more detailed and reliable results.

3.2.2 Quantitative methods

Quantitative risk analysis consists in assessing the degree of risk (the probability of occurrence of losses, as well as the amount of possible damage caused), both in relation to individual risks and the enterprise as a whole.

In total, there are four groups of quantitative risk assessment methods (Nefedova 2010):
statistical methods;

1. probabilistic and statistical methods;
2. analytical methods;
3. expert methods.

The features of the methods of quantitative risk assessment determine their application area depending on the available statistical data and the possibility of theoretical models building (Figure 1) (Romanova et al. 2019).

Statistical methods are based on the assessment of probability of a random event and on the relative frequency of this event in a series of observations. These methods are preferred in many industries because, firstly, they are simple and, secondly, their estimates are conducted on factual information. The main tools of statistical methods are: average, variance, standard deviation and coefficient of variation. According to statistical methods, of the two alternatives, the one with a better average and lower variation is considered to be the preferred one.

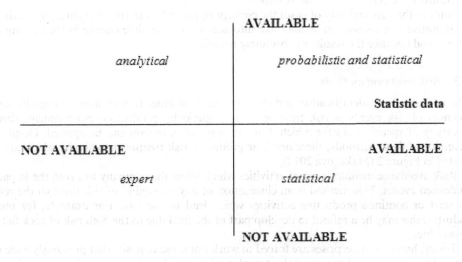

Figure 1.   Classification of methods of economic risks quantitative assessment.

The use of statistical methods is limited in those industries where there are not enough observations (Makarova 2014). To correctly assess the risks of rare events, a very large amount of statistical data is required, which is not always possible in the mining industry due to its nature. Such data as characteristics of the enclosing rocks (rock hardness, fracturing), mode of mineral occurrence (thickness, angle of slope), characteristics of water flows are unique for each specific field. Therefore, this information can only be obtained empirically and cannot be taken anywhere. In addition, the collection and processing of such amount of information can be too long and expensive.

If the available statistical information is not complete, it is sometimes possible to fill the gaps by analyzing additional indirect data or by logical reasoning. The use of a combination of statistical data and theoretical hypotheses for risk assessment is the basic idea of probabilistic and statistical methods. This extends the field of application of this group of methods, but the reliability of the results may be lower than using statistical methods.

The two previous groups of methods require sufficient or at least limited statistical data on the phenomenon under study. However, risk management has to deal with the need to assess rare events that can have very drastic consequences (Greiving 2006). In the past, these events could not occur at all because of their "rarity" or the uniqueness of the objects in question. In this case, the statistics either do not exist, or refer to other objects that are significantly different from that one under study, as in the case of consideration of risk events in the process of extraction of mineral resources. This makes impossible the use of above mentioned methods. In this case, it is possible to rely on the use of analytical methods, which are based on the building of a mathematical model of the studied risk and theoretical evaluation of its parameters. These methods are very time- and labour-consuming and have a relatively low accuracy, but in some cases are the only possible scientifically based methods of evaluation. Among the most common universal tools of this group, the use of the hypothesis of the normal distribution of the value of the analyzed risk, the application of game theory (including its section "games with nature"), etc should be mentioned (Badalova 2006).

In a situation where there is neither statistics nor the possibility of building a mathematical model, which takes place in the study of objects with uncertain parameters or unexplored properties, managers resort to the use of experience and knowledge of experts. Expert studies, despite a significant proportion of subjectivity, provide information for decision-making. The methods of expert evaluation include the organization of work with experts and the direct processing of their opinions, which can be expressed in quantitative and/or qualitative form. Among the most well-known methods of expert evaluation are Delphi method, brainstorming, scenario method. However, none of them guarantees the reliability of the results.

Since in the vast majority of cases the concept of risk refers to future events, any method of quantitative risk assessment should take into account the possible change in the existing level of risk and not take the results as absolutely reliable.

3.3 Risk treatment methods

The next step after identification and risk assessment in order to minimize the negative consequences of risk events is risk treatment. In a particular production environment, there is a variety of specific risks for which different ways of reduction can be applied. Despite the diversity of these methods, there are four groups of risk treatment methods, which are presented in Figure 2 (Makarova 2014).

Risk avoidance includes a set of activities which allow the company to avoid the impact of undesired events. This method is an elimination of any possibility of risk through the refusal to start or continue productive activities which lead to the risk. For example, for mining industry, this may be a refusal to develop part of the field due to the high risk of rock fall and loss of life.

Today, however, enterprises are forced to work out those deposits that previously were considered too dangerous because of the complex mining and geological conditions. In this case,

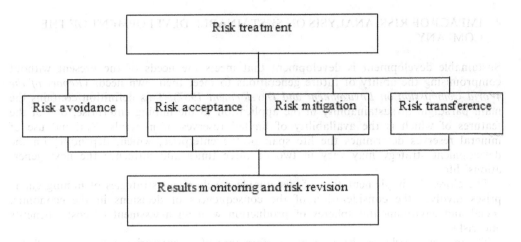

Figure 2. Main groups of risk treatment methods.

the risk avoidance will mean the cessation of activity of the enterprise, so the risk avoidance cannot be used everywhere. So, the mineral companies consciously take the risk and accept it.

Risk-taking on its own is quite rare. It can be observed in cases where the occurrence of a risk event does not depend on the activities of the company and it cannot be influenced. However, maintaining the risk at the existing level does not always mean the refusal of any action aimed at compensation of damage. The enterprise can create special funds in order to compensate the losses in case of adverse situations. The measures taken while maintaining the risk may also include obtaining loans and credits, receiving state subsidies, etc.

Risk reduction involves reducing either the possible damage or the likelihood of adverse events. In most cases it is achieved by means of implementation of preventive organizational and technical measures. These can be different ways to enhance the buildings and structures security, as well as installation of monitoring and warning systems, fire protection devices, personnel training like safety instruction etc. The mining industry is characterized by the following measures (Melnikov 2013):

- operational monitoring of the mining-geological and geomechanical state of the rock massif and forecasting of the external situation, allowing to respond quickly to any changes in the external environment;
- strategic planning of activities, which provides the development of various scenarios for the future development of mining operations and allows the managers to work out a strategy for the development of the enterprise, adapted to possible changes in the mining and geological environment;
- monitoring of the socio-economic and regulatory environment.

Measures aimed at the risk transfer mean transferring responsibility to third parties while maintaining the existing level of risk. These include insurance, which involves the transfer of risk to the insurance company for a fee, as well as various kinds of financial guarantees etc. The risk transfer can be carried out by making special comments in the text of the documents, transferring the risk to the counterparty.

The final stage is the control and adjustment of the results of selected strategy implementation based on new information. The control consists in obtaining information on losses and measures taken to minimization of them. It can be expressed in the identification of new circumstances that change the level of risk, the transfer of this information to the insurance company, monitoring the effectiveness of security systems, etc. Every few years there should be a review of data on the effectiveness of risk management measures used, taking into account information about the losses occurred during this period.

4 IMPACT OF RISK ANALYSIS ON SUSTAINABLE DEVELOPMENT OF THE COMPANY

Sustainable development is development that meets the needs of the present without compromising the ability of future generations to meet their own needs (*Report of the World Commission on Environment and Development 1987*). This definition contains the main paradigm of sustainability in the application to the mining enterprise, one of the features of which is the availability of limited reserves of minerals. Rational use of mineral reserves determines the life span of the enterprise, which, depending on the development strategy may vary in two or three times and influence the next generations' life.

Therefore, the implementation of sustainable development strategies of mining enterprises involves the consideration of the consequences of decisions in the economic, social and environmental spheres of production with an assessment of costs, benefits and risks.

The impact of risks on the economic performance of an enterprise is clear and well discussed in professional literature (Romanova 2019). Competent and forehanded risk assessment allows enterprises to take the necessary measures to reduce the likelihood of financial losses and thus keep their economic performance indicators at a high level.

As for the environmental and social component, this problem claims more attention, especially at mining enterprises. Deep and large-scale changes in the mining industry caused both by scientific and technological progress and by institutional changes that occurred during the global financial crisis and economic instability arisen because of the stagnation of the world and national economies, set the tasks for the enterprise management system. The most important of these tasks are efficient risk management and reduction of the negative impact of risks on the enterprises' activity.

Mining companies are characterized by a high degree of probability of man-made disasters, industrial accidents, natural hazards and the damage caused. The operation of such companies often has a negative impact on people and the environment. According to experts, the total losses from direct impact on the environment are estimated at 5–8% of the global domestic product, while the funds allocated for environmental protection activities do not exceed 0.6–0.8% of the global GDP (Murzin 2016). Mining production, like other branches of primary processing of natural resources, adversely affect ecosystems; therefore, the load of environmental protection measures on the economy of mining enterprises is much higher than the average indicators, both on a global scale and in the conditions of the Russian Federation.

However, unlike other industries, in mining, social and environmental aspects are inextricably linked, especially when conducting mining operations underground. The negative impact on the environment not only damages the environment, but also endangers the health and lives of people working in the mine, so proper and intelligent risk management is a much more pressing problem in the mining industry than in the majority of other sectors of the economy. For instance, for any enterprise the identification of environmental risks, such as the risk of air pollution through the release of pollutants in concentrations exceeding the maximum permissible ones, makes it possible to foresee the costs of additional treatment facilities and to avoid fines from regulatory authorities. But for a mining company the risk management is much more important because it helps not only to save money but to save human lives.

Risk management directed at ensuring the viability and efficiency of an enterprise should be a cyclical and continuous process that coordinates and directs core activities. It is advisable to do this by identifying, controlling and reducing the impact of all types of risks, including monitoring, contacts and consultations aimed at meeting the needs of the population, without prejudice to the ability of future generations to meet their needs. Risk assessment leads to a stable enterprise activity, contributing to its sustainable development.

5 DISCUSSION

The contribution of this work is the systematisation of the risk management processes and their implementation in the operation of a mining company. In Russia, where the extraction of mineral resources is of a great importance for the national economy, many enterprises just start to understand the need for the risk management and the holistic approach to its organisation. This search provides a broad picture of risk identification, risk assessment and risk treatment methods at an enterprise putting emphasis on the importance of such processes for a company in modern operating conditions.

The paper states the importance of risk management system for a company and its influence on sustainable development. Thus, risk management process helps to ensure an effective financial reporting process, as well as compliance with laws and regulations, to avoid accidents resulting in injuries and damage to the reputation of the organization and the associated consequences. Thus, the risk management process allows management to achieve its goals and therewith to avoid mistakes and unforeseen situations what also plays a significant role in the process of sustainable development of a company.

An area of further research is the development of specific indicators for the quantitative assessment of the impact of the risk management system on the sustainable development of a mining company and the testing of these indicators in terms of Russian mining enterprises. The results of this work will be presented in further scientific research.

6 CONCLUSION

Considering that high risk is the inherent characteristic of the mining industry, the company's performance depends on the efficiency and sustainability of a well-designed risk management system.

It is obvious that the risk management system should be formed on the basis of justification of measures to prevent and minimize risk, focused on ensuring the sustainability of the economic development of the enterprise.

In this paper was produced a comprehensive study of a risk-management system of a mining company, were classified and characterised methods used for risk assessment, highlighted their main features and discussed their applicability at a mining company in modern complicated operation conditions. Two groups of risk assessment methods were identified in the analysis: qualitative and quantitative ones, which, when using separately, can't offer a complete approach necessary for the risk assessment. It means that they should be used together, building a system, in order to complete each other and provide the coherent picture of risks and their relevance for the activity of a company.

The analysis carried out in the paper allows to set off the risk treatment methods applicable in mining industry due to its operating conditions and complexity of carried operations during the extraction of resources. Mining companies tend to risk more than companies of other industries due to the fact that geological conditions become more complicated while the demand on mineral products remain constant and force mining enterprises to operate constantly under risk.

Given the impact of risk on the performance of the enterprise, the concept of risk tolerance of the enterprise can be defined as the ability to manage internal risks and adapt to external risks, thereby ensuring an increase in the competitiveness of the enterprise.

That's why management risks become strategic, while enterprise competitiveness depends on the quality of the organization, on the rationality, efficiency of the structure and processes, as well as on the ability to manage internal risks (system, resource) and external risks (in consumer behavior, competitors' actions). Thus, the activity of the enterprise can act as a tool for reducing risks and harmonizing the relationship of the enterprise with the external environment The conducted research showed strong influence of the risk management results on the sustainable development of a company and proved the necessity for multi-faceted approach to mining company management.

REFERENCES

Arslanova E. R. 2010. The risk-management in crisis-management system // *Reporter of AGTU, 1*: 54-57.

Aven T. 2016. Risk assessment and risk management: Review of recent advances on their foundation // *European Journal of Operational Research* 253 (1): 1-13.

Badalova A.G. 2006. *Risk-management of industrial systems: theory, methodology, implementation mechanisms*. Moscow: Stankin, Yanus-K. 328p.

Belousov K.Yu. 2013. Actual problems of formation of sustainable development strategy in Russian companies // *Problems of modern economy, 3(47):*191-194.

Biryukova V.V. 2014. Factors of sustainable development of an oil company // Naukovedenie, 5 (24): 1-9.

Dai FC, Lee CF, Ngai YY 2002. Landslide risk assessment and management: an overview. *Engineering Geology* 64(1): 65–87.

Dyllick, T., Hockerts K. 2002. Beyond the Business Case for Corporate Sustainability // *Business Strategy and the Environment 11*, 130-141.

Greiving, S. 2006. Integrated risk assessment of multi-hazards: a new methodology. // *Schmidt-Thome' P. (ed) Natural and technological hazards and risks affecting the spatial development of European regions, geological survey of Finland*, 42, 75–81.

Ivascu L., Izvercianu M. 2011. An Approach to Identify Risks in Sustainable Enterprises // *2nd Review of Management and Economic Engineering International Management Conference, Cluj-Napoca, Romania.*

Jonkman, S. N., van Gelder, P. H. A. J. M., Vrijling, J. K. 2003. An overview of quantitative risk measures for loss of life and economic damage. // *Journal of Hazardous Materials*, A99, 1-30.

Kappes, M. S., Keiler, M., von Elverfeldt, K., Glade, T. 2012. Challenges of analyzing multi-hazard risk: a review // *Natural hazards*, 64(2),1925-1958.

Kantor O.G. 2013. Classification of quantitative estimation methods of economic risk // *Reporter of UGATU* 7 (60): 34–39.

Kashinova N. E. 2014. Identification and classification of risks as risk management tools in the anti-crisis management of a modern enterprise // *Concept, 5*: 1-7.

Kleine A., Hauff M. 2009. Sustainability-Driven Implementation of Corporate Social Responsibility: Application of the Integrative Sustainability Triangle // *Journal of Business Ethics*, *85*: 517–533.

Korchagina E.V. 2011. Sustainable development of a company: working out a model of analysis // *Problems of modern economy, 4*: 133-136.

Makarova V.A. 2014. The formation of risk management at the enterprise // *Reporter of PskovGU, 4*: 98-108.

Melnikov N.N., Kozyrev A.A., Lukichev S.V. 2013. Great depth - new technologies // *Herald of the Kola Scientific Center of the Russian Academy of Sciences, No. 4 (15)*, 58-66.

Meulbroek L. 2001. Total Strategies of Corporate Risk Control // *Mastering Risk, Vol. 1: Concepts*, London: Pearson.

Murzin M.A. 2016. Mining enterprises as a source of ecological risks // *Mining informational and analytical bulletin*. № 2: 374-383.

Nefedova T. 2010. Quantitative estimation of risks as criterion of the sustainable development of the enterprises // *Transport business in Russia* 11: 39-41.

Neil A. 2000. *Integrated Risk Management*, New York: McGraw-Hill.

Pruschak O.V. 2014. Risk management as a factor of sustainable development of innovative companies // *Reporter of the Saratov State Social and Economic University* 2: 77-81.

Pryanichnikov S.B. 2012. Characteristic of the factors, which form the level of the steady development of the industrial enterprises // *Management of economical systems, 4 (40)*.

Report of the World Commission on Environment and Development № A/42/427 from 16 June 1987 "Our common Future": Official Records of the 42nd session of General Assembly. –New York, 1987. – 383 p.

Romanova S.V., Krutova M.A. 2019. Risks as the object of crisis management: analysis and basic methods of minimization // *Journal of Economy and Business*, vol. 3-2: 82-86.

Tonelli F., Evans S., Taticchi P. 2013. Industrial Sustainability: challenges, perspectives, actions // *Int. J. Business Innovation and Research*, 7 (2): 143-163.

Vasilieva E. 2015. Actual problems of risk-management in Russia // *International scientific journal "Innovative science"* (60): 55-56.

Von Elverfeldt, K., Glade, T., Dikau, R. 2007. Naturwissenschaftliche Gefahren- und Risikoanalyse. Felgentreff, C., T. Glade (ed) Naturrisiken und Sozialkatastrophen. *Spektrum Akademischer Verlag*, Berlin, 31-46.

Zakharov V.N. 2013. The assessment and management of mining risk as a tool to increase the efficiency and safety of a mining enterprise. // *Mining informational and analytical bulletin*. № 1: 57-69.

Scientific and Practical Studies of Raw Material Issues – Litvinenko (Ed)
© 2020 Taylor & Francis Group, London, ISBN 978-0-367-86153-7

Holistik responsible mining approach

Carsten Drebenstedt
Technische Universität Bergakademie Freiberg, Germany

1 INTRODUCTION

In 1556 Agricola published the twelve books of mining and metallurgy. The first book has the title: The defense of mining against the attacks of the opposition and the evidence of their benefit. This shows that the conflict between the mining and metallurgical activities and the society has a long tradition.

Today the term "sustainable development" is often pointed out like the main guideline for the development of the modern society and for the mineral sector.

But the term is not new. The first time the term was used in 1713 by Hans Carl von Carlowitz, the chief mining inspector in Saxony, in his book "Silvicultura oeconomica" – about the economy of the forest. Because of the demand in wood e.g. for equipment construction, underground support, and char coal production for metallurgical processes, the demand in this material increased and strategies and measures were needed to reduce the consumption and to regulate the afforestation.

Initiated in 1972 by the Club of Rome with the report "The Limits to Growth" starts the modern discussion about the limit of economic growth because of the limited availability of natural resources, on special example of crude oil. The first oil crises in 1973 pushed the consideration of natural resources.

In 1987 was published the so called Brundtland Report "Our Common Future" from the United Nations World Commission on Environment and Development (WCED). An oft-quoted definition of sustainable development is defined in the report as: "development that meets the needs of the present without compromising the ability of future generations to meet their own needs."

Al least with the UNCED (United Nation Conference on Environment and Development) 1992 in Rio de Janeiro sustainable development was defined like a balanced development between economy, ecology and social needs.

However, the concept of sustainable development is to keep a system (economic, social or ecosystem) in function considering the well-being of future generations.

The mining industry may come in conflict with some of the above mentioned definitions. Can mining be sustainable? Yes, because mining is an inseparable part of the development of our society. The responsibility is, to improve the mining technologies permanent and develop science and technologies to avoid and minimize negative impacts to the environment and society.

2 SUSTAINABLE DEVELOPMENT GOALS OF UNESCO

The UNESCO submitted the sustainable development goal 2030. Analyzing the goals is becomes clear, that these goals are not in contradiction to the development of the mining industry. Up to now, the knowledge about the role of mining in the society is often understand incorrect. The picture, formed over centuries is dominates by hard and dirty work, land disturbance, negative impacts to the environment and social spheres. Looking to the Sustainable Development Goals 2030 of the UNESCO it becomes clear, that mining contributes positive to these targets.

At first there are direct links to mining products, like "Clean Water and Sanitary" (target 6), "Affordable and Clean Energy" (7), "Industry, Innovation and Infrastructure" (9).

Here contributes for example ceramics, aggregates or energy commodities. In addition, in other targets, like "Zero Hunger" (2) and "Good Health and Well-Being" (4) we find direct link to fertilizer in agriculture or metals and sources for radiation in hospital.

Other direct effects of mining to the society are employment, salaries, taxes and social responsibility. This correlates well with the target 8 "Decent Work and Economic Growth". This is the basic for targets like "No Poverty" (1), "Zero Hunger" (2), "Quality Education" (4), "Gender Equality" (5), "Reduced Inequalities" (10), "Sustainable Cities and Communities", and "Peace, Justice and Strong Institutions" (16).

The targets "Life Below Water" (14) and "Life on Land" (15) addressed to reclamation demand in mining industry, where we have today very good results in increasing biodiversity in urban areas. "Responsible Consumption and Production" (12) and Climate Action" (13) are very important issues, to which contributes mining with delivering the needed minerals for new technologies, and these are task for the mining industry by self for improvement with energy and material saving technologies, reduction of losses and waste ....

The target 17 "Partnership for the Goals" invites mining industry to be a part of the solution.

3 CONVENTIONAL AND ALTERNATIVE MINING APPROACHES

The world mining industry is characterized by several challenges. The growth of population and living standard leads to an:

– Increase of Demand in Mining Products
– Increase of Prices
– Increase of Profits (Drive of Market)
– Increase of Impacts to the Environment
– Increase of Mining Waste- Increase of Land Occupation and related conflicts
– …

Since 2004 the world mining industry develops very fast, with a growth of 2-3% per year, for several commodities faster. Nowadays the world mining production reaches more than 30 B t. In average every human being consumes 5t per year, in developed countries more than 20t per year.

Under this conditions new strategies required for covering the future raw material demand. The main directions are:

– Reducing the consumption
– Minimizing the raw material demand in products
– Use of alternative materials; material design
– Recycling (urban mining)
– Increase the recovery rates from deposits
– Exploration of new deposits
– Development of mining technologies for new deposit conditions

In the following discussed two mining approaches, the "maximum profit" driven one and the more responsible mining concept with "moderate profits". "Moderate" is a profit, when the level is attractive for investors with high environmental and social standards. Sometimes mining projects with low standards offer high (maximum) profits.

Of cause, the moderate profit concept only works, if this approach is applied for all mining project. In reality we see and today large differences in level of mining safety and the environmental and social standards in mining projects. Plus different government regulations and subsidies this differences lead to not real free market conditions.

However, the improvement of mining industry must be happening always to keep the public acceptance.

The main strategies to improve the mining activities are:

– Increase of recovery rate
– Use of environmentally friendly technologies
– Consideration of reclamation and long term environmental effects of mining

The strategies must be realized by concrete measures. In this paper presents some results of research and development projects of the Institute of mining at Technische Universität Bergakademie Freiberg.

3.1 *General mining approach*

The concept of general mining approach is maximum profit driven, which can be summarized by formula 1

$$\text{(Maximum) Profit } P_G = \text{Income } I_G - \text{Direct Costs } C_{DG} \tag{1}$$

The main characteristics of this approach are:

− not optimized processes with relatively high losses of valuable components
− in result high volume of waste material
− with negative impacts to the environment
− low consideration of indirect costs I_G

The indirect costs includes the expenditures, necessary for compensation of all negative influences of the mining project to the environment, like

− Land occupation costs
− Environmental protection costs/Health care
− Reclamation costs
− Long term costs (decrease of land value, climate change, ...)
− Monitoring costs
− Early end of deposit reserves/Early investment in new deposits
− ...

Widely exists the opinion that for all negative mining influences, especially after mining, the government is responsible, because the companies pay/paid taxes to the government budget.

In Germany the principle is, that the causer is responsible for all negative impacts and therefore he must accumulate special funds after governmental control. The experience shows, that and after decades of mine closure not all problems are solved, e.g. with slope stability and water quality. The German government paid in last 25 years approx. 11 B€ for closing coal mines, more than 5.5 B€ for closing uranium mines, and another 2 B€ for closure of other metal (Cu, Sn ...), spare, and salt mines. The mine closure was caused by the unification of Germany with the bankrupt of mining companies without financial funds for reclamation.

The economic parameters of the general mining approach are illustrated simplified in Figure 1. The general tendencies are:

− Than higher the extraction (recovery) rate of valuable components, than higher the income
− and then higher the direct costs because of additional measures
− the profits (economic zone) reaches a maximum in moderate extraction level
− considering the indirect cost, the profit reduces (ecological zone)

3.2 *Alternative mining approach*

The main characteristics of the alternative mining approach are (formula 2):

− Increase of extraction level of main product/reduce losses
− Use of byproducts and accompanying products/reduce residuals
− Use of environmental clean production methods
− Fulfill all reclamation responsibilities like integrated part of mining
− Consideration and reduction of indirect costs
− Long term production/Past investments for new deposits

$$\text{(Moderate) Profit } P_A = \text{Income } I_A - \text{Direct Costs } C_{DA} - \text{Indirect Costs } C_{IA} \tag{2}$$

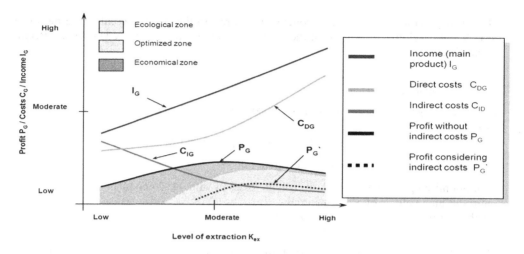

Figure 1. Economic parameters of general mining approach, depending on extraction level.

The general tendencies of economic parameters of the alternative mining approach are simplified:

– higher income because of reduction of losses and additional sellable products from bypro-ducts and accompanying raw materials
– reduce of indirect costs, because of less waste volume and contamination potential
– mainly less profit because of consideration of indirect costs and additional technical meas-ures to achieve higher extraction level and higher environmental performance

Figure 2 shows a variation of direct and indirect costs, depending of the level of additional measures. In right balanced economies, higher expenditures for environmental friendly tech-nologies result in significant decrease of direct costs, e.g. for energy and materials, and indir-ect costs.

The idea of "moderate profit" is that instead of "maximum profit" with low recovery rate and environmental and social responsibility, a part of the income is used to increase this

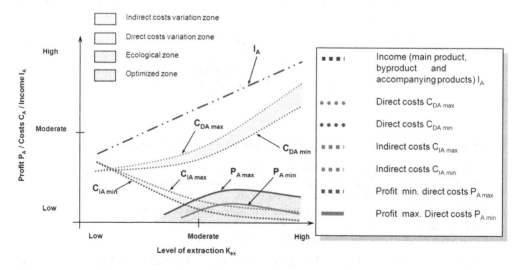

Figure 2. Economic parameters of alternative mining approach depending on extraction level.

parameters. This shall be the standard of every mining company, but in reality we see often only a minimum of realization of the requirements.

4 SUSTAINABILITY RESEARCH IN MINING

The chair surface mining at Technische Universität Bergakademie Freiberg understands sustainability in mining at first and foremost as responsible behavior of the mining companies, which takes into account the above-mentioned aspects of sustainability. Responsible for the promotion and support of this conscious action, the chair tracked in the following research strategy:

– Maximizing the use of value components of a deposit (major, minor, and accompanying minerals)
– Use of valuable minerals in the residues of raw materials extraction and processing
– Environmentally friendly mining technologies
– Ensuring public and occupational safety in all stages of the mining process
– Fast and high-quality rehabilitation of the claimed land

For this mining concept methodological basics were created, which have been published in books and international conferences. To implement the research priorities at the chair surface mining for the diverse topics are established interdisciplinary and internationally oriented research groups:

– Mine planning (operation research path)
– Mining technology (mechanical path)
– Mine water (chemical path)
– Raw material awareness/education (social path)

4.1 *Maximum utilization of the value components*

The research focus "Maximum utilization of the value components" includes the following topics:

– Construction of a rock management system starting from the deposit model to the raw material processing to optimize the mineral recovery from deposits in compliance with the environmental benefits on the example of copper mining in Chile (Drebenstedt et al. 2010a)
– Selective extraction of raw materials with high selectivity to avoid losses or dilution, e.g. hard coal mining in Vietnam, lignite mining in Germany (Xuan, Drebenstedt 2009; Pfütze, Drebenstedt 2012)
– Optimization of extraction technology for safe use of reserves, such as extraction of pillars and low-loss extraction of high rooms (chambers) in the copper mine Sheskasgan, Kazakhstan
– Development of intelligent solutions for quality control in production and storage/dumping processes (Eichler, Drebenstedt 2010)
– Targeted construction of technogenic deposits, such as for accompanying materials (Drebenstedt 2007)
– Inclusion of low-grade ores in resource recovery, such as gold in Uzbekistan (Ravshanov, Drebenstedt 2006)

4.2 *Use of valuable minerals in the residues of raw materials extraction and processing*

In continuation of the target of maximum use of valuable deposit components of cause the residuals from processing, extraction or utilization of interest. Research topics are:

- Recovery of valuable components from tailings, e.g. Phosphorite in Uzbekistan (Ishimov, Drebenstedt 2010a)
- Use of residues from treatment, e.g. Phosphorite in Uzbekistan (Ishimov, Drebenstedt 2010b)
- Use of residues of raw material use, e.g. ashes (Drebenstedt 2006c)

4.3 *Environmentally friendly mining technologies*

Under the research focus "environmentally friendly mining technologies" are investigated the following topics:

- Environmentally friendly drainage technology (HDD filter wells to reduce the demand of material, energy and land) (Struzina et al., 2011)
- Analysis of the cutting processes to reduce wear and dust as well as for energy optimization (Drebenstedt, Kressner 2008; Drebenstedt, Vorona, Gaßner 2012)
- Optimization of blasting technology to reduce emissions (vibrations, noise, gases, dust ...) with good fragmentation (Drebenstedt, Ortuta 2009, 2010, 2012)
- Use of alternative fuels for mining equipment, such as biodiesel (Drebenstedt, Jauer 2007)
- Investigations to reduce the noise on conveyor belts (Täschner 2013)
- Development of filter equipment for fine disperse sediments to reduce energy consumption (Shevshenko 2017)
- Selection of optimal technologies for the use of mining equipment and mining equipment selection for the optimal operating under given conditions (Purevsuren, Drebenstedt, Tsedendorij 2006)
- Environmental accounting for selection of environmentally friendly mining systems (example: hard rock and under water extraction) (Drebenstedt, Schmieder 2006; Gomaa, Drebenstedt 2012)
- Complex effects of small-scale mining on the mineral economy and justification of countermeasures (Grießl, Drebenstedt 2012)
- Development of holistic environmental management strategies (Drebenstedt 2008)

4.4 *Ensuring public and occupational safety in all stages of the mining process*

Environmental and aspects of health and safety are close to each other. Selected research topics under the main aspect of health and safety are:

- Development of dump monitoring systems for early detection and risk mitigation of negative impacts; use of Neural Networks (Drebenstedt 2012)
- Analysis of the cutting processes to reduce dust development (Drebenstedt, Van 2007)
- Prevention of coal fires (on example of China) (Gusat, Drebenstedt 2010; Drebenstedt et al. 2010b)
- Concept development to avoid accidences in surface mining on example of coal mining in Indonesia (Permana 2014)

4.5 *Fast and high-quality reclamation*

The research focus on "Fast and high-quality reclamation" covers the following topics:

- Developing the basics for a "dump cadaster"/"dump archive" for the evaluation of mechanical and chemical processes in dumps and appropriate countermeasures for adverse effects (Drebenstedt, Struzina 2008)
- Methods to reduce the formation of acid mine waters and effective methods for neutralization of acidic waters (Simon, Hoth, Drebenstedt, Rascher, Jolas 1012)
- Review of long-term consequences and tasks in the closure of mines (Drebenstedt 2006a)

– Optimization of mining technology to increase land reclamation area and reduce external dumps (Lozhnikov 2012, Lituchyy 2014)
– Reclamation under the aspect of post mining regional development (Toni 2015)

The stated objects of the research focus in part across themes. Further optimization of sub-processes is often only effective in the consideration of interactions with upstream and downstream processes. As an example in particular the topics selective and cutting processes as well as rock management system or optimized blasting take this in to account. In the project rock management system for example work together experts in drilling, processing, mechanical process engineering and automation processes.

The topic of mining safety is the basis and prerequisite for any of the topics e.g. coal fires or residual pillar extraction.

4.6 *Research methods*

To solve these issues following methods are used:

– Laboratory, pilot and field tests
– Mathematical modeling and simulation
– Process visualization for understanding and optimization of processes
– Artificial intelligence (neural networks)
– CAD, expert systems
– Financial-mathematical models, economic valuation
– Integration of suitable sensors for process monitoring and control
– Development of integrated mining planning and production management systems

With the selected research areas and topics the chair surface mining contribute to a responsible mining approach with balanced consideration of economic, environmental, social and safety aspects and to the responsibility, that even future generations can use the geo-materials.

In teaching this holistic approach is also taught. Serve e.g. the courses mine planning, mining abroad and mineral economics. In special courses will protected goods taken into account, for example mining water management, reclamation, and ventilation and security technology. Moreover, the aspects of sustainability are an integral part of the technical and technological courses.

5 CONCLUSIONS

To optimize the mining process under considering of both, economic and environmental/social aspects, a comprehensive investigation of all processes along the process chain over the life cycle is necessary. Positive results can be:

– minimize losses and increase the extraction rate of valuable components in the deposit
– minimize negative impacts to the environment, health and safety
– improve the acceptance and responsibility of mining projects
– optimize the economic parameters

Two mining approaches are possible: with "maximum" or "moderate" profit. To achieve the results, the maximum profit is not always the best supervisor.

For the "moderate" profit approach several solutions for more environmental and social responsibility exist. Advanced technologies and research are available and need permanent improvement.

REFERENCES

Drebenstedt C (2006a) Conditions for Mine Closure Strategy in Great European Coal Mines. In: Mine Closure Strategy (eds. A. Fourie, M. Tibbett), Perth, p. 117–1130 (ISBN 0-9756756-6-4).

Drebenstedt C (2006b) Financial Valuation of Mine Closure Alternatives. In: Mine Closure Strategy (eds. A. Fourie, M. Tibbett), Perth, p. 499–508 (ISBN 0-9756756-6-4).

Drebenstedt C (2006c) Wertstoff Braunkohleasche – Einsatzmöglichkeiten zur Verbesserung von Kippsubstraten – ein Beitrag zur nachhaltigen Rohstoffwirtschaft. In: Umweltgeotechnik und Rohstoffe, Veröffentlichungen des Instituts für Geotechnik 2006-4, Freiberg, p. 59–571 (ISSN 1611–1605).

Drebenstedt C (2007) The utilisation of accompanying mineral resources from open-cast mining – using the lusation lignite mining region as an example. Górnictwo odkrywkowe, Wroclaw, p. 13-18 (ISSN 0043-2075).

Drebenstedt C (2008) Environmental Management – from Mineral Exploration to Mine closure. In: New Challenges and Visions for Mining (eds.: Sobczyk E.J., Kicki J.), CRC Press Taylor & Francis Group, London, p. 91–101 (ISBN 978-0-415-48667-5).

Drebenstedt C (2009) State of the art and new concepts for prediction of cutting resistance on example of continuous mining equipment. In: Innovations in Non Blasting Rock Destructuring (ed.: Drebenstedt C.), TU Bergakademie Freiberg, p. 6–26 (ISBN 978-3-86012-377-5).

Drebenstedt C (2012) Schüttung von Kippen im Braunkohlentagebau, Auswirkung auf bodenmechanische Eigenschaften. In: Bodenverflüssigung bei Kippen des Lausitzer Braunkohlebergbaus, TU Bergakademie Freiberg, Freiberg, p. 82–102 (ISSN 2195-853X).

Drebenstedt C et al. (2010a) Report: RockManagementSystem (RMS), unpublished.

Drebenstedt C et al. (2010b) Risk Assessment Modeling of Coal Fires in the P. R. China Using Artifical Neural Networks and GIS. In: Latest Developments in Coal Fire Research, Bridging the Science, Economics, and Politics of a Global Disaster (Ed.: Drebenstedt et al.), Freiberg, 2010, S. 260–269 (ISBN 978-3-86012-397-3).

Drebenstedt C, Eichler R, Simon A (2013) Herausforderungen im modernen Tagebaubetrieb. Springer Verlag, Wien, p. 155–176 (ISSN 0005–8912, ISSN 1613–7531).

Drebenstedt C, Jauer J (2007) Fundamentals and Results of Use of Biodiesel in Large Mobile Mining Fleets, In: Mine Planning and Equipment selection and environmental issues in waste management in energy and mineral production (CD ROM, Eds.: Singhal et al), The Reading Matrix Inc., Irvine, p. 187–1195 (ISSN 1913–6528, ISBN 978-0-9784416-0-9).

Drebenstedt C, Kressner M (2008) Optimization of technological parameters by cutting resistance analysis on bucket wheel excavators. Metal Mine, Maanshan (China), p. 683–694 (ISSN 1001–1250, CN 34-1055/TD).

Drebenstedt C, Kressner M, Päßler S (2008) Modular concept for dynamic mine planning and production control supported by virtual reality. Metal Mine, Maanshan (China), p. 808–812 (ISSN 1001–1250, CN 34-1055/TD).

Drebenstedt C, Ortuta J (2009) New approach for energy efficient blast calculation. In: Mine Planning and Equipment Selection and Environmental Issues and waste Management in Energy and Mineral Production (eds.: Singhal, Mehrotra, Fytas, Ge), CD Rom, The Reading Matrix Inc., Irvine, p. 155–164 (ISSN 1913–6528, ISBN 978-0-9784416-0-9).

Drebenstedt C, Ortuta J (2010) Monitoring of blast-induced vibration effects on structures by fiber Bragg grating. In: Rock Fragmentation by Blasting (ed.: Sanchidrian JA), CRC Press/ Balkema, Leiden, Netherlands, p. 617–622 (ISBN 978-0-415-48296-7, 978–0–203–86291–9).

Drebenstedt C, Ortuta J (2012) Use radar reflectivity as possibility for measurements of fragmentation during blasting, In: Rock fragmentation by blasting (Eds.: Singh P, Sinha A), CRC Press/ Balkema, Leiden, The Netherland, p. 105–110 (ISBN 978-0-415-62143-4, ISBN 978-0-203-86291-9).

Drebenstedt C, Phan QV (2007) Dust Emission in Dependence on Cutting Parameters in the Process of Mechanical Rock Destruction, In: Mine Planning and Equipment selection and environmental issues in waste management in energy and mineral production (CD ROM, eds.: Singhal et al), The Reading Matrix Inc., Irvine, p. 196–205 and p. 285 – 294 (ISSN 1913–6528, ISBN 978-0-9784416-0-9).

Drebenstedt C, Schmieder P (2006) Evaluation model of alternative hard rock mining technologies. In: Mine Planning and Equipment Selection (eds.: Cardu M, Ciccu R et al.), Volume 2, p. 1330–1338 (ISBN 88-901342-4-0).

Drebenstedt C, Struzina M (2008) Overburden Management for formation of internal dumps in coal mines. In: Rock Dumps 2008 (ed.: Fourie A), Australian Centre for Geomechanics, Perth, p. 139–146.

Drebenstedt C, Vorona M, Gaßner W (2012) Improvement of cutting performance of the bucket wheel excavator ER-1250 for hard rock mining. In: Continuous Surface Mining – Latest Developments in

Mine Planning, Equipment and Environmental Protection, University of Miskolc, p. 109–118 (ISBN 978-615-5216-09-1).

Eichler R, Drebenstedt C (2010) Circular Blending Beds – New Operation Mode for Improving Continuous Homogenization. In: Continuous Surface Mining (editor: Drebenstedt C), Freiberg, p. 303–306 (ISBN 978-3-86012-406-2).

Gomaa E, Drebenstedt C (2012) A review of continuous underwater mining and the development occur on continuous dredgers. In: Continuous Surface Mining – Latest Developments in Mine Planning, Equipment and Environmental Protection, University of Miskolc, p. 41–51 (ISBN 978-615-5216-09-1).

Grießl E, Drebenstedt C (2012) Using an Expert System Approach to Identify Alternative Solutions to Environmental Degradation Caused by Artisanal Small-Scale Mining, In: Environmental Issues and Waste Management in Energy and Mineral Production 2012 (Eds.: Singhal et al.), CD ROM, The Reading Matrix, Irvine, CA, USA, p. 139–146 (ISSN 2167 3322).

Gusat D, Drebenstedt C (2010) Thermo-mechanical modelling of surface movements induced by underground coal fires. In: 12th International Symposium on Environmental Issues and Waste Management in Energy and Mineral Production (Ed.: Sklenicka P, Singhal R, Kasparova I), Publishing House of Lesnická pracá s.r.o., Prague, p. 158–163 (ISBN 978-80-213-2076-5, 978–80–87154–42–7).

Ishimov A, Drebenstedt C (2010a) Mathematical and experimental analysis of the sedimentary processes during hydraulic filling of tailings. In: 12th International Symposium on Environmental Issues and Waste Management in Energy and Mineral Production (ed.: Sklenicka P, Singhal R, Kasparova I), Publishing House of Lesnická pracá s.r.o., Prague, p. 208–213 (ISBN 978-80-213-2076-5, 978–80–87154–42–7).

Ishimov A, Drebenstedt C (2010b) Use of waste from phosphate processing for soil improvement. In: 12th International Symposium on Environmental Issues and Waste Management in Energy and Mineral Production (ed.: Sklenicka P, Singhal R, Kasparova I), Publishing House of Lesnická pracá s.r.o., Prague, p. 214–218 (ISBN 978-80-213-2076-5, 978–80–87154–42–7).

Pfütze M, Drebenstedt C (2012) Approaches for the use of contactless sensor technology for a quality-controlled selective mining in German opencast lignite mines, In: Mine Planning and Equipment Selection 2012 (Eds.: Singhal et al.), CD ROM, The Reading Matrix, Irvine, CA, USA, p. 513–520 (ISSN 2167 3322).

Purevsuren D, Drebenstedt C, Tsesendorj S (2006) Modelling of dragline excavation schemes. In: Mine Planning and Equipment Selection (eds.: Cardu M, Ciccu R, et al.), Volume 1, p. 192–201 (ISBN 88-901342-4-0).

Ravshanov D, Drebenstedt C (2006) Aspects of mineral processing of man-mad deposits. In: Mine Planning and Equipment Selection (eds.: Cardu M, Ciccu R, et al.), Volume 2, p. 891–895 (ISBN 88-901342-4-0).

Simon A, Hoth N, Drebenstedt C, Rascher J, Jolas P (2012) Practical Environmental Protection in Central German Lignite Mining New Methods for the Determination of Geogenic Acid Mine Drainage and Buffering Potentials and Applications in Mining Operation, In: Environmental Issues and Waste Management in Energy and Mineral Production 2012 (Eds.: Singhal et. Al), CD ROM, The Reading Matrix, Irvine, CA, USA, p. 343–353 (ISSN 2167 3322).

Struzina M, Müller M, Drebenstedt C, Mansel H & Jolas P (2011) Dewatering of Multi-aquifer Unconsolidated Rock Opencast Mines: Alternative Solutions with Horizontal Wells. Mine Water and the Environment, Journal of the International Mine Water Association (IMWA), Volume 30, Number 2, p. 90–104 (ISSN 1025–9112).

Xuan Nam Bui, Drebenstedt C (2009) Use of Hydraulic Backhoe Excavator in Surface Mining. In: Innovative Entwicklung und Konzepte in der Tagebautechnik (ed. Drebenstedt C), TU Bergakademie Freiberg, p. 175–189 (ISBN 978-3-86012-361-4).

Scientific and Practical Studies of Raw Material Issues – Litvinenko (Ed)
© 2020 Taylor & Francis Group, London, ISBN 978-0-367-86153-7

Modern condition and prospects for the development of forest infrastructure to improve the economy of nature management

A.A. Kitcenko, V.F. Kovyazin & A.Y. Romanchikov
Saint-Petersburg Mining University, Saint-Petersburg, Russian Federation

ABSTRACT: The article deals with the problem of improving the environmental manage-ment economy through cadastral valuation of forest lands, taking into account the degree of development of their infrastructure. The development of forest land infrastructure is an import-ant aspect of successful land policy, improving the quality of forest area assessment. This article proposes the idea of accounting for infrastructure on forest land using private infrastructure coefficients and infrastructure development coefficient in the formula for determining the cadastral value of a forest plot at the last stage of cadastral valuation of forest areas.

1 INTRODUCTION

The modern market for forest products requires compliance with the principles of environmen-tal, economic, social management and rational use of forest resources. Forest infrastructure is able to solve these problems. To date, forest land infrastructure is poorly developed and needs funding from the state. The infrastructure of forestry is a set of objects and activities used in the maintenance of forestry and creating its material, technical, technological and organizational basis. Forest land is often used inefficiently, for several reasons. Firstly, there is no register of accounting for forest infrastructure, which includes forest roads, fire fighting reservoirs, admin-istrative buildings and buildings, and much more. Control over such objects on the lands of the forest fund is not exercised. Secondly, rental payments for the use of forest land are understated and do not reflect the objective value for the use of the forest plot (Federation Council, 2018).

In addition, the main problem of inefficient use of forest land is the lack of forest roads. Transport accessibility to forest areas is a major factor in the successful development and exploitation of forests. The poorly developed road-transport infrastructure of forest use ham-pers the possibility of a more complete development of production forests and reduces the eco-nomic accessibility of woody forest resources, and also complicates control over forest areas.

2 LITERATURE REVIEW

Particular attention was paid to the consideration of factors in the cadastral assessment of for-ests in the works of the following authors Kovyazin & Romanchikov (2018), Bogomolova (2009), Bulatova (2010), Antsukevich (1998), J. Brazee (2014), R. Cademus (2015), Y. Li (2017), N. Oswalt (2016). Scientists examined the problem of cadastral valuation of forest land, taking into account the forest fund infrastructure. To date, there is no classification of infrastructure on forest land. Some authors have attempted to classify infrastructure on forest land. However, the classification of infrastructure on forest land was not taken into account in their cadastral valuation (Kovyazin & Romanchikov, 2018).

The purpose of this article is to improve the procedure for cadastral valuation of forest lands by taking into account the infrastructure on the lands of the forest fund, as well as improving the environmental management economy by establishing an objective rent for the use of forest resources. To achieve this goal, the article proposes a formula for determining the cadastral value of forest land, taking into account the development coefficient of forest area infrastructure (USSR State Committee for Construction, 1989).

The contribution of forest infrastructure to increasing the level of satisfaction of human needs through production or consumption is very high.

3 METHODOLOGY

The forest complex refers to industries that operate in sparse, inaccessible conditions. In this regard, issues of development of industrial and social infrastructure in the construction of new enterprises in areas of undeveloped forest areas remain problematic. To date, the methodology for cadastral valuation of forest land is imperfect and has several disadvantages.

To improve the cadastral valuation of forests and improve the economy of the forest fund, it is supposed to take into account the forest land infrastructures when evaluating them. To account for infrastructure in assessing forests, a classification of the infrastructure on forest land has been proposed (Figure 1).

Hydrotechnical group of infrastructure includes: dike earth locks, river passenger moorings.

The recreational infrastructure group should include the following facilities: a cable car, a complex of physical and sports facilities, a tennis court, an indoor pool, an ice skating rink, an athletics arena, a fitness center, a camping, a cycling track, sports grounds, a stadium and its stands, football and golf (USSR State Committee for Forestry,1989).

The objects of the transport and logistics infrastructure group include: highways and railways, parking lots, tunnels, bridges, overpasses, crane and access roads (rail or road).

The objects of the economic group of the infrastructure include: the bases of mechanization for the maintenance and repair of construction machinery and mechanisms, for the repair of pipes, production services, management of production and technical equipment and equipment.

Objects of the oil and gas group include such objects as field drilling bases, drill pipe bases, tubing bases, gas pipelines for various purposes, channels, trenches, holes, central collection and preparation facilities for oil, gas and water, as well as other facilities necessary for field development and oil and gas production (Forestry Code of Russian Federation, 2006).

The objects of the geological exploration group of the infrastructure should include: the base for arranging the deposit of minerals and exploration areas, a quarry, an

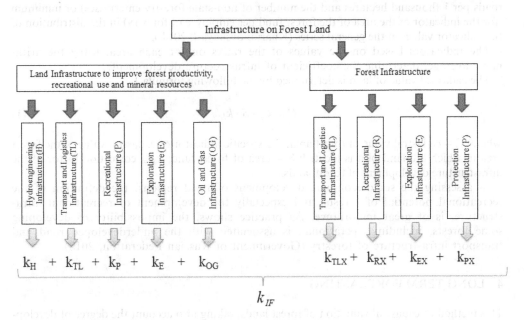

Figure 1. Types of forest land infrastructure.

201

enterprise for the extraction and enrichment of natural ores containing sulfur, boron, arsenic and barium.

The main task of the group of protective forest infrastructure facilities is to maximize forest protection efficiency with the least expenditure of funds for their construction and minimize the loss of valuable forest land. Such objects include: fire breaks, mineralized strips, platforms for water intake and others (Morkovina et al., 2016).

Conducting a cadastral assessment of forest lands, taking into account the degree of development of their infrastructure, includes the step-by-step construction of a multiple regression model. The resulting variable (cadastral value of forest land) is influenced by a large number of factor variables and the problem arises of how many and which variables should be included in the model.

The order of inclusion of variables is determined by the value of the correlation coefficient between the resulting variable and each of the factors.

Simplified scheme for constructing a model of multiple regression, which is proposed can be described as follows:

1. Analysis of factors and the degree of dependence of prices on these factors using the correlation coefficient.
2. After selecting the significant factors and building a model on their basis, this model is analyzed. For the analysis the coefficient of determination is used. It shows the percentage by which the price change is explained by all the factor variables included in the model. When selecting factors, the values of the correlation coefficients will be analyzed, rather than determination; in constructing a regression model, it will be possible to use all the factors selected for the evaluation.

The next stage of the assessment implies the establishment of the infrastructure development rate by studying the relationship between the infrastructure indicators among themselves, as well as the level of use of the calculated cutting area by region using correlation analysis. To assess the level of infrastructure development by districts in the research process, use the following indicators: the length of roads per 1 thousand hectares; infrastructure facilities on the lands of the forest fund, the area of the forest fund per one forestry worker; the number of non-state forestry enterprises. For each indicator of infrastructure, it is necessary to determine the rank calculated as a share of its maximum (for indicators of the length of roads per 1 thousand hectares and the number of non-state forestry enterprises) or minimum (for the indicator of the area of the forest fund per employee of forestry) in the distribution of the indicator values in the general series (Kharionovskaya I, 2014.).

The indicators based on the values of the ranks on for each area, using the arithmetic average, determine the coefficient of infrastructure development.

The cadastral value of land is determined by the following formula 1:

$$C = C_f \cdot S \cdot k_{IF} \tag{1}$$

where: C = cadastral value of forest area; C_f = specific indicator of cadastral value of the forest area in which the land plot is located; S = area of forest land; k_{IF} = correction factor of the infrastructure development of forest lands.

In assessing the socio-economic development of rural regions, the definition of the recreational potential of forests, and especially the development of recreational infrastructure, is of great importance. As practice shows, the impossibility of developing some forests, including operational, is associated with the underdeveloped road and transport infrastructure of forestry (Government of Russian Federation, 2018).

4 LONG-TERM FORECASTING

This method of cadastral valuation of forest lands, taking into account the degree of development of their infrastructure, can be applied on the territory of the Leningrad Region and in the North-West region of Russia. In this part of the country, the forest land infrastructure is most developed. The influence of indicators of forestry infrastructure on the level of territorial economic development should be assessed on the basis of indicators of the level of use of the calculated cutting area and the overall infrastructure development rate for each district. Environmental economics involves the internal and external economic assessment of natural resources and the contribution of each natural resource to environmental, health, social and other spheres. Based on the obtained indicators, it will be possible to obtain an objective picture of the development of forest infrastructure within the boundaries of a specific territory. In addition, an objective cadastral assessment is the main factor stimulating the development of infrastructure on forest land by the state and industrial companies. Consequently, the size of payments for the use of forest land will correspond to real time and bring large investments in the forest complex (Federation Council, 2018).

The economic assessment of natural resources is a very complex scientific and practical problem. The fact is that the cost of any thing is determined by the labor costs for its production, and natural resources are products of nature, not of man. However, labor is invested in their exploration, development, protection and reproduction, that is, value is created. Currently, there are two main concepts for assessing natural resources: cost and rental. When the cost concept takes into account the cost of the development of natural resources, and the quality of natural goods, their usefulness act as an additional factor in the measure of value. In the rental concept, differential rent is calculated, that is, a different amount of income obtained from the exploitation of natural resources of different quality and location (for example, lands located far or close to transport routes, etc.) (Kovyazin & Romanchikov, 2018).

5 DISCUSSION

Thus, the specific cadastral indicator of the cadastral value of forest lands will reflect their quality, taking into account the degree of infrastructure development. Based on the proposed indicators, a comparative and integrated assessment of the development of forestry infrastructure by region will be given (National Duma Committee for Natural Resources, Property and Land Use, 2017).

Based on the assessment results, groups of areas with different levels of infrastructure development are identified. Also, on the basis of the data used for cadastral valuation, it is possible to determine the main problems and possible directions for the development of the forestry infrastructure within a specific territory. Today, it is necessary to modernize all the activities of the timber industry, the introduction of modern innovative and scientific technologies (Morkovina et al., 2016). Technically and technologically modernize the timber industry complex through measures aimed at enhancing the attraction of investments in the development of the transport and production infrastructure of the forest complex. In the economic assessment of natural resources, it is very important to find a compromise solution taking into account both approaches. Accordingly, the formula for determining the cadastral value of forest land, proposed in this article takes into account both approaches and takes into account the following factors: damage assessment from their irrational use of environmental protection efficiency; the effectiveness of measures to stimulate the rationalization of environmental management, the reasonableness of the size of fees for the use of natural resources; export profitability of wood resources; fair distribution of profits from joint ventures (National Duma Committee for Natural Resources, Property and Land Use, 2017).

6 CONCLUSION

There is a lag in the technological development of the forest complex at the global level. Forest infrastructure influences technological growth and product upgrades. In accordance with international standards, the pace of technological renewal in the Russian timber industry is several times less than the competitiveness of domestic forest products. Today, it is necessary to modernize all the activities of the timber industry, the introduction of modern innovative and scientific technologies. To realize investments, to increase the development of transport infrastructure on forest land leased for development of forests (Kharionovskaya, 2014). Cadastral valuation of forest land is one of the priorities of the state land policy. Based on the cadastral value of the forest area, the amount of rent for the use of forest land and the amount of tax are determined. To obtain the exact value of the cadastral value of the forest area, it is necessary to take into account all factors affecting the total value of the cadastral value, including the need to take account of the infrastructure on the forest fund lands.

In addition, based on the data of the state cadastral assessment, the amount of rent for the use of forest land, which is currently unreasonably undervalued, is determined. The cadastral assessment of forest lands, taking into account the degree of development of their infrastructure, will help to contribute to the economics of environmental management, preserve and increase the main social, economic and political resource for the state, which is the forest.

REFERENCES

Government of Russian Federation 2018 On Approval of 2030 Forestry Development Strategy (Decree N 1989 20 September 2018 (ed. 28 February 2019) [Moscow: Government of Russian Federation].

Federation Council. 2018. Condition and Prospects of Forestry Transport and Industrial Infrastructure [Moscow: Federation Council].

Kharionovskaya I. 2014. Features of development of forestry infrastructure in the republic of Komi Journal Izvestiya Komi Science Center № 1 (25): 104-111.

Kovyazin V and Romanchikov A. 2018. The problem of cadastral valuation of forest land, taking into account forest fund infrastructure Journal of Mining Institute 229: 98-104.

Morkovina S, Vasilyev O and Ivanova A. 2016. Innovative infrastructure of the forestry system: forest selection-seed-crop centers Journal of Forestry 4:221-230.

National Duma Committee for Natural Resources, Property and Land Use. 2017. Round-table conference materials: "Exercise of Authority in the Field of Forestry by Government Bodies of Russian Regions: Issues, Aims, and Prospects" [Irkutsk: National Duma Committee for Natural Resources, Property and Land Use].

National Duma of Russian Federation. 2006. Forestry Code of Russian Federation [Moscow: National Duma of Russian Federation].

USSR State Committee for Forestry. 1989. Industrial Construction Standards. Standard for Forest Roads Designing VSN 7-82 [Moscow: USSR State Committee for Forestry].

USSR State Committee for Construction. 1989. Construction Rules and Regulations. Industrial Transport SNiP 2.05.07-85*[Moscow: USSR State Committee for Construction].

Scientific and Practical Studies of Raw Material Issues – Litvinenko (Ed)
© 2020 Taylor & Francis Group, London, ISBN 978-0-367-86153-7

Responsible mining

O. Langefeld & A. Binder
Clausthal University of Technology, Institute of Mining, Clausthal-Zellerfeld, Germany

ABSTRACT: Mining influences the natural and social environment largely. More than other businesses, the mining industry is aware of this impact, especially regarding their historical legacy. This awareness and the dealing with the impact characterize responsible mining. The common definition of sustainable development reflects the combination of that awareness and the handling of development in present and future. Refusal of mining does not represent a sustainable development because the present generation needs raw materials for their living. Rather, mining needs to be performed with regard to the responsibility for the present and the future. Therefore, mapping of needs and impacts represents an essential task for sustainable activities. Taking up the definition of sustainable development, responsible mining should shape the presence, add value, keep doors open for future generation, and provide them a plurality of options without comprising future abilities. Hereafter, thoughts and ideas for those areas are presented for sustainable mine practice. Afterwards, possibilities are shown to integrate discussed aspects into mining engineering education.

1 INTRODUCTION

Mining influences the natural and social environment to a great extent. More than other industries, the mining industry is aware of this impact, especially regarding their historical legacy. Responsible mining is characterized by this awareness and the dealing with the impact. Also, mining companies take on the responsibility for the mineral supply and support the economic development by the mineral extraction. (Mirande, Chamber, Coumans, 2005) (Klein, 2012)

The common definition of sustainable development according to the Brundtland-Commission reflects the combination of that awareness and the handling of development in present and future It is a "development that meets the needs of the present without comprom-ising the ability of future generations to meet their own needs." (World Commission on Environment and Development, 1987)

Refusal of mining does not represent a sustainable development because the present gener-ation needs raw materials for their living. Rather, mining needs to be performed with regard to the responsibility for the present and the future. Therefore, the mapping of needs and impacts represents an essential task for sustainable activities. Due to the numerous affected areas and stakeholders, the mapping is complex and the task cannot be solved completely. By using the triple-bottom-line concept, which is considering the areas of economy, environment and society, a consciousness of affected sectors can be created by focusing on three areas. The Australian Centre for Sustainable Mining Practices (ACSMP) expands the concept for sus-tainable mine practice by introducing two important points for the mining industry: resource efficiency and safety. The deposits are the basis for the activity. They are limited and predeter-mined but dictate the approach. Resource efficiency represents a responsible exploitation. Furthermore, safety is an important focus of the mining practice. Often, approaches and measures differ from other industries because of the operational characteristics, the accessibil-ity as well as the dimensions (Laurence, 2011).

Taking up the definition of sustainable development, responsible mining should shape the presence, add value and keep doors open for future generation and provide them a plurality of options without comprising future abilities. Hereafter, thoughts and ideas for those areas are presented for sustainable mining practice. Afterwards, possibilities are shown to integrate discussed aspects into the mining engineering education.

2   KEEP DOORS OPEN: PERSPECTIVES ON TAILINGS

Relating to the exploitation of mineral resources, resource efficiency is discussed less often than relating to materials or energy. Besides present demands, a sustainable design must consider the future needs. The exclusive mining of high-grade areas is not sustainable if the subsequent extraction is hindered. At best, deposits can get fully exploited. (Laurence, 2011)

Though, the definition of waste and value is changing depending on the available technology and the demand at the date of mining. Hence, waste can become value and non-profitable areas can be exploitable or a reprocessing of tailings is considered.

Since 2015, the recycling of tailings is investigated at the former ore mine Rammelsberg in Goslar funded by the federal ministry of education and research. (Technische Universität Clausthal, 14 May. 2015) Until its exploitation, the deposits has been one of the biggest deposits for indium in Europe. During the operation until 1988, indium wasn't extracted. After the closure, the demand increased significantly due to the usage of indium-tin-oxide (ITO) for solar panels and flat screens. The remnants of the deposits are the tailing ponds in which the waste material was disposed in the last fifty years of operation. Under the new circumstances, the potential of the tailings is investigated. For this purpose, the deposit was modelled using historic and newly gained data (Figure 1). Furthermore, approaches for mining and processing are planned. Entrusted with these tasks, the viewpoint of a future generation opened. The research shows which steps in the past could have eased the resumption of activities.

For the deposit modelling, historic documentation, data of previous investigations and assays of new sampling were used. The data situation was complicated due to the transformation of the operating company. Hence, only fragments of the operational documentation were available and published data needed to be used. Thus, the request can be formulated to keep and store all captured data to provide the opportunity of usage for future generation. Public archives are important to support the preservation of the knowledge.

The possibility of reprocessing should be considered in the selection of approaches in the design and construction of tailings ponds. Often, the dam is supported by the material. This can lead to huge harm in case of a lack of knowledge about the rheology as in the case of the dam failure in the Brazilian state Minas Gerais in November 2015 (Samarco, 2016). Also, this kind of construction leads to difficulties in dam reconstruction. The investigations showed

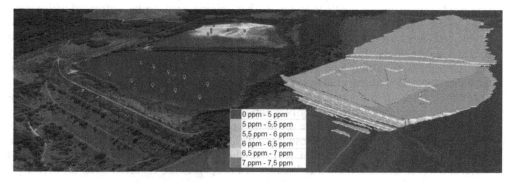

Figure 1.   Modelled deposit with indium grades and sampling points.

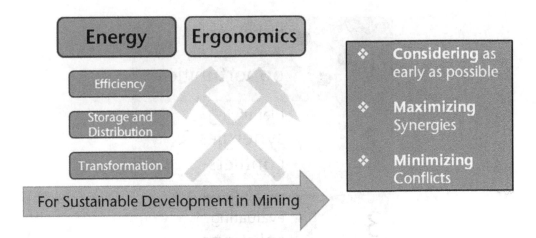

Figure 2.   Areas and principles of Blue Mining.

a slight weathering caused by the water coverage abetting the reprocessing. Therefore, an isolation of
   the material relating to surroundings fostering weathering is preferable.

   In conclusion, a subsequent usage is supported by an early awareness of such possibilities if those thoughts lead to an adjusted planning. To integrate those subsequent and multiple usages into the planning process, represents a goal of the Blue Mining approaches, which focuses on ergonomics and energy. Three main principles frame the approach, which are shown in Figure 2 on the right side.

   Mine planning should be performed with regards on future possibilities and awareness of impacts and consequences for the future. That way, present and future interest are considered and doors are kept open to add value as described in the following section. (Langefeld, Binder, 2017)

3   ADD VALUE: MORE THAN A MINE

Besides surface installations, an underground mine consists of cavities which provide an access to the deposits, facilitate extraction, haulage and processing. Furthermore, the cavities enables the access to the rock mass which can be a source for heat and gases. Value can be added to operation by using this access. Besides, areas in a new depth are constructed. The height difference can be used for pump storage facilities. The restructuring of surface areas provides further possibilities for energy transformation installations.

   The Blue Mining concepts aims to harmonize those possibilities and the extraction of raw materials during and after the production stage. Assessing only the subsequent usage of abandoned mines, the modification has drawbacks based on previous decisions leading to advantages of new constructions. (H.-P. Beck, 2011) This can be prevented by an early consideration as claimed by the Blue Mining concept.

   To implement the concept in the planning, three steps are arranged shown in Figure 3. The accuracy of planning results increases by multiple runs at different stages of mine planning. In the first stage, opportunities are identified. Based on the condition and the demand, opportunities are evaluated. The development of the environment initiated by opening and closure of the mine needs to be considered. For chosen opportunities, conflicts and synergies with the existing plans are identified. In the third stage, solutions are developed and evaluated. Those should maximize synergies and minimize conflicts. The factors for assessing the solution are chosen with consideration of different dimensions of sustainable mine practice.

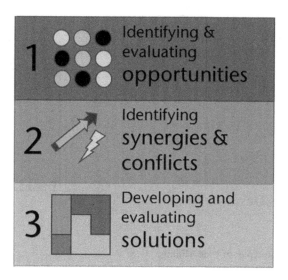

Figure 3. Stages in Blue Mining.

Results of the process are implemented into the technical planning. An example are the orientation of heaps or roof areas as well as the planning of cavities close to the shafts for the operation of pump storages. Additional measures are developed parallel and extend the benefits of the mining operation outlasting the production phase and leading to long-term positive impacts. Synergies can improve the balance of the whole operation. For measuring and communicating the impacts, the approach of Life Cycle Assessment can be used. Different solutions can be assessed and strength and weakness can be investigated. However, the approach depends on data, which is just slightly available for mining operations. The existing data is based on few case studies which incorporate only environmental but not social impacts. (Awuah-Offei, Adekpedjou, 2011) For a satisfying application, further development in this field is needed.

4   SHAPE THE PRESENCE: RESPONSIBILITY TOWARDS PEOPLE

Responsibility towards people is an essential topic talking about responsible mining, particularly towards the organizations own employees. The responsibility can be divided into three areas: technical, organizational and personal.

The technical responsibility can be assigned to the manufacturers and to the management. Manufacturers are intended to design and produce their products according to usual standard. In Europe, it means the compliance with CE standards and in particular with ATEX standards. In other countries further standards are established for example by the American National Standards Institute (ANSI) since 1918. It aims to strengthen the position of the US-market with the global economy and ensure at the same time the safety and health of consumers as well as the protection of environment. Besides the exclusive usage of standardized and certified machines and equipment, the maintenance and repair must be compliant.

In future, manufacturer will produce increasingly more automated equipment and machines making the workers tasks easier and safer. At this stage, fully automated machines move in dangerous areas of mines. Eventually, automation is the solution. Thus, nobody can be harmed where no people are. (Langefeld, 2017).

The mining company has the organizational responsibility. It is necessary to ensure that equipment and machines are used rule-consistent and safe. Furthermore, the employees need to be informed about the machine control and related hazards. Additionally, the exposure of the employees needs to be monitored to protect them from environmental impacts. Workplace control can be carried out when the workplaces and their exposure for example to noise, heat and dust as well as the work time at each place is registered. An integrated approach including appropriate medicals survey was conducted for instance over decades.at the Ruhrkohle AG at its several mines

A safe use of machines can be achieved by using a risk assessment. It is conducted in several steps. First of all, the risks must be investigated by determination of working areas and tasks in the operational structure, hazardous substances table as well as tables of installations and equipment. Secondly, hazards resulting from used machines and substances are investigated and assessed. During the third step, required measures are determined to reduce the hazards. In the fourth step, measures are collected in a table of measures, which can be used for the control of implementation. In the last step, measures are checked for efficiency. (Sächsisches Staatsministerium für Wirtschaft, Arbeit und Verkehr, 2011) For this process, it is important to establish an operational structure including an appropriate documentation using an organisation chart. Visualizing all working areas of the organisation, tasks can be assigned to the areas. The organisational structure is the foundation for the systematic determination of risks and hazards. Specialist create risks assessments and control the implementation of measures. Employees get briefed systematically in seminars, so that each employee can detect and handle hazards at his or her workplace. These measures do not only protect the employees but also the social environment having the "Vision Zero".

In this context, Zero Vision means that no accidents should happen. To this, the German Social Accident Insurance Institution for the raw materials and chemical industry (BG RCI) has introduced the concept „Vision Zero and the 7 Golden Rules in Mining your first steps to Success in Prevention" (Figure 4) (Meesmann, 2017). The prevention strategy defines goals for the year 2024. Quantitative goals are chosen to sensitize all stakeholders to make further efforts and achievements in prevention. Therefore, milestones must be defined and a reporting system has to be installed. For all prevention measures, the effectiveness, perception and acceptance of the social environment must be examined systematically and regularly.

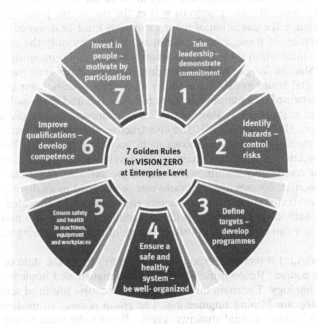

Figure 4.   Vision Zero by BG RCI (Meesmann 2017).

The establishment of standby force for emergencies is very important for the social environment. Potential incidents, which are affecting not just the mine itself, can be localized and mitigated in a safe way.

The International Council on Mining and Metals (ICMM), which incorporate already six of the ten biggest mining companies, aims to have a central role in the promotion of social and economic development. Encouragement of collaboration amongst the industry leaders should contribute to alleviate poverty and give people access to a better life. Ten principles are defined and all members are committed to them. (International Council of Mining & Metals, 2015). The principles describe responsibility towards the social environment as addressed by this paper especially towards the employees. For instance, the principles refer to respect human rights and interest, cultures, traditions and values of employees and affected communities. Furthermore, the implementation of risk management systems and strategies based on scientific evidence as well risk perception of stakeholders must be considered. By continuous improvement of health and safety efforts, zero vision is pursued as an ultimate goal in principle five. The ninth principle targets the continuous improvement of social services to contribute to the social, economic and institutional development of countries and communities. The topics of water, tailings and post-mining are also focused.

The personal responsibility is situated in the social environment. It means for the employees to wear the personal protection equipment provided by the mining company as well as regular medical examinations. By constructive collaboration with the mining company, people from the farer social environment take their responsibility. The mining company contributes by direct and indirect employment of community member as well as by the local business development, the economic diversifications and strategic social investments. (International Council of Mining & Metals, 2015)

5 EDUCATION FOR RESPONSIBLE MINING

The responsible mine planning and practice is mandatory for the future of mining. A transparent stakeholder communication is required to prove taking-over of responsibility and ensure the acceptance for the activity. The field of work for future mining engineers includes the communication with different groups as well as the sustainable planning and realisation of mining projects. Hence, the education of mining engineers must be designed for this purpose.

The implementation of those aspects cannot be realized by only the extension of existing classes. A holistic integration into the curriculum is required. To integrate sustainability into the curriculum, (Shields, Verga, Andrea Blengini, 2014) describes four ways: a limited coverage of environmental issues in existing modules or courses, specific courses on sustainability, sustainability intertwined as an approach for regular courses and the possibility for specialization in the area of sustainability. Possibilities for implementation are presented by (Binder, Langefeld, Clausen, Hutwalker, 2017). The constructive alignment of intended learning outcomes, teaching-learning activities and assessment according to (Biggs, 1996) is an essential foundation. Formulated as learning outcomes, the development of skills is the objective of the courses. By the successful and continuous linkage of professional, technical and social skills as well as system expertise, courses and programs can be designed more diverse in the given temporal and personnel framework and aware acting engineers can be educated. But, appropriate teaching-learning-activities are required differing from common frontal monologues. Experiental learning spaces can enhance the positive effects of activate learning methods. (Clausen, Binder, 2017)

Based on the relevant topics on Responsible mining as well as the state of modern teaching and learning, the course "Responsible Mining" was designed and implemented at Clausthal University of Technology. The main target group of the course are third semester Master students from the program Mining Engineering. The group is small to medium with 10-15 students and with 86 % international students diverse. Besides the main target group, interested students from other programs and Erasmus students join the course. The intended learning outcomes are spread over different competency areas and the lectures cover all stages of the lifecycle of the mine:

1. Introduction to Responsible Mining, the future of Mining and its role in the circular economy
2. Sustainable Development in Mining
3. Pre-Mining: Planning for Responsible Mining
4. Impacts of Mining in Production
5. Responsible Mining methods of the future
6. Shaping the footprint of Mining: Mine closure

The Teaching and learning activities support the development of skills through activating methods and an early integration of project work, which is part of the final assessment. In the project, the students show their communication skills on Responsible Mining by choosing a critical mining topic related to the lecture and preparing a communication strategy to the wider society. The objective, methodology, concept and action is presented by a report while the topic selection is presented during the first seven minutes of the oral exam, which is followed by questions and answers on responsible mining and reflection. The first implementation showed a great success based on the evaluation of teachers, learners and experts.

6 CONCLUSION

Mining must take over responsibility for human and environment by actions shaped in regards to sustainability, by the development of technologies and decision paths and by the education of engineers, who can look at the bigger picture besides their area and spot the social responsibility. Therefore, all parties involved must collaborate in industry, science, education and community.

REFERENCES

AWUAH-OFFEI, K. and A. ADEKPEDJOU, 2011. Application of life cycle assessment in the mining industry [online]. *The International Journal of Life Cycle Assessment*, **16**(1), 82-89. Available from: 10.1007/s11367-010-0246-6

BIGGS, J., 1996. Enhancing teaching through constructive alignment [online]. *Higher Education*, **32**(3), 347-364. Available from: 10.1007/BF00138871

BINDER, A., O. LANGEFELD, E. CLAUSEN, and A. HUTWALKER, 2017. Linking Sustainability and Underground Mining: Course development in the Master Mining Engineering. In: M. CARDU, ed. *28th SOMP Annual Meeting and Conference. Proceedings- Papers.*

CLAUSEN, E. and A. BINDER, 2017. Innovative learning spaces for experiental learning: Underground mines. In: R. BRENNAN, K. EDSTRÖM, R. HUGO, J. ROSLÖF, R. SONGER, and D. SPOONER, eds. *The 13th International CDIO Conference. Proceedings Full Papers*, pp. 595-604.

H.-P. BECK, M.S., 2011. *Windenergiespeicherung durch Nachnutzung stillgelegter Bergwerke*. Abschlussbericht. Goslar [viewed 22 September 2017]. Available from: http://www.gbv.de/dms/clausthal/E_BOOKS/2011/2011EB1130.pdf.

INTERNATIONAL COUNCIL OF MINING & METALS, 2015. *ICMM 10 Principles* [online]. 2018 [viewed 3 January 2018]. Available from: https://www.icmm.com/en-gb/about-us/member-commitments/icmm-10-principles.

KLEIN, P., 2012. *Why the Future of Mining Depends on Social Change* [online]. 23 February 2012, 12:00 [viewed 4 January 2018]. Available from: https://www.forbes.com/sites/csr/2012/02/23/why-the-future-of-mining-depends-on-social-change/#4a56e9e936f9.

LANGEFELD, O. and A. BINDER, 2017. Blue Mining - Planning future mines today. *World of Mining - Surface & Underground*, **69**(2), 109-112.

LANGEFELD, O., 2017. Fully Automated Underground Mine. *Mining Report*, **153**(5), 429-435.

LAURENCE, D., 2011. Establishing a sustainable mining operation [online]. An overview. *Journal of Cleaner Production*, **19**(2-3), 278-284. Available from: 10.1016/j.jclepro.2010.08.019.

MEESMANN, U., 2017. *Vision Zero and the 7 golden rule in Mining. Your first steps to sucess in prevention* [viewed 3 January 2018]. Available from: http://visionzero.global/sites/default/files/2017-09/Your%20First%20Steps%20to%20Success%20in%20Prevention.pdf.

MIRANDE, M., D. CHAMBER, and C. COUMANS, 2005. *Framework for Responsible Mining. A Guide to Evolving Standards* [viewed 3 January 2018]. Available from: http://www.frameworkforresponsiblemining.org/pubs/Framework_20051018.pdf.

SÄCHSISCHES STAATSMINISTERIUM FÜR WIRTSCHAFT, ARBEIT UND VERKEHR, 2011. *Arbeits- und Gesundheitsschutz in kleinen Unternehmen.* 2.th ed. [viewed 3 January 2017]. Available from: https://publikationen.sachsen.de/bdb/artikel/11228.

SAMARCO, 2016. *one year after the fundao dam failure* [online]. 2016 [viewed 21 December 2017]. Available from: http://samarco.com/wp-content/uploads/2017/01/Book-Samarco_Ingles_v1.pdf.

SHIELDS, D., F. VERGA, and G. ANDREA BLENGINI, 2014. Incorporating sustainability in engineering education [online]. *International Journal of Sustainability in Higher Education*, **15**(4), 390-403. Available from: 10.1108/IJSHE-02-2013-0014.

TECHNISCHE UNIVERSITÄT CLAUSTHAL, 14 May. 2015. *Wieder Rohstoffe aus dem Harz: Neues Forschungsprojekt läuft an.* Goslar.

WORLD COMMISSION ON ENVIRONMENT AND DEVELOPMENT, 1987. *Our common future.* Repr. Oxford: Oxford Univ. Press.

Scientific and Practical Studies of Raw Material Issues – Litvinenko (Ed)
© 2020 Taylor & Francis Group, London, ISBN 978-0-367-86153-7

Legal issues of the use of digital technology in mining - data ownership and liability

Marian Paschke
University Hamburg, Manager of the workgroup digitalisation at DRRF, Germany

1 INTRODUCTION

As many other industries, mining has arrived in the digital future. Technologies are constantly being optimized and further developed. Especially, artificial intelligence and robotics play a decisive role in the digitalization of mining. Cyber Physical Systems (CPS) are used increasingly, which include automated, highly automated, remote-controlled and autonomous systems. These systems are expected to enable and implement a more economically efficient and reliable and, last but not least, safer handling of mining tasks.

While the technical and economic know-how for the use of digital technology in mining is well advanced, this does not apply in the same way to the legal framework. In particular, two fundamental legal issues are in focus: Firstly, the question of the assignment of data used and generated in digitized mining to ownership and disposal rights. Secondly, the question of personal and material liability for damages caused by the use of digital technology. The following article focuses on these two issues.

2 RIGHTS OF DISPOSAL OF DATA

Every person communicates in public and discloses personal data about themselves in the context of such communication. Without communication, participation in social life would be neither conceivable nor possible. A long going discussion about the protection of data revolves around the question of whether the idea of a right to limit access to data is legally feasible and meaningful. Today, as far as personal data is concerned, this is commonly denied, since every person communicates in public and discloses personal data about themselves in the context of such communication. Granting such a right would render participation in social life inconceivable and impossible. Notwithstanding the justified exclusion of personal data from the allocation of exclusive rights of disposal, personal data constitute a high level of protection in any legal system.

2.1 Rights of disposal of machine data

However, concerns about "data ownership" do not apply to the economically significant area of "machine data" or anonymized data. These data, which have no personal reference, are of sole importance to the use in the mining sector. They have a considerable economic value and are the basis for the reliable, efficient and safe use of digital technology.

In the economically significant processes of generating and using machine data, however, the focus is not on the legal assignment of such data to a specific asset as property. Rather, the companies are concerned with the question of whether exclusive rights of use or exclusion can be established for the respective data. Accordingly, the legal discussion is less focused on the legally defined concept of "ownership of machine data" rather than on a more open concept of "data ownership" of machine data.

2.2.1 *Legally significant interests*

The interests of concerned parties emphasize a number of arguments for and against the recognition of data ownership. The concept is supported in particular by the fact that a data owner would be strengthened by the recognition of a legal position, which should in particular contain defensive and surrender claims. There would be additional incentives for companies to develop a data market in the economy. Also, ownership of data could create investment incentives. Integrity protection would protect data from changes, and confidentiality protection would ensure that only "authorized" persons have access to the data. Last but not least, legally secured ownership of data makes it easier to make economic allocations that promote the use of the data.

On the other hand, the discussion about the legal recognition of ownership of machine data refers to a number of possible disadvantages. For example, it may be argued that the "owner" would regularly be not concerned with the protection of the data as such, but with the protection of the information represented by the data or the knowledge represented by this information. Exclusive rights to data were rights of domination or even exclusivity over information and this is ultimately accompanied by a restriction of freedom of opinion, information and economic activity. Further, new technologies require an open society. Information would therefore be too important for the information society to be assigned to an ownership position that entails exclusive rights. Also, practical problems would arise during implementation, especially with regard to the proof of "data ownership" and the overlapping of assigned rights between different parties. In mining, for example, such overlapping would occur between the owner of mining machines, the manufacturer of these machines, the operating or service personnel as well as external controllers. These references parallel the discussion concerning data collected by vehicles in private ownership. In that regard, it is disputed whether the owner, driver or passenger or other persons (e.g. insurers or service providers) are to be assigned the data collected by the vehicle.

2.2.2 *Legal basis*

In positive terms, ownership of data is out of the question, since there is no generally binding legal regulation that assigns exclusive rights or rights to use a person's data. The traditional linking of the concept of property to the physicality of the object of reference stands in the way of a recognition of data ownership consolidated by property law.

According to a widely accepted technical definition by the Association for Information and Image Management, data are "representations of facts, concepts or instructions in a formal manner suitable for communication, interpretation or processing by humans or computers" and are machine-readable information encoded and stored in binary code. The essential elements for the term digital data are the coding of the information as a binary character set (syntactic level) and the storage (structural level), while the so-called semantic level, i.e. the content of the digital data, is not relevant for the technical data term.

The legal challenge of the data concept conceived in this way arises because property in the legal sense is linked to an element of corporeality or a material substrate. This, however, does not apply to data. As long as data is not embodied by storage on data carriers, the concept of ownership in this traditional legal constitution is not capable of including data as an object of ownership.

Some voices, therefore, want to pave the way for a functional orientation of the concept of capability of ownership in order to extend the legal concept of property and thus speak out in favor of a modification of the traditional concept of property solely applicable to corporal objects. This, however, is a legal-political demand, which is presented with the legal-political intent to facilitate electronic traffic and to promote the implementation of digital technology.

This legal view remains unrestricted in practice to this day. It has been expressly stated by the highest court in Germany, that data not embodied on a suitable data carrier are not objects that are accessible for property attribution (Federal Court of Germany, 21.9.2017, I ZB 8/17, para. 15, MDR 2018, p. 227). This jurisprudence confirms the prevailing assessment in the legal literature. If (digital) data are not objects in the legal sense, then there is also no original right of ownership to such data and, accordingly, a legal "owner" of the data cannot claim ownership as a legally secured position against third parties.

This legal starting position does not mean that digital data have no legal significance or enjoy no legal protection. On the contrary, the legal system provides for regulations relating to data in several areas.

Existing provisions which protect against the unauthorized reading of data by overcoming technical security mechanisms are widespread under national criminal law, e.g. in Germany the regulation of § 202a (1) German Penal Code, while in the regulation of § 303a (1) German Penal Code the integrity of digital data is sanctioned against unauthorized modification by third parties. It is recognized in copyright law that databases are protected. A database is a collection of works, data or other independent elements which are arranged systematically or methodically and are individually accessible by electronic means or by other means (definition according to Art. 1 para. 1 of Directive 96/9/EC); however, this requires the selection and/or arrangement of the collected data to be based on a special intellectual achievement which goes beyond the usual (representation/preparation) criteria.

Business secrets are protected data in many jurisdictions. Accordingly, it is a criminal offence to fulfil the conditions of the facts mentioned therein and to disclose the information previously kept secret by a company to third parties without authorization. However, these protection standards do not apply to data per se. All data which can no longer be regarded as secret in the sense of this provision is withdrawn from the scope of the protection standards.

The subject of data protection law is limited to personal data. According to German and European data protection law, personal data is only data relating to natural persons. All other data, in particular data generated by machines without personal reference, are not covered by the scope of protection of data protection law.

In 2019, an EU regulation for machine-generated data came into force. This was the "Free Flow of Data" regulation 2018/1907 (Official Journal of the EU of 28.11.2018, No. L 303/59). It regulates how machine-generated data should be handled. The purpose of this Regulation is to establish a framework for the free movement of non-personal data within the Union and a basis for the development of the data economy and the improvement of the competitiveness of the Union industry for the processing of data other than personal data within the internal market of the Union. This Regulation neither establishes nor intends to establish the right of ownership for the protection of machine data. Rather, the regulation lays down the principle of free movement within the Union for non-personal data, unless a restriction or prohibition is justified on grounds of public security.

2.2.3 *Protection of machine data under current law*

Currently, there is no right of ownership of digital data under the current legal situation. However, this does not mean that the interests of data producers and other market participants cannot be adequately protected. The starting point is that the data stored in or generated by a mining machine belongs to the owner of the machine. In that regard, the mining company may "own" all data stored on its computers and other machines at its disposal if those machines are its sole property.

However, the increasing networking of the Cyber Physical Systems in mining through IoT technology entails that machine data does not remain on the respective machines, but may be transferred to other machines and computers. Physical, chemical or vehicle data, for example, may be transmitted to control centers or workshops, fed to a logistics center or forwarded to an external service provider. If the data is evaluated by control centers, workshops, logistics centers and third-party providers, the question arises as to who is entitled to participate in the associated added value and who has the rights to use the data obtained.

Companies that create value through digital services or launch intelligent products on the market can legally protect their business models and investments in the collection of data, in order to protect their economic interests. In particular, this may be done through contractual agreements on access and usage rights to data. Companies that generate data in the course of mining activities can prevent unauthorized access to "their" data by third parties by means of technical and legal protective measures. In each case, the companies are in a position to conclude contracts with third parties on access to and handling of data due to their de facto control over the data. In conjunction with contractual agreements, the factual ownership of the data forms the basis for proper protection and market-driven handling of machine data.

According to today's state of knowledge, sufficient legal protection of machine data is provided by contractual agreements as well as factual control over such data through the use of "smart products" equipped with sensors as well as the operation of other "smart machines" and their digital networking via the Internet of Things. Just as in the case of self-propelled vehicles in mining, the contract law system is set up and applicable in such a way that the data can be controlled in a legally secure manner without hindering the mining processes and applications.

On the basis of proper contracts, the data owner has the right to demand from his contractual partners the surrender of digital data not stored on a special data carrier. A contract which deals with the exchange of data may include regulations which regulate the surrender of the data at least at the time of termination of the contractual relationship. In addition, such contracts should contain deletion obligations on the part of the contractual partner, in order to guarantee data protection. However, this does not mean, that the original data owner may demand the contractual partner to surrender all storage media on which the data made available by him is contained.

2.2.4 *Conclusion*

The applicable law does not recognize ownership of machine data. The discussion on this is expected to continue in the future. At present, however, a need of a further development of the traditional understanding of ownership beyond the area of embodied objects is not recognizable. Appropriate data ownership can be achieved within the framework of the applicable legal system by drawing up appropriate contracts with market participants. In the field of mining, i.e. the use of digital sensor technology, the linking of machines within the framework of IoT equipment, the use of digital services from third-party suppliers and other applications of digitization in mining, a proper order of the interests of market participants by means of contractual regulations currently appears to be sufficient and reliable.

3 LIABILITY

3.1 *Basic principles*

With the increasing influence of digital technology on mining activities, the liability responsibility of the manufacturer or operator of automated, highly automated or autonomous control technology is gaining in importance. The current legal system is essentially unprepared for this. There is currently no uniform regulation, either at national or international level, establishing legal principles of manufacturer liability. National or regional regulations currently shape the legal situation, but likewise have no overriding liability principles. The call for a globally uniform regulation as far as possible therefore encounters particular implementation difficulties in the area of digital control systems; it is, however, entirely justified in the interest of harmonizing the initial liability situation affecting the position of the players in global economic competition.

Currently, a discussion revolves around objective justification for introducing a manufacturer's hazard liability for remote-controlled or autonomous systems. As long as the guiding principle for strict liability is primarily based on the idea of improving the acceptance of digital systems in this way, no legally viable basis for such producer liability has yet been found. In the future, the discussion should be conducted with greater consideration of the interests involved. With such an approach, it should be borne in mind that, with an automation level that continues to have "human in the loop", unrestricted liability on the part of the system manufacturer independent of fault cannot always be justified.

Overall, it can be stated that the growing responsibility of the manufacturer of digitally networked and controlled systems has not yet been reflected in the liability bases of the applicable law. For fully autonomous systems with the highest degree of automation, strict product liability on the part of the system manufacturer appears to be the appropriate "price" for the systemic risk introduced with autonomous control, which only the manufacturer is able to influence.

3.2 Liability for digital failure

According to conventional legal control technology, liability for damage caused in the mining industry by the use of digital technology can initially be considered a liability for fault. However, such a liability approach proves to be unsuitable in so far as the human influence on the control of machines and processes in the liability situations to be assessed is no longer decisive. It is precisely the use of digital technology, digital sensor, control and monitoring technology that should push back the human factor in mining and exclude or at least qualitatively and quantitatively minimise cases of damage caused by fault. Only in those cases in which culpable human errors lead to a damaging event in the mining context is contractual or tortious liability ultimately called upon to ensure proper responsibility for the damage.

In the course of increasing digitalization, the cause of damage events traced back to human fault will decrease; liability based on fault will then no longer make it possible to assign liability risks and responsibilities properly. This does not prevent the software programmer from being held responsible for culpable behavior. It is generally accepted that from a certain degree of complexity it is hardly possible to program software without errors. Software errors are often systemic and less caused by individually caused failures. The "classic" legal questions of attributing "behavior and fault" then ignore the causes of malfunctions and damage. Liability based on fault no longer forms an appropriate basis for coping with such causes of damage that are no longer caused by fault.

It is therefore considered to separate liability for digital technical failure from fault in future. Liability could therefore be conceived as a liability regardless of further fault. Such would derive its justification from the fact that the software programmer uses technical aids instead of acting himself and this use leads to damage. Such liability is known under French law as "responsabilité du fait des choses" (cf. Art. 1242 para. 1 Civil Code). This is a strict tortious liability of the owner of an object. The owner is liable for damages caused by the object; he may not exonerate himself by proving that he was not at fault.

This concept of "liability for technical failure" is convincing insofar as liability is linked to the cause to be sanctioned: namely technical failure instead of human fault. At the same time, however, it runs the risk of going beyond the objective, since circumstances such as material fatigue could give rise to liability, which ultimately could hardly be regarded as technical failure. The definition of technical failure would become the lynchpin of such liability, and it is easy to see that such a definition would encounter considerable difficulties.

A "liability for technical failure" is in fact close to a strict liability if the legal concept of technical failure cannot be precisely defined. It then shares the criticism levelled against the concept of the introduction of strict liability as a whole and against liability for digital systems in particular. This criticism centers on the inappropriateness of strict liability to encourage the person affected by the regulation to behave carefully. The addressee of the regulation is neither given an incentive to behave as carefully as possible nor can such a liability concept contribute to improving and optimizing digital technology.

However, it should be borne in mind that a concept of strict liability can certainly provide an incentive for the optimization of the technology, since the owner of the technology would endeavor to minimize the probability of the damage occurring. This, however, will only succeed if the regulation is constantly adapted to "best available technology". The mining entrepreneur will opt for the use of digital technology even under the conditions of strict liability if the benefits outweigh the risks.

The current design of strict liability in the rules of product liability forms a possible blueprint for future legal developments in the technical and mining fields. As of today, however, it is not tailored to the use of digital technology in the corporate sector because the applicable rules are limited to damage to property for private purposes and actually private use.

The fact that in the transport sector there are already a number of rules on liability based on fault indicates that liability in the event of a failure of digital technology is not unimaginable from a legal policy point of view. The liability facts affecting the owner or operator of motor vehicles, aircraft or railways are capable of recording the facts of traffic damage cases due to the failure or malfunction of digital systems. For the use of digital technology in mining, liability solutions based on these models can be created in the future.

4 FINAL REMARK

The increasing use and importance of digital technology in mining requires a legal order appropriate to the new technical standards. At present it is not clear whether the legal order is conceptually oriented towards the phenomenon of digital technology or that of mining technology. If mining takes place in an increasingly digitally networked environment, the mining of the future will not only undergo ever more extensive automation pushes, but will itself become part of a digitized business environment. From this perspective, it seems appropriate that the legal order to be created for mining under the conditions of digitization should at least take into account the governance developing for the digitization of the law as a whole. This would be the best way to take into account the networking of both areas in economic life, the development of mining and the development of digitization. In this respect, a legal order for the digital mine would be a partial aspect of a new order to be created for the use of digital technology in the economy as a whole.

As long as the applicable national and international legal systems recognise neither ownership of data nor a right of disposal over data ownership, it is up to the market participants themselves to manage the opportunities and risks of the use of digital technology in mining through appropriate contractual arrangements. Under liability law, however, the opportunities and risks of the use of digital technology in mining will not be manageable without proper further development of the applicable rules on liability for culpability and hazards.

REFERENCES

Amstutz, Dateneigentum, AcP 218 (2018), S. 438 ff.
Dorner, Big Data und „Dateneigentum", CR 2014, 617 ff.
FZI Forschungszentrum Informatik (Hrsg.), Daten als Wirtschaftsgut, Europäische Datenökonomie oder Rechte an Daten?, Smart-Data-Begleitforschung 2017.
Grützmacher, Dateneigentum – ein Flickenteppich, CR 2016, 485 ff.
Heymann, Rechte an Daten, CR 2016, 650 ff.
Hilty/Drexl/Harhof, Ausschließlichkeits- und Zugangsrechte an Daten, Positionspapier des Max-Planck-Instituts für Innovation und Wettbewerb, vom 16. August 2016.
Paschke/Lutter, Zur künftigen Rechtsordnung der unbemannten Schifffahrt, 2018.
Soebbing, Fundamentale Rechtsfragen zur künstlichen Intelligenz: (AI Law), 2019.
Wildhaber (Hrsg.), Leitfaden Information Governance, 2015.
Zech, Daten als Wirtschaftsgut, CR 2015, 137 ff.

Scientific and Practical Studies of Raw Material Issues – Litvinenko (Ed)
© 2020 Taylor & Francis Group, London, ISBN 978-0-367-86153-7

Optimization model for long-term cement quarry production scheduling

Trong Vu & Carsten Drebenstedt
TU Bergakademie Freiberg, Freiberg, Germany

ABSTRACT: Production scheduling optimization techniques have not become common in cement quarry mining. This research sought to present and implement a new optimization model based on Mixed Integer Linear Programming (MILP) along with an efficient solution method to address the Long-term Cement Quarry Production Scheduling Problem (LCQPSP). A multi-step method was applied to solve the LCQPSP including block clustering, and finding a starting integer feasible solution (SIFS) by a heuristic technique. The implementation of the MILP model and its solution method at a cement quarry show their ability to generate practical schedules in a reasonable time while decreasing the cost for developing the raw mix by $6.2-7.8 million when compared with a solution generated using a common industry practice.

Keywords: cement quarry, raw mix, long-term cement quarry production scheduling, mixed integer linear programming, clustering, block, optimization

1 INTRODUCTION

For the production of cement it is important to make the raw mix whose chemical compositions are within certain limits. Frequently, limestone mined from quarry is mixed together or with limestone from other quarries, or additives purchased from the market. The success of cement plant operation is possible only if the raw mix with optimum composition at desired quantity is continuously supplied. This requires a long-term view in production planning of cement quarries. Nowadays, modern cement plants apply the latest technology integrated with computer controlled processing to monitor and make decision through every stage of plant operation. Manual planning and scheduling using computer aided design (CAD) programs with little knowledge about objectives make little coordination between planning and operation. This approach locks mine planners to existing short term plan, and long-term plan is ignored.

Mine planning and production scheduling is hierarchy works from long-term to short-term. Mine planners must answer two questions which blocks to remove and when to remove them in order to maximize the discounted value of project. Those questions could not be easy to answer globally because of the large size of mathematical optimization problem in real world, despite of the advances in computer and software technology throughout many decades. Optimization techniques in open pit mine planning are still not widely used in cement quarry mining. The impossible establishment of economic block model based on cement market value makes the solution to the cement quarry production planning problem different from the open pit mining problem (S.U. S. Srinivasan *et al*, 1996; K. Dagdelen *et al*, 2002;). Most proposed solutions to the LCQPSP are heuristics with no indication of solution quality. A sequencing algorithm was presented by (M.W.A. Asad, 2011) to solve the long-term production planning of cement quarry operations. However, the optimality of the solution can be difficult to achieve and the model loses flexibility. A MILP for short-term production scheduling of cement quarry operations was proposed and used by (S.U. Rehman *et al*, 1996). A long-term production planning model was developed by (D. Joshi *et al*, 1996) to supply consistent quantity and quality of limestone to a cement plant. To reduce the problem size, they applied the block aggregation method to group the blocks with similar location and lime saturation factor (LSF) to form the aggregates and then discretized the main problem to a number of small sub-problems. Nevertheless, the quality of the solution is unknown.

The aim of this research is to develop a new MILP optimization model along with an efficient solution method to address the LCQPSP. The objective of the model is to minimize the cost for developing the raw mix in cement manufacturing while simultaneously considers the mine operations, raw mix blending requirements, and amount of additives purchased. To achieve the research aim, a multi-step method is developed to solve the LCQPSP. Firstly, blocks are aggregated into mining- cuts using hierarchical clustering algorithms in order to reduce the size of problem and to provide practical schedules. Secondly, the global LCQPSP is decomposed into smaller sub-problems, each associated with a period t (t =1,...,T), and solve them sequentially. The solution found is provided for the solver as a SIFS to solve the global problem.

2 BLOCK CLUSTERING

In this research, blocks are aggregated in the same level or mining bench into mining-cuts (MCs) based on attributes: location and grade distribution. These MCs are then fed into the production scheduling model. Hence, instead of solving the production scheduling problem at block level, we did at mining-cut (MC) level. This technique decreases the problem size, which allows generating the optimum solutions in a reasonable time, and creates a mining schedule at a selective mining unit (SMU) that is practical from mining operation perspective. However, it is important to note that solving the production scheduling problem at MC level reduces the resolution of the problem resulting in a reduced NPV values or increased cost values in comparison with that at block level.

The clustering algorithms work normally based on the similarity or dissimilarity index which is measured by the distance between two objects. There are two main categories of clustering algorithms: partitioning and hierarchical clustering. In this research, the general procedure of clustering algorithm is hierarchical and performed as follows (S.C. Johnson, 1967):

(i) Start by considering each block as a MC and calculate the similarities between blocks in the same bench based on location and grade distribution
(ii) Link pairs of MCs into single MC that are in close proximity
(iii) Compute similarities between the newly formed MC and the rest of the MCs
(iv) Repeat steps (ii) and (iii) all blocks are merged together into MCs.

Besides, to be used as SMUs, the MCs must have a minable shape and a reasonable size (number of blocks per a MC) (M. Tabesh et al, 2011). A computer code is developed and implemented in Matlab (MATLAB Software) to aggregate blocks into MC as well as refine the MC's shape and size.

3 MILP MODEL FOR LCQPSP

3.1 Blending requirements in cement manufacturing

The key driver for the LCQPSP is the consistent provision of blend of raw mix at desired quality. Hence, blending plays a crucial role in guiding both mining and processing operations. The primary requirement for developing an acceptable raw mix is the existence of calcium oxide (CaO), silica (SiO2), alumina (Al2O3), iron (Fe2O3), magnesia (MgO), potassium oxide (K2O), etc. within the allowable range. For practical purposes, the raw material composition requires a balance of these oxides which can be guided by indicators including silica ratio (SR), lime saturation factor (LSF) and alumina ratio (AM), and clinker minerals including alite (3CaO.SiO2) represented as "C3S", belite (2CaO.SiO2) represented as "C2S", aluminate (3CaO.Al2O3) represented as "C3A", and brownmillerite (4CaO.Al2O3.Fe2O3) represented as "C4AF". Equations from (1) to (7) (Rehman *et al*, 2008) are used to express these indicators:

$$SR = SiO_2/(Al_2O_3 + Fe_2O_3) \tag{1}$$

$$LSF = CaO/(2.8SiO_2 + 1.18Al_2O_3 + 0.65Fe_2O_3) \tag{2}$$

$$AM = Al_2O_3/Fe_2O_3 \tag{3}$$

$$C_3S = 4.071 \times CaO - 7.60 \times SiO_2 - 6.78 \times Al_2O_3 - 1.43 \times Fe_2O_3 \tag{4}$$

$$C_2S = -3.071 \times CaO + 8.6 \times SiO_2 + 5.068 \times Al_2O_3 - 1.079 \times Fe_2O_3 \tag{5}$$

$$C_3A = 2.65 \times Al_2O_3 - 1.692 \times Fe_2O_3 \tag{6}$$

$$C_4AF = 3.043 \times Fe_2O_3 \tag{7}$$

In this research, the raw mix stockpile is the combination of the raw materials from multiple parts or benches of the quarry and the additives from outside sources. This framework is consistent with practical cement manufacturing, and used to illustrate how the MILP model works.

3.2 MILP model for LCQPSP

The mathematical formulation of the model is presented as follows:

Indices and sets

$i \in I$	set of block i
$j \in N_i \subset I$	predecessor set of block j which must be extracted before block i
$t, t' \in T$	set of scheduling periods
$a \in A$	set of additives from outside
$k \in K$	set of K chemical indices

Parameters

B_i	Quantity of material within a block i
C_{it}	Mining cost of block i during period t
C_{at}	Purchasing cost of additive a during period t
$minMC_t, maxMC_t,$ $maxMC_t$	Minimum and maximum quantity of raw materials to be extracted from the quarry in period t
$minA_{at}, maxA_{at}$	Minimum and maximum quantity of additive k purchased from the market in period t
g_{ki}	Average grade of chemical k in block i
g_{ka}	Average grade of chemical k in additive a
$minG_k, maxG_k$	Minimum and maximum allowable content of chemical k in the raw mix

Decision variables

X_{it}	Equal to 1 if block i is scheduled in period t; 0 otherwise.
Y_{at}	Quantity of additive a from outside sources

Objective function:

$$\text{Minimize} \sum_{t \in T} \left(\sum_{i \in I} C_{it} X_{it} + \sum_{a \in A} C_{at} Y_{at} \right) \tag{8}$$

Subjects to:
Precedence constraints

$$N_i.X_{it} - \sum_{t' \in t} \sum_{j \in N_i} X_{jt} \leq 0, \qquad \forall i \in I, \forall t \in T \tag{9}$$

221

Mining capacity constraints

$$minMC_t \leq \sum\nolimits_{i\in I} X_{it}\cdot B_i \leq maxMC_t, \qquad \forall t \in T \qquad (10)$$

Additive capacity constraints

$$minA_{at} \leq A_{at} \leq maxA_{at}, \qquad \forall t \in T, \forall k \in K \qquad (11)$$

Resource constraints

$$\sum\nolimits_{i\in I} X_{it} \leq 1, \qquad \forall t \in T \qquad (12)$$

Raw mix blending constraints

$$minG_k \leq \frac{\left[\sum_{i\in I} X_{it}\cdot g_{ki}\cdot B_i + \sum_{a\in A} Y_{at}\cdot g_{ka}\right]}{\left[\sum_{i\in I} X_{it}\cdot B_i + \sum_{a\in A} Y_{at}\right]} \leq maxG_k, \qquad \forall t \in T, \forall k \in K \qquad (13)$$

$$minG_k \leq \frac{\left[\sum_{i\in I} X_{it}\cdot g_{(CaO)i}\cdot B_i + \sum_{a\in A} Y_{at}\cdot g_{(CaO)a}\right]}{\left[\begin{array}{l} 2.8\left[\sum_{i\in I} X_{it}\cdot g_{(SiO_2)i}\cdot B_i + \sum_{a\in A} Y_{at}\cdot g_{(SiO_2)a}\right] + \\ 1.18\left[\sum_{i\in I} X_{it}\cdot g_{(Al_2O_3)i}\cdot B_i + \sum_{a\in A} Y_{at}\cdot g_{(Al_2O_3)a}\right] + \\ 0.65\left[\sum_{i\in I} X_{it}\cdot g_{(Fe_2O_3)i}\cdot B_i + \sum_{a\in A} Y_{at}\cdot g_{(Fe_2O_3)a}\right]\end{array}\right]} \leq maxG_k, \quad \forall t \in T \qquad (14)$$

$$minG_k \leq \frac{\left[\sum_{i\in I} X_{it}\cdot g_{(SiO_2)i}\cdot B_i + \sum_{a\in A} Y_{at}\cdot g_{(SiO_2)a}\right]}{\left[\begin{array}{l}\left[\sum_{i\in I} X_{it}\cdot g_{(Al_2O_3)i}\cdot B_i + \sum_{a\in A} Y_{at}\cdot g_{(Al_2O_3)a}\right] + \\ \left[\sum_{i\in I} X_{it}\cdot g_{(Fe_2O_3)i}\cdot B_i + \sum_{a\in A} Y_{at}\cdot g_{(Fe_2O_3)a}\right]\end{array}\right]} \leq maxG_k, \quad \forall t \in T \qquad (15)$$

$$minG_k \leq \frac{\left[\sum_{i\in I} X_{it}\cdot g_{(Al_2O_3)i}\cdot B_i + \sum_{a\in A} Y_{at}\cdot g_{(Al_2O_3)a}\right]}{\left[\sum_{i\in I} X_{it}\cdot g_{(Fe_2O_3)i}\cdot B_i + \sum_{a\in A} Y_{at}\cdot g_{(Fe_2O_3)a}\right]} \leq maxG_k, \quad \forall t \in T \qquad (16)$$

$$minG_k \leq \frac{\left[\begin{array}{l} 4.071\left[\sum_{i\in I} X_{it}\cdot g_{(CaO)i}\cdot B_i + \sum_{a\in A} Y_{at}\cdot g_{(CaO)a}\right] \\ -7.6\left[\sum_{i\in I} X_{it}\cdot g_{(SiO_2)i}\cdot B_i + \sum_{a\in A} Y_{at}\cdot g_{(SiO_2)a}\right] \\ -6.718\left[\sum_{i\in I} X_{it}\cdot g_{(Al_2O_3)i}\cdot B_i + \sum_{a\in A} Y_{at}\cdot g_{(Al_2O_3)a}\right] \\ -1.430\left[\sum_{i\in I} X_{it}\cdot g_{(Fe_2O_3)i}\cdot B_i + \sum_{a\in A} Y_{at}\cdot g_{(Fe_2O_3)a}\right]\end{array}\right]}{\left[\sum_{i\in I} X_{it}\cdot B_i + \sum_{a\in A} Y_{at}\right]} \leq maxG_k, \quad \forall t \in T \qquad (17)$$

$$minG_c \leq \frac{\left[\begin{array}{l} 2.650\left[\sum_{i\in I} X_{it}\cdot g_{(Al_2O_3)i}\cdot B_i + \sum_{a\in A} Y_{at}\cdot g_{(Al_2O_3)a}\right] \\ -1.692\left[\sum_{i\in I} X_{it}\cdot g_{(Fe_2O_3)i}\cdot B_i + \sum_{a\in A} Y_{at}\cdot g_{(Fe_2O_3)a}\right]\end{array}\right]}{\left[\sum_{i\in I} X_{it}\cdot B_i + \sum_{a\in A} Y_{at}\right]} \leq maxG_c, \quad \forall t \in T \qquad (18)$$

$$\min G_c \leq \frac{\begin{bmatrix} -3.071\left[\sum_{i\in I} X_{it}\cdot g_{(CaO)i}\cdot B_i + \sum_{a\in A} Y_{at}\cdot g_{(CaO)a}\right] \\ +8.6\left[\sum_{i\in I} X_{it}\cdot g_{(SiO_2)i}\cdot B_i + \sum_{a\in A} Y_{at}\cdot g_{(SiO_2)a}\right] \\ +5.068\left[\sum_{i\in I} X_{it}\cdot g_{(Al_2O_3)i}\cdot B_i + \sum_{a\in A} Y_{at}\cdot g_{(Al_2O_3)a}\right] \\ -1.079\left[\sum_{i\in I} X_{it}\cdot g_{(Fe_2O_3)i}\cdot B_i + \sum_{a\in A} Y_{at}\cdot g_{(Fe_2O_3)a}\right] \end{bmatrix}}{\left[\sum_{i\in I} X_{it}\cdot B_i + \sum_{a\in A} Y_{at}\right]} \leq \max G_c, \quad \forall t \in T \tag{19}$$

$$\min G_c \leq \frac{3.043\left[\sum_{i\in I} X_{it}\cdot g_{(Fe_2O_3)i}\cdot B_i + \sum_{a\in A} Y_{at}\cdot g_{(Fe_2O_3)a}\right]}{\left[\sum_{i\in I} X_{it}\cdot B_i + \sum_{a\in A} Y_{at}\right]} \leq \max G_c, \quad \forall t \in T \tag{20}$$

Decision variable definition constraints:

$$X_{it}, X_{jt} \in \{0,1\}; Y_{at} 0, \forall t \in T, \forall i \in I, \forall a \in A \tag{21}$$

The objective function (8) seeks to minimize the mining cost for developing the raw mix of cement plant. Constraints (9) ensure that block i is not extracted by time period t unless every block j in predecessor set of block i is also extracted by time period t. Constraints (10) and (11) enforce minimum and maximum capacity of raw materials extracted from the quarry and additive purchased from outside sources, respectively. Constraints (12) ensure that a MC is extracted in one period only. Constraints (13)-(21) define the raw mix requirements. Constraints (13) require the average grade of CaO, SiO2, Al2O3, Fe2O3, and MgO, etc. to be between a minimum and maximum in each time period. To maintain the balance of oxides in the raw mix, constraints (14)-(19) keep indicators LSF, SR, AM, C3S, C3A, and C4AF in the desired limits. Finally, constraints (21) enforce internality and non-negativity of the decision variables.

4 SOLUTION METHODOLOGY

In solving the MILP model proposed in the previous section, the mining capacities defined by constraints (10) present common properties of the Knapsack Problem (KP). Then, the precedence constraints (equation (9)) turn the KP into a Precedence Constrained Knapsack Problem (PCKP). Both KP and PCKP themselves are NP-hard that means the optimal solution of the LCQPSP cannot be obtained within a practical computation time. Therefore, variable elimination and heuristic methods have to be utilized to reduce the solution time and get near-optimal solutions to the LCQPSP.

4.1 *Decision variable elimination*

We reduce the problem size through applying the concepts of earliest and latest start times for each MC to eliminate the binary decision variables in the model (Dciooer, 2003; N. Boland et al, 2007; M.P. Gaupp, 2008). The earliest start time implies how long it takes to reach MC i based on the MC precedence and maximum mining capacity. The MC precedence determines a predecessor cone above MC i that must be exacted before extracting MC i. The cone tonnage and the cumulative maximum mining capacity from the first period up to period t are calculated and compared against each other. If the cone tonnage exceeds the cumulative maximum mining capacity, the earliest start time for MC i is period t. All decision variables corresponding to extracting MC i during the periods prior to period t must be equal 0 in the optimal solution, and so, we can eliminate these variables from the problem.

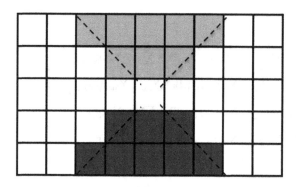

Figure 1.   Predecessor (light grey) and successor (dark grey) cones of MC *i*.

Whereas, the latest start time for MC *i* reflects the possible longest time to reach MC *i* based on the MC precedence in conjunction with minimum mining capacity. In this case, the MC precedence determines a successor cone below MC *i* that can only be extracted after MC *i* is completely extracted. The cone tonnage is used to calculate the tonnage of the leftover MCs in the pit which is the largest extracted tonnage to reach MC *i*. The cumulative minimum mining capacity from period *t* to the end of mine life is calculated and compared against the tonnage of the leftover MCs. If the tonnage of the leftover MCs below the cumulative minimum mining capacity, the latest start time for MC *i* is period *t+1*. All decision variables corresponding to extracting MC *i* at or after its latest start time can be set to 1 in the optimal solution. The predecessor and successor cones are shown in Figure 1.

4.2   *Decision variable elimination*

Another method to reduce the solution time in solving the MILP model is supplying the solver with a SIFS. Such a solution may preclude evaluation of dominated solutions and enable more aggressive use of the optimizer's local search heuristic (*IBM, ILOG CPLEX,* 2009). Some heuristic approaches are employed to identify SIFS such as greedy heuristic (R. Chicoisne *et al,* 2012*)* or a sliding window heuristic (C. Cullenbine *et* al, 2011). In this research, we determined the SIFS by discretizing the LCQPSP into smaller sub-problems according with a period t (t∈T) and solving them sequentially in increasing order of t. The MCs assigned to each sub-problem are eliminated from the block model after solving and the remaining MCs are assigned to the next period. The size of each sub-problem is smaller considerably than the global LCQPSP and hence, an exact method can be used to solve the sub-problems. Finally, the partial integer feasible solutions of sub-problems are linked together to generate the SIFS. If the value of the SIFS is close to the upper bound (< a pre-determined MILP gap) derived from the linear relaxation in solving the LCQPSP, the solution returns to the SIFS. Otherwise, the solver continues to improve the quality of the SIFS.

5   MODEL IMPLEMENTATION

The implementation of the MILP model is demonstrated on a cement quarry deposit which belongs to Cement Company in Central Vietnam. The study area represents a gentle slope downward from West to East. The main rock formations encountered in the area include low grade marl, marginal grade limestone and high grade limestone. Marl with low CaO grade (CaO <36%) and marginal grade limestone (36%≤CaO≤40%) cannot be used independently for cement manufacturing and can be sent to the dump, resulting in an increase of mining cost and loss of raw materials. The company applied Surpac software to construct a block model which has 22216 blocks with block size of 25×25×10m using ordinary kriging technique following the standard procedure. Table 1 presents the descriptive statistics of the attributes in the quarry block model. CaO grade distribution at the 145m bench is shown in Figure 2(a).

Table 1. Basic statistics of the grade attributes in the quarry block model.

	CaO	SiO2	Al2O3	Fe2O3	LOI	MgO	Na2O	K2O
Mean	46.01	5.25	1.30	0.44	39.49	1.67	0.08	0.21
Median	49.26	5.67	1.42	0.45	40.53	1.68	0.07	0.18
Variance	65.40	4.45	0.28	0.02	44.27	0.19	0.01	0.01

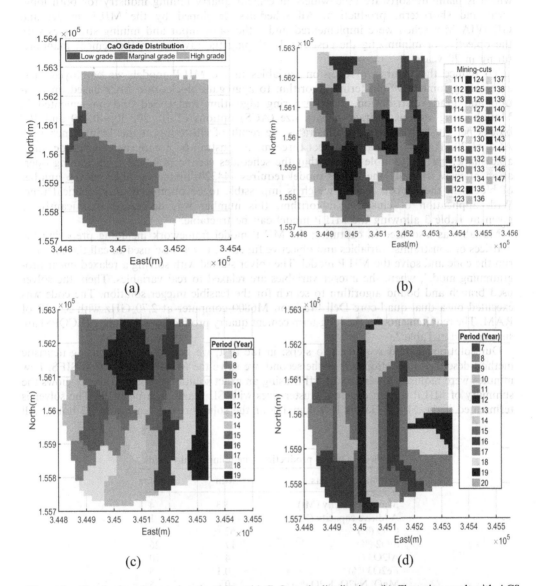

Figure 2. Bench plan view at elevation 145m: (a) CaO grade distribution; (b) Clustering result with ACS of 30; (c) MILP production schedule with ACS of 30; (d) GEOVIA Minesched production schedule.

The cement operation utilizes clay, sand, high grade limestone, and iron ore purchased from the different market suppliers as additives to blend with the raw materials from the quarry to achieve the quality and quantity requirements in cement manufacturing.

Matlab was also used in capturing the MILP model framework including preparing the matrices of constraints, variables and objective function. It was also used to call the solver to run the code and solve MILP model. The solver started with solving a relaxed linear programming model, where the integer variables are relaxed to real variables. Then, the solver used branch and bound algorithm to search for the feasible integer solution. The code was executed on a dual quad-core Dell Precision M6800 computer at 2.70 GHz with 32 GB of RAM. The production schedules generated by the MILP model were compared against the production schedule generated by GEOVIA Minesched which is planning software used widely in cement quarry mining industry for both long-term and short-term production. All schedules developed by the MILP model and GEOVIA Minesched were implemented under the same input and mining strategies with the objective of minimizing the cost for developing the raw mix in the cement manufacturing in 20 years.

To reduce the number of decision variables in the MILP model, we employed hierarchical agglomerative clustering algorithm to aggregate blocks into MCs based on location and grade distribution. The clustering algorithm was coded and implemented in Matlab using various average cluster size (ACS) options of 10, 20, 30, 40, 50 and 100 blocks per cluster. Figure 2(b) illustrates the result of clustering at the 145m bench with ACS of 30. The MCs created respect three main constraints: homogeneity in grade, minable shape, and reasonable size so that the schedules can be practical from mining operation of view. Initially the MILP model requires 444320 binary variables as the number of blocks in the deposit model, which is impossible to be optimized in a practical time. With application of clustering algorithm, this number was decreased significantly as given in Table 3 allowing the MILP model can be tractable.

Matlab was also used in capturing the MILP model framework including preparing the matrices of constraints, variables and objective function. It was also used to call the solver to run the code and solve the MILP model. The solver started with solving a relaxed linear programming model, where the integer variables are relaxed to real variables. Then, the solver used branch and bound algorithm to search for the feasible integer solution. The code was executed on a dual quad-core Dell Precision M6800 computer at 2.70 GHz with 32 GB of RAM. The requirements for the long-term cement quarry production scheduling (LCQPS) are summarized in Table 2.

Our solution method requires two steps: in the first, we generate a SIFS using heuristic method described in Section 4.2; in the second, we solve the MILP model with a SIFS. Few minutes were required to obtain the SIFS using an exact solution method. Table 3 shows the summary of MILP run on different cluster sizes with SIFS and with no SIFS. The solver is terminated if the MILP Gap is less than 1% or the solution time reaches to 12 hours. All

Table 2.   Quarry production scheduling requirements.

Production targets	Min	Max
Mining capacity (Mt)	3.3	4
Additive capacity (Mt)	0	1
CaO (%)	62	69
SiO2 (%)	17	26
Al2O3 (%)	4	10
Fe2O3 (%)	0.1	5
MgO, Na2O, K2O (%)	0	5
SR	2	2.6
LSF	0.9	1
AM	1	3
C3S (%)	40	60
C3A (%)	5	15
C4AF (%)	10	18

Table 3. Summary of MILP run on different MC sizes with SIFS and with no SIFS.

		No. of MCs	No.of binary vars	MILP with no SIFS			MILP with SIFS		
ID	ACS			Objective value ($)	MILP Gap (%)	Solution time (s)	Objective value ($)	MILP Gap (%)	Solution time (s)
I	10	2446	48920	N/A	N/A	N/A	190053700	0.05	3133
II	20	1276	25520	N/A	N/A	N/A	190120000	0.09	777
III	30	960	19200	N/A	N/A	N/A	190149000	0.07	418
IV	40	648	12960	N/A	N/A	N/A	190250000	0.14	167
V	50	526	10520	N/A	N/A	N/A	190330000	0.17	119
VI	100	268	5360	N/A	N/A	N/A	191610000	0.84	103

*N/A = not available

scenarios first employed the decision variable elimination technique. It can be seen that the MILP model with no SIFS felt to solve in all scenarios within the time limit. Meanwhile, the MILP model with SIFS was solved successfully in all scenarios within the gap of 1% indicating that the heuristic method generated a very good solution. The longest solution time was only nearly one hour. As we expected, the larger sizes of MCs were, the larger objective values as well as the smaller solution times were. On the other hand, multiple runs with various MC sizes proved the ability of the MILP model in controlling the number of MCs generated and solution time.

The production schedules generated by the MILP model were compared against the production schedule generated by GEOVIA Minesched which is planning software used widely in cement quarry mining industry for both long-term and short-term production. All schedules developed by the MILP model and GEOVIA Minesched were implemented under the same input and mining strategies with the objective of minimizing the cost for developing the raw mix in the cement manufacturing in 20 years. The schedules generated by the MILP model using ACS of 30 blocks and GEOVIA Minesched at bench 145m can be seen in Figure 2(c) and 2(d), respectively. Figure 3(a) and 3(b) illustrates South-North cross sections 156142.5 m of the schedules, indicating the differences in their physical sequence of extraction.

In Figure 4, the MILP production schedules yielded 3.17-3.96% cost of raw mix smaller than GEOVIA Minesched in all experiments, establishing a cost saving equivalent of nearly $6.2-7.8 million. This is considerable amount with a fairly small quarry. The costs required to produce one ton of raw material in the MILP production schedules are smaller than that in GEOVIA Minesched schedule, except in the scenario V. Figure 5 (a), (b) shows the annual quarry production and amount of additives purchased for 20 periods in the MILP production schedule with ACS of 30 and GEOVIA Minesched production schedule, respectively. The quarry production and additive capacity as set up in Table 2 are met in both schedules. This is especially notable that GEOVIA Minesched required purchasing all kinds of additives for 20 periods while MILP model only needed clay and iron ore for the same time horizon. The comparisons of quality of raw mix between MILP production schedule with ACS options and GEOVIA Minesched production schedule are given in Figure 6. The ability of MILP model in meeting all raw mix quality requirements was well proven in all ACS options. Meanwhile, GEOVIA Minesched method shows violations of the quality targets in most cases, except of the case of MgO grade and C4AF.

6 CONCLUSION

Production scheduling optimization techniques have been not applied widely in cement quarry mining. Implementation of long-term cement quarry production planning and scheduling using economic block values are not reliable and impractical. MILP have been applied

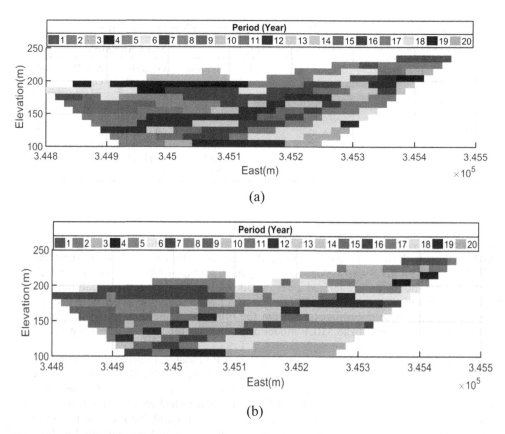

(a)

(b)

Figure 3. South-North cross section 156142.5 m: (a) MILP production schedule with ACS of 30; (b) GEOVIA Minesched production schedule.

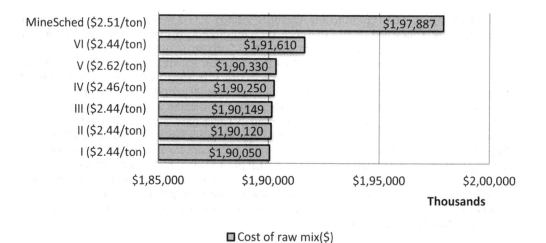

Figure 4. Cost of raw mix of MILP production schedules and GEOVIA MineSched schedule.

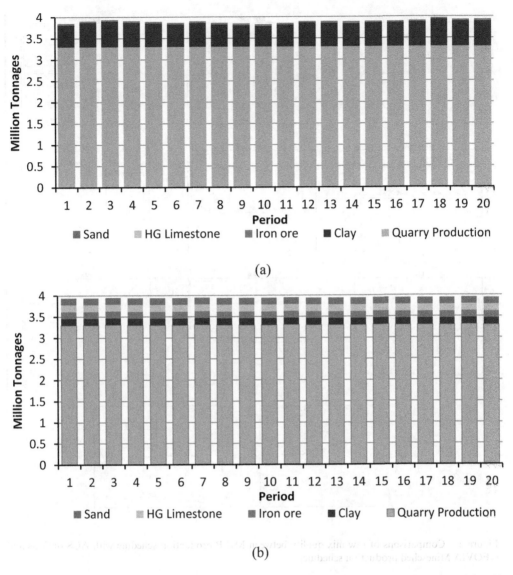

Figure 5. Annual quarry production schedule and amount of additives purchased generated by: (a) MILP production schedule with ACS of 30; (b) GEOVIA Minesched production schedule.

effectively to tackle mine planning and production scheduling problems. However, the size of the problem in real-life is still a long-standing challenge. This research develops a new MILP model that integrates the LCQPSP in an optimization framework to provide sustainably raw materials for cement manufacturing at minimum cost. To reduce the problem size, the blocks were aggregated into MCs, using clustering algorithms. The MCs are considered as an input for the MILP model. For solving the MILP model, two major steps are taken including: generation of a SIFS using heuristic method and providing the solver with a SIFS.

The case study in a cement quarry deposit showed that application of the clustering algorithm was able to not only reduce considerably the number of binary variables in the MILP model but also generate practical schedules based on SMUs. The heuristic method based on decomposition technique quickly identified a SIFS. It is observed from the case study that solving the MILP model with SIFS can guarantee to produce a better solution or return the SIFS, and with the

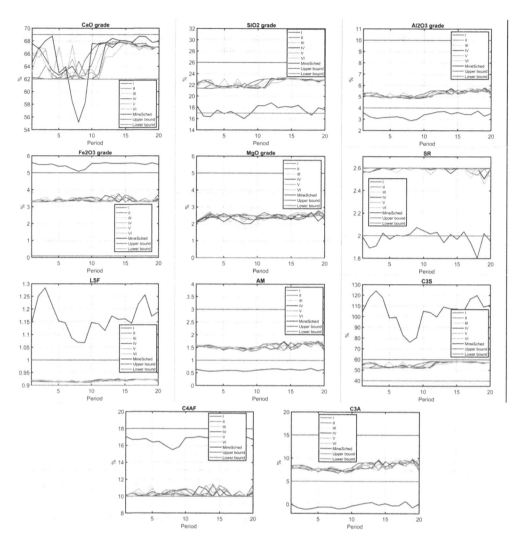

Figure 6. Comparisons of raw mix quality between MILP production schedule with ACS options and GEOVIA Minesched production schedule.

indication of solution quality. In case of solving the MILP model with no SIFS, the problem was intractable, whereas with SIFS the problem became simpler and therefore, yielded faster solution times. These observations demonstrate that the solution method proposed in this paper is able to overcome the challenge of problem size of MILP model in solving LCQPSP.

The MILP model was verified and validated by a comparative study against one of the common commercial software in quarry mining – GEOVIA Minesched. The analysis showed that application of MILP model resulted in substantially higher cost saving than GEOVIA Minesched that is not based on mathematical programming methods. Besides the ability of optimality of the objective, the MIP model also considered simultaneously the raw mix blending requirements, whereas GEOVIA Minesched aimed to meet capacity targets but violated the quality targets in most cases. This highlights the importance of application of optimization techniques in solving the LCQPSP.

In some practical cases, the raw mix is supplied by multiple quarries and some operations stores raw materials in stockpiles. Therefore, the next of this research is to develop alternative model that considers multi-cement quarrying complexes and stockpiles.

REFERENCES

D. Joshi, S. Chatterjee, and S.M. Equeenuddin, *Limestone quarry production planning for consistent supply of raw materials to cement plant: A case study from Indian cement industry with a captive quarry*, Journal of Mining Science 51 (2015), pp. 980-992.

IBM, ILOG CPLEX. 2009, Incline Village, NV;

I. uublicatim Dciooer, *Production scheduling at LKAB s Kiruna Mine using mixed-integer programming*, Mining engineering (2003), p. 35.

K. Dagdelen, and M.W. Asad, *Optimum cement quarry scheduling algorithm*, APCOM 2002: 30 th International Symposium on the Application of Computers and Operations Research in the Mineral Industry (2002), pp. 697-709.

MATLAB Software.. MathWorks Inc. 9.3 (R2017b);

M.P. Gaupp, *Methods for improving the tractability of the block sequencing problem for open pit mining*, COLORADO SCHOOL OF MINES GOLDEN, 2008.

M. Tabesh, and H. Askari-Nasab, *Two-stage clustering algorithm for block aggregation in open pit mines*, Mining Technology 120 (2011), pp. 158-169.

M.W.A. Asad, *A heuristic approach to long-range production planning of cement quarry operations*, Production Planning & Control 22 (2011), pp. 353-364.

N. Boland, C. Fricke, and G. Froyland, *A strengthened formulation for the open pit mine production scheduling problem*, Available at Optimization Online (2007).

Rehman, S, MWA Asad and I Khattak (2008), '*A Managerial Solution to Operational Control of the Raw Material Blending Problem in Cement Manufacturing Operations*, Proceedings of the COMSATS International Conference on Management for Humanity and Prosperity, Lahore, Pakistan.

R. Chicoisne, D. Espinoza, M. Goycoolea, E. Moreno, and E. Rubio, *A new algorithm for the open-pit mine production scheduling problem*, Operations Research 60 (2012), pp. 517-528.

S.C. Johnson, *Hierarchical clustering schemes*, Psychometrika 32 (1967), pp. 241-254.

S. Srinivasan, and D. Whittle, *Combined pit and blend optimization*, PREPRINTS-SOCIETY OF MINING ENGINEERS OF AIME (1996).

S.U. Rehman, and M.W.A. Asad, *A mixed-integer linear programming (milp) model for short-range production scheduling of cement quarry operations*, Asia-Pacific Journal of Operational Research 27 (2010), pp. 315-333.

Scientific and Practical Studies of Raw Material Issues – Litvinenko (Ed)
© 2020 Taylor & Francis Group, London, ISBN 978-0-367-86153-7

The application of MPC to improve the efficiency of an AC electric drive vector control

S. Erokhov
Mining University, St. Petersburg, Russia

ABSTRACT: The article discusses the electric motor control system using the MPC speed controller. A model of the asynchronous electric motor is shown; a number of assumptions are made to simplify the model. A brief description of a fuzzy PID controller is given. The principle of operation and the block diagram of the MPC controller are considered. The simulation modeling of the electric drive vector control system in the "MatLab" environment was carried out. The modeling results are given. The results prove that the control system with an MPC controller provides better transient characteristics compared to a fuzzy PID controller.

1 INTRODUCTION

Currently, due to the decrease of hydrocarbon reserves in traditional producing regions, offshore reserves are of the greatest interest. For their development, drilling platforms and ships, which include technological complexes, are used. The technological complex provides drilling, production and transportation of hydrocarbons. The economic performance of a drilling platform or a ship largely depends on the effective operation of this complex (Kozyaruk: 2016). The drive of mechanisms, included in the technological complex, is a complex dynamic system. Complex dynamic systems are objects with non-linear static characteristics, i.e., objects that are described by differential equations with time-varying parameters. As practice has shown, control of such objects using traditional proportional-integral-derivative (PID) controllers does not allow achieving the required control quality. To improve the efficiency of automatic control systems of complex dynamic objects, researchers around the world (Guzelkaya & Eksin & Yesil: 2003, Siddique: 2014) are trying to combine a standard PID controller with a fuzzy adaptive controller. Currently there are various types of fuzzy controllers, but the ones based on PID controllers are the most common (Fereidouni & Masoum & Moghbel: 2015, Ang & Jafar: 2014).

For this study fuzzy PID controllers is used (Manenti & Rossi & Goryunov & Dyadik & Kozin, Nadezhdin & Mikhalevich: 2015), but without a procedure of identifying the parameters of the controlled object. A distinctive feature of the fuzzy PID controller is that it determines the PID controller coefficients based on expert estimates. Also recently controllers based on the predictive model - model predictive control (MPC) controllers - have been extensively used for complex dynamic objects control (Manenti: 2011, Nayhouse & Tran & Kwon & Crose & Orkoulas & Christofides: 2015). To synthesize such a controller, it is necessary to make a mathematical model of the controlled object on the basis of which the controller will predict the change of the controlled variable for a certain period of time and calculate the optimal control action to ensure the best trajectory of the controlled variable. The purpose of this study is a comparative analysis of automatic control systems with a fuzzy PID controller and a controller based on a predictive model (MPC controller).

2 MODEL OF ASYNCHRONOUS ELECTRIC MOTOR

For correct operation of the controller an adequate model of the object is necessary, in this case an asynchronous electric motor with squirrel-cage rotor acts as an object. The controlled value is the rotation frequency of the electric motor output shaft.

When modeling an asynchronous electric motor, a number of generally accepted assumptions were made (Sipajlov & Loos: 1980):

- machine magnetic system is not saturated;
- there are no steel losses;
- machine phase windings are symmetric and shifted strictly by 120°;
- magnetomotive winding forces and magnetic fields distributed along the circumference of the air gap according to a sinusoidal law;
- air gap is constant;
- machine rotor is symmetrical;
- real distributed winding is replaced by the equivalent concentrated one, creating the same magnetomotive force.

When modeling an asynchronous motor, the choice of the coordinate system is not fundamental and depends on the considered drive control system, but it is more convenient to describe it in a fixed relative to the stator coordinate system $(\omega_k = 0)$, axes in which are designated (a, b).

In this case, the system of equations of the asynchronous motor in the operator form will take the following form:

$$
\left.
\begin{aligned}
I_{S\alpha} &= \tfrac{\frac{1}{R_S}}{T_S \cdot p + 1} \cdot \left(U_{S\alpha} - \tfrac{L_m}{L_r} \cdot p \cdot \psi_{r\alpha} \right) \\
I_{S\beta} &= \tfrac{\frac{1}{R_S}}{T_S \cdot p + 1} \cdot \left(U_{S\beta} - \tfrac{L_m}{L_r} \cdot p \cdot \psi_{r\beta} \right) \\
\psi_{r\alpha} &= \tfrac{1}{T_S \cdot p + 1} \cdot \left(L_m \cdot I_{S\alpha} - T_r \cdot \omega \cdot \psi_{r\beta} \right) \\
\psi_{r\beta} &= \tfrac{1}{T_S \cdot p + 1} \cdot \left(L_m \cdot I_{S\beta} - T_r \cdot \omega \cdot \psi_{r\alpha} \right) \\
M_e &= \tfrac{3 \cdot L_m}{2 \cdot L_r} \cdot p_n \cdot \left(\psi_{r\alpha} \cdot I_{S\beta} - \psi_{r\beta} \cdot I_{S\alpha} \right) \\
\omega &= \tfrac{1}{J_\Sigma p} \cdot \left(M_e - M_r \right) \\
\sigma &= 1 - \tfrac{L_m^2}{L_S \cdot L_r}
\end{aligned}
\right\}
\tag{1}
$$

where: p_n= number of poles pairs; L_m= mutual inductance; L_S = stator windings inductance; L_r= rotor windings inductance;
electromagnetic permanent stator circuits:

$$
T_S = \frac{L_S}{R_S}
\tag{2}
$$

electromagnetic permanent stator and rotor circuits:

$$
T_r = \frac{L_r}{R_r}
\tag{3}
$$

R_S = stator winding active resistance; R_r = rotor winding active resistance; $I_{S\alpha}$, $I_{S\beta}$ = components of the motor stator current vector in axes (α, β); $\psi_{r\alpha}$, $\psi_{r\beta}$ = components of the motor rotor flux linkage vector in axes (α, β); ω, M_e = rotor speed and electromagnetic moment of asynchronous squirrel-cage motor; M_r = static load moment; J_Σ = total moment of the motor and mechanism inertia; σ = motor dispersion coefficient.

3 CONTROL SYSTEM WITH FUZZY PID CONTROLLER

A diagram of the proposed automatic control system with a fuzzy PID controller is shown in Figure 1.

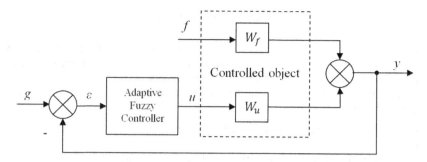

Figure 1.   Block diagram of the control system with a fuzzy PID controller.

Shown in Figure 1 variables have the following meaning: g = set point; f = measured perturbation; u = control action; W_U = controlled object channel; W_f = perturbation object channel; y = controlled variable; ε = control error defined as $\varepsilon = g - y$

The adaptive fuzzy controller consists of the following blocks: a fuzzy rule base generator, a Mamdani-type fuzzy controller and a system for calculating Jn, Je, and Ju. The optimization task is to minimize the functional, which plays a key role in setting up adaptive and optimal control systems. Optimized functional has the following form:

$$\min(Je_k + Ju_k + Jn_k) \tag{4}$$

$$Je_k = \sqrt{\frac{\sum_{j=k}^{k+he} (\varepsilon_j)^2}{he - 1}} \tag{5}$$

$$Ju_k = \sqrt{\frac{\sum_{j=k}^{k+hu} (u_j - u_k)^2}{hu - 1}} \tag{6}$$

where: Jn_k = number of control errors fluctuating in the interval, he, ε_j = control error, u_j = controlled variable, he = control error interval, hu = control interval, j = sampling time index.

The controller is described in more details in the study (Manenti: 2011).

4   CONTROL SYSTEM WITH MPC CONTROLLER

The automatic control system with MPC controller is similar to the circuit with a fuzzy controller (Figure 2).

The main idea of control with a predictive model can be represented as follows, there are u and y inputs of the controlled objects, output controlled variable, g is the desired value (dependence) of the change of the controlled variable.

Consider the system in discrete time, i.e., only at time points $t = k \cdot \Delta T$, where in ΔT = some quantization period, and k = integer. For convenience, the graphical representation will be considered as $\Delta T = 1$.

The main feature of control using the MPC controller is the presence of mathematical model of the controlled object (process, apparatus), which describes its behavior quite accurately. The presence of an adequate mathematical model of the controlled object allows us to predict the values of the controlled variable at a certain number of steps forward (Figure 3).

Values of the controlled variable $y(t)$, predicted at some point in time t, in Figure 3 shown as follows - $\hat{y}(t)$. The prediction horizon is built on a certain number of cycles. The predicted trajectory of the controlled variable will depend on the future values of the control action $u(t)$.

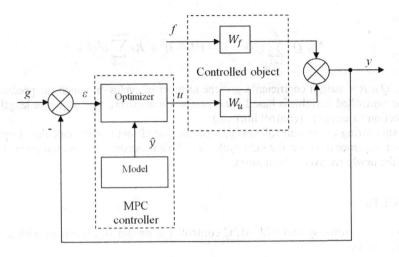

Figure 2.   Block diagram of the control system with MPC controller.

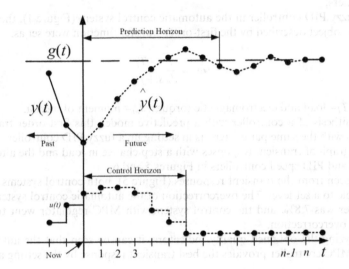

Figure 3.   Graphic representation of the control concept using the MPC controller.

The method consists in finding a certain sequence of values of a controlled variable $u(t)$, which will provide the best predicted trajectory for the controlled variable $y(t)$. The length of the calculated control actions sequence $u(t)$ is a fixed value and is called "the control horizon". The desired sequence of control action values is determined by solving some optimization problem. The choice of the best trajectory of the controlled variable is determined by the quality control indicators.

The study applies a quality indicator, which contains the square of the mismatch between the predicted output variable of the controlled object $y(t)$ and the desired trajectory $g(t)$. When choosing the optimal values of the controlled variable $u(t)$, the controller aims to minimize the function (Zaki Diab & Kotin & Anosov & Slepcov: 2015):

$$J = Q \cdot \sum_{i=1}^{p} (y(k+i) - r(k+i)^2 + R \cdot \sum_{i=0}^{m=1} u^2(k+i) \qquad (7)$$

where: Q и R = weight coefficients, p = the number of cycles on which the predicted behavior of the controlled variable is based $y(t)$ (prediction horizon), m = sequence length of future control action values $u(t)$ (control horizon).

After submitting to the controlled object of the first element of the calculated optimal control action sequence $u(t)$, on the next cycle the whole procedure is repeated anew, taking into account the newly received information.

5 RESULTS

The automatic control system with MPC controller is similar to the circuit with a fuzzy controller (Figure 4).

where: RPS = Rotor position sensor, RFLO = Rotor flux linkage observer.

For comparison, consider the transient responses of the control system with a PID and MPC controllers.

To set a fuzzy PID controller in the automatic control system (Figure 1), the parameters of the controlled object described by the first-order transfer function were set as:

$$\omega_r = \frac{T_e - T_i}{J_m s + f_d} \qquad (8)$$

where: T_e, T_l= load and electromagnetic torques; J_m= moment of inertia.

For the synthesis of a controller with a predictive model, this first-order transfer function was also used with the same parameters as in setting up a fuzzy PID controller.

There is a graph of transient responses with a step change in load and the alternate application of MPC and PID speed controllers in Figures 5 and 6.

As can be seen from the transient responses (Figure 5), both control systems bring the controlled variable to a set level. The overcorrection of the automatic control system with a fuzzy PID controller was 7.8%, and the control system with MPC regulator went to the specified level without overcorrection.

After analyzing the obtained quality indicators, it can be seen that the automatic control system with MPC controller provides the best transient response by the setting action.

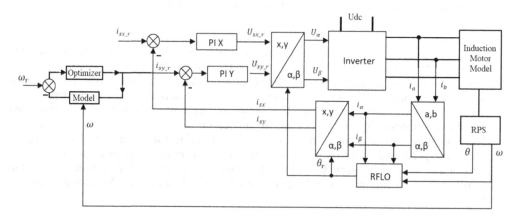

Figure 4.   Block diagram of the electric drive vector control with the MPC controller.

236

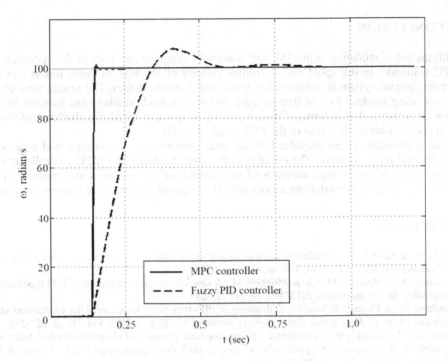

Figure 5. Graph of rotational speed changes when using MPC and PID-speed controllers.

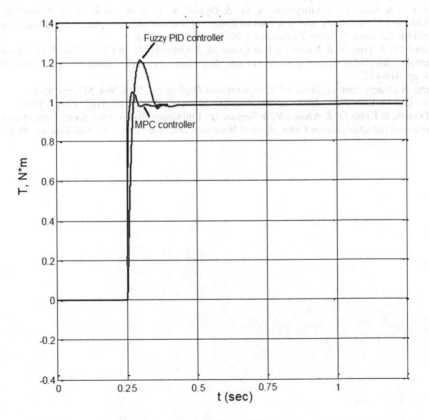

Figure 6. Graph of torque changes when using MPC and PID-speed controllers.

6 CONCLUSION

With the use of modeling in the "Matlab-Simulink" program the study of the operation of the MPC controller in the speed vector control systems of the asynchronous motor was made. The synthesized system is researched in static and dynamic modes. The results were obtained for operating modes: start at free running and with a load, sudden load increase in steady mode. It is shown that in terms of motor current, torque and speed, the efficiency of the MPC controller was better than that of the PID speed controller.

The synthesized system provides high dynamic performance, smoothness and a deep range of speed and torque control. The use of negative speed feedback and MPC controller in electric drives allows achieving high accuracy of stabilization of the set speed and moment, a quick response to external perturbation actions with the required quality of the transient response.

REFERENCES

Ang L. Y. & Jafar F. A.: Simulation analysis of non-linear fuzzy PID temperature controller, *Applied Mechanics and Materials*, 2014, Vol. 465-466, pp. 677-681.

Fereidouni A. & Masoum M.A.S. & Moghbel M.: A new adaptive configuration of PID type fuzzy logic controller, *ISA Transactions*, 2015, Vol. 56, pp. 222-240.

Guzelkaya M. & Eksin I. & Yesil E.: Self-tuning of PID-type fuzzy logic controller coefficients via relative rate observer, *Engineering Applications of Artificial Intelligence*, 2003, Vol. 16, pp. 227-236.

Kozyaruk A.E.: Development experience and development prospect of electromechanical technological complexes of movement and positioning of technic shelf development equipment, *Journal of Mining Institute*, 2016, Vol. 221, pp. 701-705.

Manenti F.: Considerations on nonlinear model predictive control techniques, *Computers and Chemical Engineering*, 2011, Vol. 35, pp. 2491-2509.

Manenti F. & Rossi F. & Goryunov A. G. & Dyadik V. F. & Kozin K. A. & Nadezhdin I. S. & Mikhalevich S. S.: Fuzzy adaptive control system of a non-stationary plant with closed-loop passive identifier, *Resource-Efficient Technologies*, 2015, Vol. 1, pp. 10-18.

Nayhouse M. & Tran A. & Kwon J.S.I. & Crose M. & Orkoulas G. & Christofides P.D.: Modeling and control of ibuprofen crystal growth and size distribution, *Chemical Engineering Science*, 2015, Vol. 134, pp. 414-422.

Siddique N.: Fuzzy control, *Studies in Computational Intelligence*, 2014, Vol. 517, pp. 95-135.

Sipajlov G.A. & Loos A.V.: *Mathematical modeling of electrical machines*, High school, 1980.

Zaki Diab A. & Kotin D. & Anosov V. & Slepcov O.: Full Order Observer for Speed Estimation of Vector Controlled Induction Motor Drive, *Applied Mechanics and Materials*, 2015, Vol. 698, pp. 40-45.

Features of the walking mechanism of a floating platform autonomous modular complex for the extraction and processing of peat raw materials

D. Fadeev & S. Ivanov
St. Petersburg Mining University, Saint-Petersburg, Russia Federation

ABSTRACT: Intensification of peat production sets a priority task to re-equip the peat industry with new and modernized equipment based on energy efficiency and energy saving. The solution may be an Autonomous floating complex for the extraction and processing of peat raw materials from undrained deposits.

1 INTRODUCTION

Thousands of settlements in the Russian Federation are far from centralized sources of energy supply. Delivery of fuel (coal, fuel oil, gas) to these regions requires enormous costs, at the same time they have significant reserves of local fuel (Lazarev & Korchunov,1982). Strategic objectives of the use of local fuels according to the Draft energy strategy of Russia (izm. to № 35-FZ "On electric power industry" from 03.07.2016) for the period up to 2030 diversification of fuel and energy balances and improvement of the level of energy security and reliability of energy supply of the subjects of the Russian Federation and the country as a whole, while reducing the cost of fuel transportation.

A return to peat is not a return to the past, but a reasonable approach in Economics and ecology. Peat is one of the forgotten, but the most important and promising local fuel sources (Timofeeva & Mingaleeva, 2014). Total reserves of peat in Russia – more than 175 billion tons, which is more than 40% of the world's reserves. Local peat resources of a number of regions of the country allow for the long term to ensure the implementation of socially import-ant tasks of stable heat and energy supply of both social and industrial facilities.

Providing the needs of municipal and industrial consumers with both electric and thermal energy on the basis of mini-thermal power plants using peat as a fuel is a real alternative to the existing situation, which can also significantly improve the environmental situation by reducing the anthropogenic impact on the environment (Shtin,2012).

The field of application of peat raw materials is very wide, primarily it concerns energy (direct burning of peat, gasification, pyrolysis, coking and semi-coking, production of energy-dense fuel – briquettes and granules) and agriculture (fertilizers, peat litter, substrates, micro-particles, fodder yeast, peat pots, substrate peat slabs). Also peat raw materials are actively used in the chemical industry (wax, hubic acids, glycerin, alcohol, filter elements, sorbents, polymers, furfural, dry ice, etc.).); construction (construction peat briquettes, peat blocks, additives in concrete, filtering and sorbing materials for the collection and disposal of indus-trial and storm drains), medical and cosmetic industries (therapeutic muds, extracts, medi-cines, balms, etc.) (Bondarev & Zvonarev,2015).

The most modern and promising method for the extraction and processing of peat raw materials is the use of an Autonomous modular complex, for example, according to patents of the Russian Federation № 2655235 and № 2599117.

2 THE DESIGN OF THE PLATFORM

The floating platform contains tanks consisting of four side faces, the upper face and the bottom, which form a hollow body together. The tank body has a multilayer structure, where the shells are sealed. There are hermetic hatches on the outer sides of the tanks. On the side faces of the tanks are installed specialized fixing elements that serve to connect into a single platform. Above the tanks, a frame-type carrier frame is installed, on which there is a deck flooring for placing the modules of the equipment of the Autonomous complex for peat extraction (Fadeev & Khudyakova & Ivanov, 2018). Also, the tank's outer perimeter is equipped with pile-anchors, with devices for installation and extraction of piles. In addition, there are technological holes, closed hatches for ballasting products of technological processing of peat raw materials and means of ballasting.

For the successful operation of the complex of mining equipment for the extraction and processing of peat raw materials, it is necessary to consider its movement through the developed space filled with water. In contrast to the widely used on the waters and floodplains of the dredgers for the extraction of peat and sapropel (Khudyakova & Vagapova,2018), moving across boggy fields using the screw unsafe, additionally, the dredger near impossible to get on the field, in contrast to the platform (Kokonkov & Liakh,2018).

3 FORCE CALCULATION

In order to determine what forces, act in the supports with a statically indeterminate loading scheme, conditionally leave only 2 supports (support A and support E), successively moving from one pontoon to another. Write the equation of the force load given the reactions in supports and the equations of moments closer relative to the point A.

$$YR_A + YR_{E(DCB)} = P$$
$$YR_{E(DCB)}l_{E(DCB)} = l_P P \tag{1}$$

where $YR_{(ABCDE)}$ - reaction force in the support; P - total load in the frame; l_E - distance between the reaction forces of the supports; l_P - shoulder from the reaction force of the support A to the load point P;

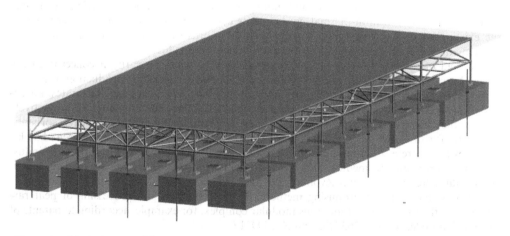

Figure 1. 3d platform model.

Solving together a system of equations (1) sequentially defining conditional reactions in supports. As a result of the calculation of the expression (2) we obtain the values of the conditional reactions.

$$
\begin{cases}
YR_E = \frac{l_p P}{l_E} & YR_E = P - YR_{A1} \\
YR_D = \frac{l_p P}{l_D} & YR_D = P - YR_{A2} \\
YR_C = \frac{l_p P}{l_C} & YR_C = P - YR_{A3} \\
YR_B = \frac{l_p P}{l_B} & YR_B = P - YR_{A4}
\end{cases}
\tag{2}
$$

To identify the coefficients of influence, we add the conventional values

$$
YR_{A1} + YR_{A2} + YR_{A3} + YR_{A4} + YR_B + YR_C + YR_D + YR_E = 4P
\tag{3}
$$

using (2) replace in (3) the corresponding values:

$$
\begin{cases}
YR_{A1} = f(P) \\
YR_{A2} = f(P) \\
YR_{A3} = f(P) \\
YR_{A4} = f(P)
\end{cases}
\begin{cases}
YR_E = f(P) \\
YR_D = f(P) \\
YR_C = f(P) \\
YR_B = f(P)
\end{cases}
\tag{4}
$$

$$
K_E = \frac{YR_E}{P}; \; K_D = \frac{YR_D}{P}; \; K_C = \frac{YR_C}{P}; \; K_B = \frac{YR_B}{P}; \; K_A = \frac{\sum YR_{Ai}}{4P}
\tag{5}
$$

$$
\begin{cases}
R_E = K_E P; \\
R_D = K_D P; \\
R_C = K_C P; \\
R_B = K_B P; \\
R_A = K_A P;
\end{cases}
\tag{6}
$$

4   THE PRINCIPLE OF OPERATION OF THE AUTONOMOUS MODULAR COMPLEX

The platform works as follows. In the initial position, the floating platform is fixed on the surface by piles-anchors, buried in the bottom. The tank outer perimeter spaced a maximum telescopic housing paluszak are in the pushed state. For the stepper movement of the floating platform, the anchor piles are removed from the bottom by the devices for the installation and extraction of the piles, freeing the entire row of tanks of the outer perimeter from the intended direction of movement of the floating platform. Then linear motor locking elements of the coupling move the whole row of tanks on the outer perimeter of the step. The specified direction of movement is provided by simultaneous movement of the extreme row of tanks with the linear and angular guides fixed on their top faces and invariable position of the external racks of the bearing frame established in grooves of these guides. At the end of the step, the value of which is due to the given course of the telescopic linear motors of the fixing elements of the couplings, the extended series of pontoons is fixed by devices for installing and removing piles into the bottom. The piles-anchors of the floating platform, which fixed it in the process of walking, are removed, and the linear motors belonging to the fixing elements of the couplings of the extended series are switched to the reverse, which ensures the tightening of the floating platform to the extended row of tanks of the outer perimeter. At the new standing point, the floating platform is fixed by the installation of all piles-anchors, returning to the original position. If necessary, the pacing process is repeated in any of the possible directions of movement and the required step value (Bessonov,2005).

The principle of operation of the walking pile move is carried out when two piles are installed on the stern of the platform and which can only move vertically (Figure 2).

With winches lowered piles on the bottom of the reservoir and introduced into the ground. The mechanism of the walking pile stroke with fixed guides is installed along the perimeter of the platform to ensure movement in any of the parties. On the frame 8, a winch 7 for lifting the pile and a portal 3 are mounted, the racks of which are located symmetrically relative to the diametrical plane of the dredger. Structurally sheptytskyi winch not differ from ramapajama. On the racks of the portal, unfastened struts, on the transverse beams at a certain distance in pairs, in one plane vertically fixed (two at the top and bottom) four guides 2 with folding clips. In the guides, piles 1 with a tip 9 are moved. The figure shows the lower suspension of the pile when the rope from the drum through two double blocks 4 and a single roller 5 passes through the blocks and is attached to the lower guide. Limits the course of the pile limit switch 6 – a special device that disconnects the motor of the pile winch from the network when the polispast stroke is selected. Friction grip for lifting and lowering piles, consists of two pads, interconnected by plates on horizontal hinges. One block is connected to the pile lifting pulley. When the pulley moves up, the gripper grips the pile due to the frictional forces her up. If the winch is turned on for unwinding the rope, the polispast together with the capture and the pile begins to fall down. As soon as the grip comes into contact with the guide, the rope will weaken, the pads will open, and the pile will begin to fall under its own weight (Mikhailov & Ivanov,2015).

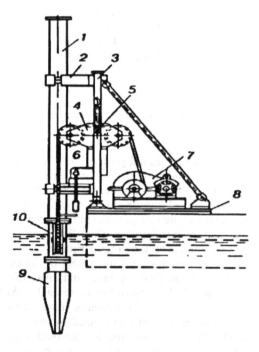

Figure 2. Walking pile stroke design scheme; 1 – pile; 2 – guide piles; 3 – portal; 4 – double block; 5 – roller; 6 – switch; 7 – winch; 8 – frame; 9 – pile tip; 10 – block.

5 CONCLUSIONS

The proposed algorithm allows to achieve an accurate result of loads on the platform in order to avoid overturning or breaking the connections of the pontoon base. The proposed method of calculation complements and allows a more rational use of the resources of the platform strength and is recommended for use in the design of this structure and the preservation of operational properties. The absence of the need for measures for preliminary drainage of the field during its development with the use of an autonomous complex has the following positive aspects: a) improvement of the environmental situation (refusal to drain the swamps leads to minimizing the emission of greenhouse gases, as well as reducing the anthropogenic impact on the hydrosphere, reducing the risks of peat fires); b) increase in seasonality (ensuring year-round raw flow); c) accelerated restoration of the ecosystem of the field; d) increase in economic efficiency; e) providing energy to remote areas and hard-to-reach areas of the Russian Federation.

The technical result is the expansion of operational capabilities of transportation and relocation of the modular floating platform to the place of operation.

The use of this kind of walking mechanism has the following positive aspects:

- increased cross-country ability in marshland;
- ability to move from a place in any direction;
- point support walking machine can improve the environmental situation by reducing the anthropogenic impact on the environment.

REFERENCES.

Bessonov E. A. encyclopedia of hydro-mechanized works. Dictionary-reference / E. A. Bessonov // M.: 2005.– 520 p.

Bondarev Y.Y., Zvonarev I.E., Ivanov S.L. Valkovo-disk separator of Autonomous modular complex of extraction and processing of peat raw materials for fuel. Vestnik pnipu. Geology. Oil and gas and mining. - 2015.- №14. – P. 72-81.

Fadeev D.V., Khudyakova I.N., Ivanov S.L RF patent № 2655235 Floating platform Publ. 24.05.2018.

Khudyakova I.N., Vagapova E. A., Ivanov S.L. Raw peat production and processing from flooded fields and approaches to maintain dehydration// - IOP Conf. Series: Earth and Environmental Science 194 (2018) 032010.

Kokonkov A.A., Liakh D.D., Ivanov S.L. Autonomous complex module for peat development on watered deposits - IOP Conf. Series: Earth and Environmental Science 194 (2018) 032011.

Lazarev A.V., Korchunov S.S. Handbook of peat. – Moscow: Nedra, 1982. –760 p.

Mikhailov A.V., Ivanov S.L., Gabov V.V. Formation and effective use of the machine Park of peat mining companies. Vestnik pnipu. Geology. Oil and gas and mining. - 2015.- №14. -P. 82-91.

Shtin S.M. Hydro-Mechanized peat extraction and production of peat products for energy purposes / – M.: Mountain book, 2012. – 357s.

Timofeeva S.S., Mingaleeva G.R. prospects for the use of peat in the regional power industry. Collection of materials of the all-Russian peat forum. – 2014. No 4. -P. 46-55.

Tomlinson M., Woodward J. Pile Design and Construction Practice. New York, Taylor & Francis, 2008, 566 p.

Scientific and Practical Studies of Raw Material Issues – Litvinenko (Ed)
© 2020 Taylor & Francis Group, London, ISBN 978-0-367-86153-7

Simulation of combined power system with storage device

B. Garipov & D. Ustinov
Saint-Petersburg Mining University, Saint-Petersburg, Russian Federation

ABSTRACT: The article rates the effect of combined power grid consisting of the steam-turbine and wind-powered plants upgrading with electric power storage device. The study is based on the mathematical modeling and simulation of the system in the Simulink MatLab software. The article describes the used model blocks. The results show a positive effect from the use of energy storage devices in power supply systems.

1 INTRODUCTION

The growth of production volumes, powers of electric machines, energy prices and changes in production technology lead to the development of engineering solutions aimed at more efficient use of energy (Skamyin et al., 2019) from the energy consumption point of view.

Nowadays traditional sources of electricity (generators of thermal power plants, gas turbine power plants, nuclear power plants, hydroelectric power plants, etc.) are complemented with alternative ones that run on solar or wind power. Wind power plants are the most widely used in the world (Murphy-Levesque et al., 2017), however, such sources are characterized by a variable level of generated energy, which depends on weather conditions. In turn, traditional generators can have rather high inertia (Piriz et al., 2012) (high resistance to any changes of energy consumption). This can lead to excessive generation of electricity (Khodayar et al., 2016), and, consequently, to the loss of electricity. Thus, one urgent task is to accumulate of excess electricity for using in the future.

Storage devices are of interest to scientists around the world. In (Díaz-González et al., 2016) modeling, control, and imitation of accumulation systems performed on various components are considered. In (Chen et al., 2009) and (Rampazzo et al., 2018) the authors model Li-ion energy storage devices. In (Maia et al., 2017), the author uses modeling to study the behavior of energy storage systems and their lifespan. However, in all these works attention was not paid to modeling the joint operation of storage devices with wind generators and traditional generators.

2 RESEARCH METHODS

The methods of this work include literature review, mathematical modelling, simulation using MatLab and experiment. This set of research methods is determined by the fact that it is not actually possible to study the object of research (Kabakov & Kabakov, 2005).

3 THE SUBJECT OF RESEARCH AND SOURCE DATA FOR SIMULATION

The electric power system with a combined source of energy consists of steam-turbine plant, wind-powered electrical generator and electricity consumers. The energy storage is included in the system during the simulation too.

Table 1. Variables used in equations (1) - (5).

Variable	Meaning/Equivalent
$P_{TOT}(t)$	Total power production
$P_{ST}(t)$	Steam-turbine output power
$P_{WP}(t)$	Wind-powered output
$P_{EXC}(t)$	Excess power
$P_{CON}(t)$	Power consumption
$W_{ACC}(t)$	Accumulated energy
η	Energy storage device efficiency
$P_{ACC}(t)$	Energy storage output power
$P_{LOSS}(t)$	Power losses

Table 1 shows the variables used in all equations. The equation to compute the power generated by the sources of electricity is as follows:

$$P_{TOT}(t) = P_{ST}(t) + P_{WP}(t). \tag{1}$$

The difference between the generated power and the power consumed by the power receivers determines the excess generation:

$$P_{EXC}(t) = P_{TOT}(t) - P_{CON}(t). \tag{2}$$

To simulate a system without an energy storage device is of interest to know the amount of excess generation, which in this case is wasted. The amount of stored energy is (in the case of modeling a system with an energy storage device):

$$W_{ACC}(t) = \eta \cdot \int P_{EXC}(t)dt. \tag{3}$$

Instantaneous power of energy storage device is as follows:

$$P_{ACC}(t) = \frac{dW_{ACC}(t)}{dt}. \tag{4}$$

It is worth noting that the electric power storage device has a certain amount of capacity, therefore it is necessary to estimate the amount of excess generated and non-accumulated power (power losses):

$$P_{LOSS}(t) = P_{EXC}(t) - P_{ACC}(t). \tag{5}$$

Equations (1) - (5) include mathematical description of a power system with a combined power source and storage device. A block diagram of the simulation model based thereon is constructed and shown in Figure 1.

4 DESCRIPTION OF THE MODEL'S BLOCKS

In this study simulation model was built in Simulink MatLab software as the most common in engineering practice (Egorova et al., 2015; Tariq Ahmed Hamdi & Abdul Hussein, 2017) corresponding to the structural scheme in Figure 1.

A Constant block, a Step block and a Transfer Fcn block of the first-order aperiodic link with time constant 3 hours describe the steam-turbine plant. The Constant block sets the initial

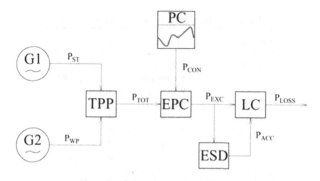

Figure 1. Block diagram of combined power supply system simulation model: G1, G2 – steam-turbine and wind-powered generators respectively; TPP – total power production calculator; PC – power consumption; EPC - excess power calculator; ESD – energy storage device; LC – losses calculator.

power of 110 MW generated by the source. The Step block simulates an adding 40 MW of the generated power at 17 h. The Transfer Fcn simulates the inertia of the generator when the input signal changes. Figure 2 shows the oscillogram of the output power of the steam-turbine plant.

The wind power plant is based on the Random Number block, the Saturation block and the Repeating Sequence Stair block. This model allows us to simulate the energy production of wind turbines depending on random weather conditions (Tavner et al., 2013) (see Figure 3). Figure 4 shows the total power produced by the generators.

The energy storage unit is made using a Gain block as an amplifier, a Switch block as an ideal switch, an Integrator and a Derivative blocks. The value of gain in the Gain block is defined as an efficiency of the energy storage device (Mizutani et al., 2016). The key is intended for simulating the condition of energy buildup (when the input value is ≥ 0) and the condition of energy return to the system (when the input value is > 0). Integration and Derivative blocks are needed to compute the amount of stored energy and the instantaneous power of the energy storage device respectively.

The daily energy needs are simulated with a Repeating Sequence Stair block, which describes typical consumption during the day (Kasaeian et al., 2014). Peak consumptions occur in the morning and evening while low consumption occurs during the day (see Figure 5).

Scope blocks are used to display simulation results over the entire simulation period.

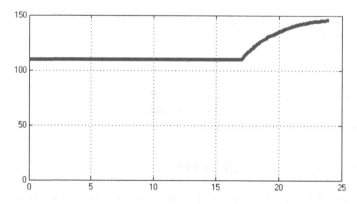

Figure 2. Steam-turbine power output.

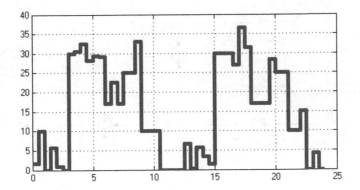

Figure 3.　Wind-powered output.

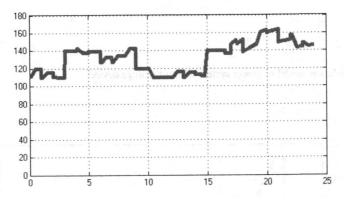

Figure 4.　Total power production.

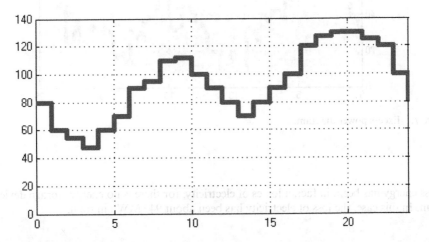

Figure 5.　Daily energy needs diagram.

5　MODEL SIMULATION WITHOUT ENERGY STORAGE DEVICE

The simulation model of a power system without an energy storage device is shown in Figure 6.
Model debugging has resulted in a graph of excess energy production (shown in Figure 7).

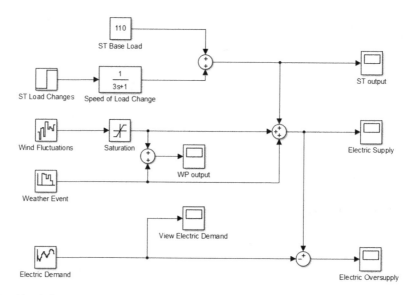

Figure 6.   Simulation model of power system without storage device.

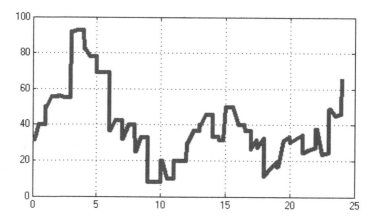

Figure 7.   Excess power diagram.

Excess energy has been, in fact, a losses of electricity, for there is no energy storage device in the system. In this case, the loss of electricity has been about 940 MWh in total.

6   MODEL SIMULATION WITH ENERGY STORAGE DEVICE

The simulation model of a power system with a storage device is shown in Figure 8. Model debugging has resulted in the graphs of excess energy production and electric power losses (shown in Figure 9).

As can be seen from the graphs, a significant portion of the electricity was accumulated in the storage device. There is a graph of the charge of the storage device in Figure 10. The

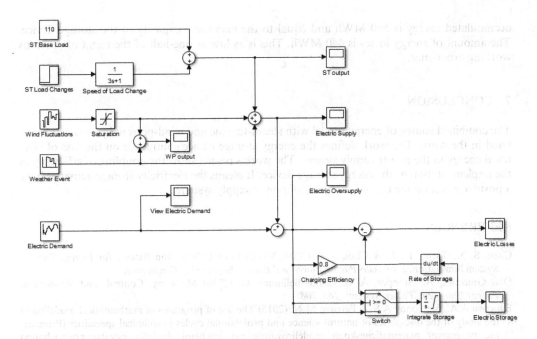

Figure 8.   Simulation model of power system with storage device.

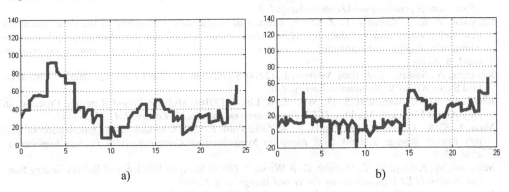

a)

b)

Figure 9.   Diagrams of: a) excess power; b) power losses.

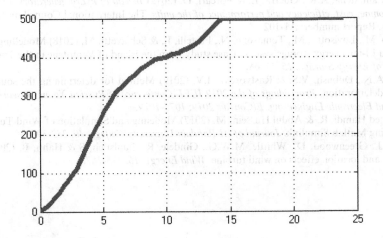

Figure 10.   Energy storage device charge diagram.

accumulated energy is 500 MWh and equal to the maximum capacity of the storage device. The amount of energy losses is 440 MWh. This is as low as one-half of the result of previous working experience.

7 CONCLUSION

The combined source of energy model with steam-turbine and wind-powered plants was simulated in the work. The work defines the energy storage device's influence on the loss of electrical energy in the power supply system. The work's results show the doubling of efficiency in the implementation of the energy storage device. It means the electricity storage activities have a positive effect on the energy efficiency of power supply systems.

REFERENCES

Chen, S. X., Tseng, K. J., & Choi, S. S. (2009) Modeling of Lithium-Ion Battery for Energy Storage System Simulation. *2009 Asia-Pacific Power and Energy Engineering Conference.*

Díaz-González, A., Sumper, A. & Gomis-Bellmunt, O. (2016) Modeling, Control, and Simulation. *Energy Storage in Power Systems. 163–208.*

Egorova A.A., Semjonov A.S. & Petrova M.N. (2015) The use of programs of mathematical modeling in the study of the disciplines of natural science and professional cycles in technical specialties [Primenenie programm matematicheskogo modelirovanija pri izuchenii disciplin estestvennonauchnogo i professional'nogo ciklov u tehnicheskih special'nostej]. *Modern problems of science and education [Sovremennye problemy nauki i obrazovanija], 2.*

Kabakov, Z. K. & Kabakov P. Z. (2005) The use of mathematical modeling in the development of new technologies and in education [Primenenie matematicheskogo modelirovanija pri osvoenii novyh tehnologij i v obrazovanii]. *Successes of modern science [Uspehi sovremennogo estestvoznanija], 6: 85-86.*

Kasaeian, A., Sameti, M. & Razi Astaraei, F. (2014) Simulation of a ZEB Electrical Balance with aHybrid Small Wind/PV. *Sustainable Energy, 2. 5.*

Khodayar, M. E., Manshadi, S. D., Wu, H., & Lin, J. (2016). Multiple Period Ramping Processes in Day-Ahead Electricity Markets. *IEEE Transactions on Sustainable Energy, 7(4),1634–1645.*

Maia, L. K. K., Güven, Z., La Mantia, F., & Zondervan, E. (2017). Model-based Optimization of Battery Energy Storage Systems. *27th European Symposium on Computer Aided Process Engineering, 2563–2568.*

Mizutani, M., Kobayashi, T., Watabe, K. & Wada, T. (2016) Study of Efficiency of Battery Energy Storage System. *IEEJ Transactions on Power and Energy. 136. 824-832.*

Murphy-Levesque, C., Donnan, C., Rossman, J., & Clee, P. (2017) IEA Wind TCP Annual Report. *PWT Communications Inc., September 2018: 4.*

Piriz, H., Cannatella, A.R., Guerra, E. & Porcari, D. (2012) *Inertia of hydro generators. Influence on the dimensioning, cost, efficiency and performance of the units.* The International Council on Large Electric Systems. Report number: A1-102.

Rampazzo, M., Luvisotto, M., Tomasone, N., Fastelli, I., & Schiavetti, M. (2018) Modelling and simulation of a Li-ion energy storage system: Case study from the island of Ventotene in the Tyrrhenian Sea. *Journal of Energy Storage, 15, 57–68.*

Skamyin, A.N., Dobush, V.S. & Rastvorova, I.V. (2019) Method for determining the source of power quality deterioration. *Proceedings of the 2019 IEEE Conference of Russian Young Researchers in Electrical and Electronic Engineering, ElConRus 2019: 1077-1079.*

Tariq Ahmed Hamdi, R. & Abdul Hussein, M. (2017) Modeling and Simulation of Wind Turbine Generator Using Matlab-Simulink. *Journal of Al-Rafidain University College. 40. 282-300.*

Tavner, P.J., Greenwood, D., Whittle, M.W.G., Gindele, R., Faulstich, S & Hahn, B. (2013) Study of weather and location effects on wind turbine. *Wind Energy. 16. 175-187.*

Scientific and Practical Studies of Raw Material Issues – Litvinenko (Ed)
© 2020 Taylor & Francis Group, London, ISBN 978-0-367-86153-7

Arc steel-making furnaces functionality enhancement

E. Martynova, V. Bazhin & A. Suslov
Saint-Petersburg Mining University, Saint-Petersburg, Russian Federation

ABSTRACT: The problems of arc steel-making furnaces parameters controlling at existing production facilities are revealed. Current situation in the departments of automation is analyzed. A mathematical model of arc steel-making furnace is developed. The control algorithm of arc furnaces for steel-making is developed. A method for controlling the parameters of arc furnaces is developed. A draft solution for controlling the thermal state of an electric arc furnace at a modern electrometallurgical plant is proposed.

1   INTRODUCTION

Currently, in Russia and abroad, electric arc steel-making furnaces are widely used as the main units for steel production. The advantages of the electric arc method are high productivity, as well as the ability to use scrap metal, liquid iron melt, and metallized pellets as raw materials. Electric arc furnaces are considered as the most common and environmentally friendly equipment for steel-making, but there is a number of lacks and problems of their sustainable operation.

The main source of heat energy in an electric arc furnace is the electric arc. At that, a power of about 75 MW is released in the arc column, and problems of uniform heating arise for the melting of charge materials loaded into the furnace (Abramovich & Sychev, 2016).

The uniform heating of the furnace and the charge material depends on the control of the arc. Therefore, there is a need to maintain a certain electric mode of the furnace operation with the given values of current, voltage and power. The load in the furnace is changing when the arc length changes from zero (short circuit; length and arc resistance equal to zero; maximum load) to infinity (arc breakage; load equal to zero).

In the process of melting in an electric arc furnace, the length of the electric arc is constantly changing, therefore, the power of this arc changes proportionally, and this directly affects the thermal condition of the furnace shaft and its elements and structures when the charge materials are heated to melt.

The existing methods of control of arc furnaces and the level of automation do not correspond to the state-of-the-art trends in the development of technology and the physicochemical conditions of multicomponent alloys melting processes, which leads to premature failure of the electrodes and lining of furnaces, as well as a large number of defects with low production of suitable products (Kolesnichenko & Afanasieva, 2018).

Furthermore, the control of the thermal state during melting and the effective management of the electric arc furnace during the process of the source metallurgical raw material melting is an unsolved problem to the date in full and remains one of the urgent tasks to be solved in the metallurgical industry.

It is advisable to characterize the melting of steel in an electric arc furnace with the use of three modes: electric, thermal and technological. Electric melting mode is the nature of the change in power supplied to the furnace over the time. The temperature mode includes a change in temperature of the metal, slag and lining along the melting process. The technological mode of melting includes change the chemical composition of metal and all the technological operations performed over the time (slag discharge, feeding of alloying additives, etc.).

All the modes, of course, are interconnected, but for a better understanding of automation systems description, they are usually considered individually (Lapshin, 2014).

Despite the fact that there is quite a lot of liquid metal temperature sensors developed, measuring the temperature of a metal in arc furnace is a difficult task. This is primarily due to the large temperature gradient over the surface and depth of a bath. In this regard, it is necessary to ensure the correct determination of sensor insertion point, namely, a representative point, which is characterized by the fact that the temperature measured in it is equal to the average temperature in the bath, and approximately equal to the temperature of the steel of the melt in the bucket after discharge from the furnace (Kurbatov & Vasilenko, 2013).

2 MATHEMATICAL MODEL OF SCRAP METAL MELTING PROCESS IN ELECTRIC ARC FURNACE

Currently, thanks to the development of computational technologies, a method of mathematical simulation has become the main method of research. It is based on the concept of a mathematical model of the process.

Mathematically, the process of steel melting in the arc furnace can be described by the heat balance equation obtained based upon data:

$$T = \int_0^t \left(\frac{3 \cdot I \cdot (a + b \cdot L) \cdot \cos(\phi) - P_{losses}}{M_m \cdot C_m} \right) dt + T_{init} \tag{1}$$

where T_{init} is the temperature in the arc furnace at the initial moment of time. Model assumptions:

1. Ambient temperature and heat capacity of ferroalloys and coke are not taken into account;
2. Control is performed by the entire electrode plug at the same time, but not by each individual electrode;
3. There is no arc ignition system;
4. There are heat losses in the model, indicated by one factor depending on the period of steel production;
5. The arc current used in the model is equal to the output current of the transformer;
6. The power of the model is equal to the active component of the transformer power; the reactive component resulting from inductance coils in the transformer is not taken into account;
7. The resistance of the short network is not taken into account; the resistance in the model is equal to the resistance of the arc;
8. The possibility of a short circuit in the circuit is not taken into account.

This model is used to study the influence of the main parameters (impacts) on the temperature of the system, since the basic physical laws of the processes occurring in the object are taken into account.

To simulate the steelmaking process in arc furnace, Matlab R2014b process model was compiled (Figure 1).

Control object model includes subsystems calculating the voltage values of the electric arc, based on the length of the arc; arc power using the arc voltage value, arc current value and installation power factor value; taking into account the number of electrodes in the arc furnace (3), the temperature in the arc furnace, calculating it on the vase of arc power value, taking into account heat losses during the melting process, the mass of the metal and its heat capacity, as well as the initial temperature of the metal in the furnace. Band-Limited White Noise unit is used to simulate the electrical noise of a thermocouple. In order to eliminate the electrical noise of the thermocouple, a smoothing filter in the form of a first-order aperiodic link is used. A detailed description of each subsystem is presented further (Fedorova & Firsov, 2018).

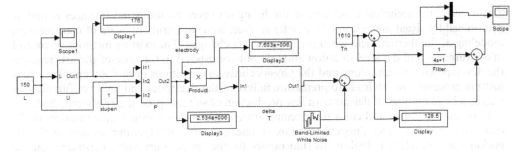

Figure 1. Control object model.

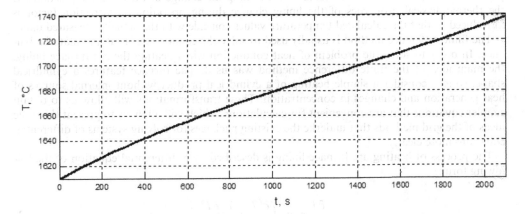

Figure 2. Temperature change over the melting period in arc furnace.

As a result of the work of the arc furnace, data were obtained on the change of steel temperature over the entire period of melting (Figure 2). In this figure, the abscissa is the time in seconds (s), and the ordinate is the temperature in the arc furnace in degrees Celsius (°C).

3 FURNACE TEMPERATURE STATE CONTROL ALGORITHM

Considering the above reasons, it is necessary to constantly monitor the thermal state during melting and effectively control the arc steel-making furnace during the process of the initial metallurgical raw materials melting. This problem is still unsolved to the full and remains one of the urgent tasks in the metallurgical industry.

Arc steel-melting furnaces are the batch-type furnaces. Such mode of operation features a significant impact on the lining materials of the furnace shaft. The typical arc steel-melting furnace DSP-90, for example, is operated at the metallurgical enterprises of Russia. The steel-making cycle in this arc furnace is 52 minutes, while the temperature of the melt reaches 1,620-1,630°C. After pouring the metal, the lining temperature decreases down to 700-800 °C. Such temperature drops negatively affect the condition of the furnace lining: microcracks arise, a change in the thickness of the seams occurs, chips and fractures are formed). At the same time, the features of arc furnace operation affect the thermophysical properties of the lining elements themselves (metallization and slagging). All this changes the heat balance of the unit. A need to continuously adjust the arc power arises, since the efficiency of the furnace and the amount of energy consumed are directly dependent on the heat loss that occurs through the furnace lining.

Currently, the technical condition of the lining on operated industrial furnaces is mainly determined by visual inspection of the outer surface, while not using the data of mathematical modeling of the thermal field when the arc power is changed. The existing methods of control of arc furnaces and the level of automation do not correspond to the state-of-the-art trends in the development of technology and the physicochemical conditions of multicomponent alloys melting processes, which leads to premature failure of the electrodes and lining of furnaces, as well as a large number of defects with low production of suitable products (Zatsepin, 2009).

In order to monitor and control the temperature mode and its thermal state throughout the entire volume of the metallurgical unit (arc furnace), a software algorithm as part of the SE package was developed. Built-in thermocouples fix the temperature value depending on the change of the arc length and the degree of heating of the charge and melt in the furnace shaft. The adaptive regulator matches the values of the lining temperature from all the measurement points and sends a signal to the actuator controlling the change of the electrode positions. If the thermophysical properties of the lining change during operation, a correction factor is introduced so that the calculated temperature values coincide with the actually measured data.

The finite difference method is usually used to solve the equations of heat and material balance. In order to solve the problem of heat conduction when heating the arc furnace lining, the main idea of the finite difference method was used. The furnace features a cylindrical shape, so let's consider a simple heating of a cylindrical product without internal sources of heat generation and changes in concentration. These transformations will allow us to obtain simpler equations that are convenient for calculations, which will allow us to present the basic ideas of the grid methods that underlie the existing packages for solving systems of differential equations more clearly.

The process of heating arc furnace lining is described by a differential equation of the following form:

$$\frac{\partial T}{\partial t} = a \cdot \left(\frac{\partial^2 T}{\partial r^2} + \frac{1}{r} \cdot \frac{\partial T}{\partial r} \right) \tag{2}$$

Initial and boundary conditions for this task:

$$T(0,r) = T_H(r); \quad \frac{\partial T(t,0)}{\partial r} = 0; \quad -\lambda \cdot \frac{\partial T(t,R)}{\partial r} = \alpha(T(t,R) - Too) \tag{3}$$

The initial condition expresses the temperature distribution in the furnace lining at the initial moment of time; the 1st boundary condition expresses the symmetric temperature distribution along the heating radius of the arc furnace, and the 2nd boundary condition expresses the equality of the heat flow from the environment (electric arc and metal) and the heat flow inside the lining due to thermal conductivity.

In this case, $0 \leq r \leq R$ is the radial coordinate; R is the furnace radius, m; a is factor of thermal diffusivity of the heated material, m²/s; α is the factor of heat transfer from the environment, W/(m²·K); λ is the factor of thermal conductivity of the heated material, W/(m·K). For convenience of the analysis, a dimensionless radius is entered into the equation:

$$\rho = \frac{r}{R}; r = \rho \cdot R; dr = R \cdot d\rho; \ dr^2 = R^2 \cdot d\rho^2 \tag{4}$$

By substituting the values of the radial coordinate and its derivatives from equation (4) into equations (2) and (3), we obtain the following equations:

$$\frac{\partial T}{\partial t} = \frac{a}{R^2} \cdot \left(\frac{\partial^2 T}{\partial \rho^2} + \frac{1}{\rho} \cdot \frac{\partial T}{\partial \rho} \right) \tag{5}$$

$$T(0,\rho) = T_H(\rho); \quad \frac{\partial T(t,0)}{\partial \rho} = 0; \quad -\lambda \cdot \frac{\partial T(t,1)}{\partial \rho} = \alpha(T(t,1) - Too) \qquad (6)$$

Continuous space in T (ρ, t) coordinates, which is a plane, can be represented as a set of discrete points with T (i, j) coordinates (Figure 3).

The continuous derivatives for a point with coordinates (i, j) will be replaced by their finite-difference equivalents expressed through the function values at the points $(i-1, i, i+1)$ with respect to the spatial coordinate and $(j-1, j, j+1)$ by time coordinate.

Let's denote the distance between the points in space and time variables as follows:

$$\Delta\rho = \frac{1}{N}; \rho = i \cdot \Delta\rho; \text{ and } t = j \cdot \Delta t \qquad (7)$$

Since the melting time is limited, it is necessary to specify Δt for the reasons of accuracy. Derivative values through the differences:

$$\left.\frac{\partial T}{\partial \rho}\right|_i \cong \frac{T_i^{(j-1)} - T_{i-1}^{(j-1)}}{\Delta\rho}; \left.\frac{\partial c}{\partial \rho}\right|_{i+1} \cong \frac{T_{i+1}^{(j-1)} - T_i^{(j-1)}}{\Delta\rho};$$

$$\frac{\partial^2 T}{\partial T^2} \cong \frac{\frac{T_{i+1}^{(j-1)} - T_i^{(j-1)}}{\Delta\rho} - \frac{T_i^{(j-1)} - T_{i-1}^{(j-1)}}{\Delta\rho}}{\Delta\rho} = \frac{T_{i+1}^{(j-1)} - 2 \cdot T_i^{(j-1)} + T_{i-1}^{(j-1)}}{\Delta\rho^2}; \qquad (8)$$

$$\frac{\partial T}{\partial t} \cong \frac{T_i^{(j)} - T_i^{(j-1)}}{\Delta t}$$

$$\left.\frac{\partial T}{\partial \rho}\right|_i \cong \frac{T_i^{(j-1)} - T_{i-1}^{(j-1)}}{\Delta\rho}; \left.\frac{\partial c}{\partial \rho}\right|_{i+1} \cong \frac{T_{i+1}^{(j-1)} - T_i^{(j-1)}}{\Delta\rho};$$

$$\frac{\partial^2 T}{\partial T^2} \cong \frac{\frac{T_{i+1}^{(j-1)} - T_i^{(j-1)}}{\Delta\rho} - \frac{T_i^{(j-1)} - T_{i-1}^{(j-1)}}{\Delta\rho}}{\Delta\rho} = \frac{T_{i+1}^{(j-1)} - 2 \cdot T_i^{(j-1)} + T_{i-1}^{(j-1)}}{\Delta\rho^2}; \qquad (9)$$

$$\frac{\partial T}{\partial t} \cong \frac{T_i^{(j)} - T_i^{(j-1)}}{\Delta t}$$

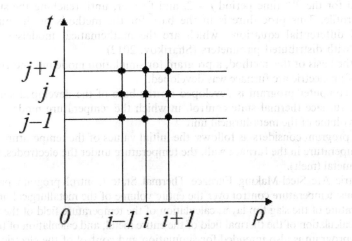

Figure 3. The layout of continuous coordinates dividing into discrete points for the issue of thermal conductivity of arc furnace lining.

By substituting the difference values of the derivatives into the original equation (5) and into the boundary conditions equations (6), we get:

$$\frac{T_i^{(j)} - T_i^{(j-1)}}{\Delta t} = \frac{a}{R^2} \cdot \left(\frac{T_{i+1}^{(j-1)} - 2 \cdot T_i^{(j-1)} + T_{i-1}^{(j-1)}}{\Delta \rho^2} + \frac{1}{i \cdot \Delta \rho} \cdot \frac{T_i^{(j-1)} - T_{i-1}^{(j-1)}}{\Delta \rho} \right);$$

$$T_i^{(j)} = T_i^{(j-1)} + B \cdot \left(T_{i+1}^{(j-1)} - 2 \cdot T_i^{(j-1)} + T_{i-1}^{(j-1)} \right) + \frac{B}{i} (T_i^{(j-1)} - T_{i-1}^{(j-1)}) \qquad (10)$$

$$B = \frac{a \cdot \Delta t}{R^2 \cdot \Delta \rho^2}$$

The recurrence relation (10) allows us to calculate all the points from $i = 1$ to $i = N-1$ by the spatial variable on the j-th time layer. The value of the zero point and the N-th points are calculated on the base of the boundary conditions.

$$T_0^{(j)} = T_1^{(j)} \qquad (11)$$

$$-\frac{\lambda}{R} \cdot \frac{T_N^{(j)} - T_{N-1}^{(j)}}{\Delta \rho} = a(T_N^{(j)} - T_{oc});$$

$$T_N^{(j)} \left(1 + \frac{a \cdot R \cdot \Delta \rho}{\lambda}\right) = T_{N-1}^{(j)} + \frac{a \cdot R \cdot \Delta \rho}{\lambda} \cdot T_{oc} \qquad (12)$$

$$T_N^{(j)} = \frac{T_{N-1}^{(j)} + \frac{a \cdot R \cdot \Delta \rho}{\lambda} \cdot T_{oc}}{\left(1 + \frac{a \cdot R \cdot \Delta \rho}{\lambda}\right)}$$

Consequently, the relations (10-12) for calculating all the temperatures on the spatial and temporal coordinates are obtained. In this case, the following calculation procedure is used. All the temperature values from the initial conditions are recorded for selected points of the zero-time layer in accordance with the selected number of points for i = 0, 1, ..., N. Then, using the equation (10), the temperatures are calculated for all the points from $i = 1$ to $i = N-1$ on the 1st time layer $j = 1$. Then, according to the equation (11), the temperature value is calculated in the center of the furnace (for $i = 0$), and the temperature on the outer surface of the lining is calculated by equation (12). Thus, the entire first period of time is covered. After that, in the same order, all the calculations are performed for the 2nd time period $j = 2$, and further, until reaching the steady-state temperature profile. This procedure is in the base for the methods of the numerical solution of partial differential equations, which are the mathematical models of technological processes with distributed parameters (Sharikov, 2012).

On the basis of this method, a program for simulation modeling and control of the thermal state of an electric arc furnace was developed.

The computer program is developed on the base of the developed software algorithm for the arc furnace thermal state control, in which the temperature mode is controlled over the entire volume of the metallurgical unit.

The program considers as follows: the initial values of the temperature of the furnace bath, the temperature in the furnace wall, the temperature under the electrodes, and the temperature of the metal (melt).

Electric Arc Steel-Making Furnace Thermal State Control program provides the following functions: temperature control over the entire volume of the metallurgical unit; calculation of the temperature of the slag top layer; calculation of the temperature field of the metal under the electrode; calculation of the thermal field of the entire metal; and calculation of the wall temperature.

The program is also intended for simulation and control of the electric arc furnace thermal state.

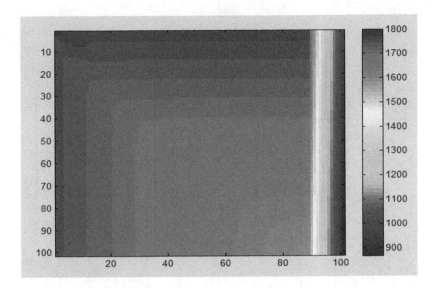

Figure 4. Temperature field of electric arc furnace.

The use of this control system will allow considering the temperature measurements at several points and more accurately estimating the length of the arc, which will directly help to control the heat state of the furnace and use the results obtained in the SCADA system control, which will make it possible to predict and evaluate the control efficiency of the melting process in the arc furnace at the one of the enterprises (Martynova et al., 2018).

4 ELECTRIC ARC FURNACE PARAMETERS CONTROL METHOD

In order to reduce energy consumption and improve the quality of the produced melt, to reduce the number of prematurely failing electrodes and the bottom and side lining through the increase of the reliability of estimation of the thermal condition of the furnace, a method for the arc furnace parameters control was developed.

The method is implemented as follows. Thermocouples (TPP (S)) are installed at the control points to measure the temperature and control the thermal state of the arc furnace. Control points are assumed on each electrode of a three-electrode plug directly below the electrode holder mounting, between the furnace lining and the casing (at six points along the perimeter of the unit), and in the furnace bottom (at six points opposite to each electrode and between them). Thermocouples arrangement is presented in Figure 5. The numbers in the figure indicate as follows: 1 – thermocouple; 2 – electrode; 3 – electrode holder; 4 – furnace roof; 5 – furnace lining; 6 – furnace bath; 7 – furnace casing.

In this case, the melting time in the DSP-90 furnace is approximately 52 minutes. After the start of melting, the current melting time (t) is determined and the temperature is measured by the specified thermocouples; then the data is transferred to the controller for calculation.

The program in the controller operates according to the following algorithm (Figure 6).

According to the formulas, the average temperature of the electrodes (T_E), the average temperature of the molten metal (T_M), and the average temperature of the lining (T_L) are calculated.

$$T_E = \frac{\sum_{1}^{i} T_{Ei}}{i} \tag{13}$$

257

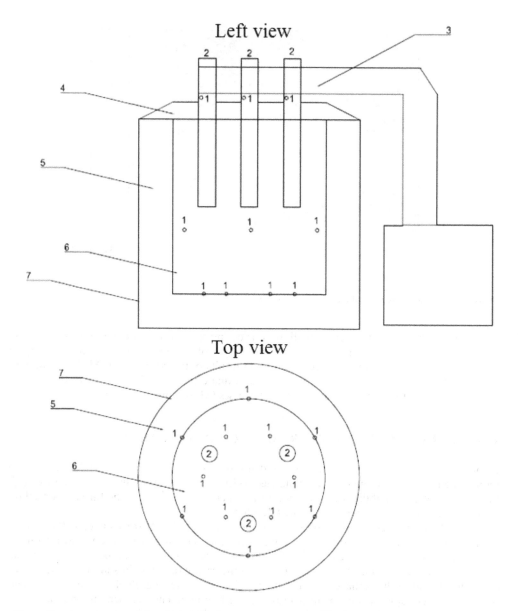

Figure 5.   Arrangement of thermocouples for temperature control in the arc furnace.

$$T_M = \frac{\sum\limits_{1}^{n} T_{Mn}}{n} \tag{14}$$

$$T_L = \frac{\sum\limits_{1}^{m} T_{Lm}}{m} \tag{15}$$

The current value of the melting time is determined. If this value of the melting time is less than 10 minutes, then the factor of temperature change at the electrode is calculated (C_E).

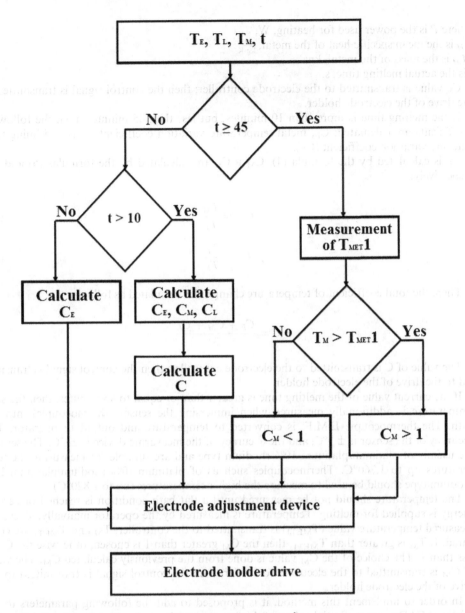

Figure 6. Program algorithm for arc furnace parameters control.

$$C_E = \frac{T_E}{T_P} \qquad (16)$$

where

$$T_P = \frac{P}{M_M \cdot C_H} \cdot t \qquad (17)$$

where P is the power used for heating, W;
C_H is the mean specific heat of the metal, $\frac{J}{kg \cdot K}$;
M_M is the mass of the metal, kg;
t is the actual melting time, s.

C_E value is transmitted to the electrode controller; then the control signal is transmitted to the drive of the electrode holder.

If the melting time is more than 10 minutes, but less than 45 minutes, then the following coefficients are calculated: C_E, metal temperature variation coefficient (C_M) and lining temperature variation coefficient (C_L).

C_E is calculated by the formula (4), C_M и C_L are calculated by the formulas (5) and (6), respectively.

$$C_M = \frac{T_M}{T_P} \tag{18}$$

$$C_T = \frac{T_L}{T_P} \tag{19}$$

Then, the total coefficient of temperature change C is calculated as follows:

$$C = \frac{C_E + C_M + C_L}{3} \tag{20}$$

The value of C is transmitted to the electrode controller; then the control signal is transmitted to the drive of the electrode holder.

If the current value of the melting time is greater than or equal to 45 minutes, then the steel temperature is additionally measured when immersing the sensor (thermocouple) into the bath. The thermocouple E.M.F. is converted to temperature and output to operator. The accuracy of the sensor is ± 3°C, and the accuracy of the measuring device is ±2°C. The sensors are usually of platinum/platinum-10% rhodium type and are suitable for measuring the temperatures up to 1,760°C. Thermocouples such as of platinum-10% rhodium/platinum-13% rhodium type should be used to measure the higher temperatures (up to 1,820°C).

The temperature should not be measured until a flat bath condition is reached or certain energy is supplied for melting. Temperature is measured by the operator manually. Then, the measured temperature value (T_{MET1}) is transmitted to the controller. T_M and T_{MET1} are compared. If T_M is greater than T_{MET1}, then the C_M greater than 1 is chosen, otherwise the C_M is less than 1. The choice of the C_M value is done from the previously calculated C_M. The value of C_M is transmitted to the electrode controller; then the control signal is transmitted to the drive of the electrode holder.

In order to implement this method, it is proposed to add the following parameters to the existing SCADA system: initial values of the furnace bath temperature, temperature in the furnace wall, temperature under the electrodes, and metal (melt) temperature. This will improve the management of the thermal state of the electric arc furnace and optimize the existing parameters to narrower values by expanding the information flow. The project of entering new parameters into the current SCADA system is presented in Figure 7.

The developed arc furnace control system will make it possible to consider the temperature measurements at several points and more accurately estimate the arc length, which will directly help controlling the thermal state of the furnace and using the results obtained in the SCADA control system. This will provide an opportunity to predict and estimate the efficiency of the melting process control in the arc furnace at metallurgical enterprises.

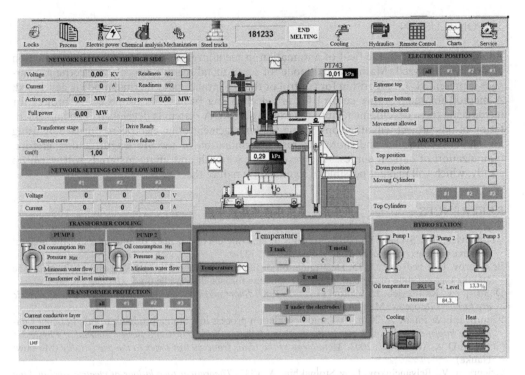

Figure 7.   Project of entering new parameters into the current SCADA system.

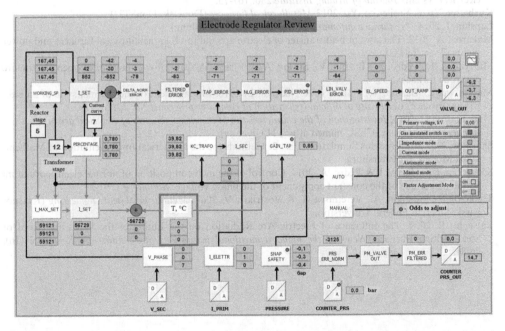

Figure 8.   The project of new parameters entering into the existing SCADA system (into the electrode controller).

5   CONCLUSION

The main problems in the field of automation in metallurgical enterprises for the production of steel are analyzed. Mathematical model of scrap metal melting process in electric arc furnace is presented. An algorithm for control of the arc furnace bath thermal mode is developed. A method for temperature control for arc steelmaking furnaces is proposed. A project for introducing these developments into the existing SCADA system is developed. The implementation of this project will allow optimizing the management of an electric arc furnace by using an adaptive controller and more accurate determination of the arc length, and, therefore, minimization of current surges. As a result, furnace bath and lining materials will warm up evenly, which will make it possible to significantly reduce their wear.

REFERENCES

Abramovich, B & Sychev, Yu. 2016 Problems of ensuring energy security of enterprises of the mineral resource complex. *Journal of Mining Institute* 217: 132-9.

Beloglazov, I. & Pedro, A. 2015. Determination of the temperature of the working end of the electrode and the thermal state of the bath of the ore-smelting furnace during idle time. *Metallurg* 6: 66-9.

Dantsis, Ya., Kantsevich, L., Zhilov, G., *et al.* 1987. *Short networks and electrical parameters of electric arc furnaces* (Moscow: Metalurgy).

Fedorova, E. & Firsov A. 2018. Red mud flocculation process in alumina production. *International Conference Information Technologies in Business and Industry 2018. IOP Conf. Series: Journal of Physics.*

Gavrilov, D. & Alekseev V. 2008. *Metallurgy of the Urals from ancient times to the present day* (Moscow: Nauka).

Grigoryan, V., Belyanchikov, L. & Stomakhin, A. 1987. *Theoretical foundations of electric steel-making processes* (Moscow: Metalurgy).

Kolesnichenko, S. & Afanasieva, O. 2018. Theoretical aspects of the evaluation of the technical level of electrical systems *Journal of Mining Institute* 230: 167-75.

Kurbatov, Yu. & Vasilenko, Yu. 2013 *Metallurgical furnaces* (Donetsk: DonSTU).

Lapshin, I. 2014. *Arc furnace automation* (Moscow: MSU Publishing).

Makarov, A. 2012. The laws of heat transfer of electric arc and torch in metallurgical furnaces and power plants (Tver: TSTU).

Martynova, E., Pedro, A., Bazhin, V., Suslov, A. & Firsov, A. 2017. Power control of open electric arc electrode furnace. *Steel* 7: 21-3.

Martynova, E., Bazhin, V., Petrov, P. & Nikitina, L. 2018. Management of the thermal state of an electric arc furnace. *Certificate of official registration of computer programs No. 2018614175.*

Mikhadarov, D. 2015. *Investigation of the characteristics of the characteristics of electrical arc in arc furnaces* (Cheboksary: ChSU named after I N Ulyanov).

Sharikov, Yu. 2012. Systems Simulation. Part 2: Methods for the Numerical Implementation of Mathematical Models. St. Petersburg.

Suslov, A., Beloglazov, I. & Pedro, A. 2014. Control of the melt composition of normal electrocorundum by the magnitude of the constant component of the phase voltage. *Metallurg* 1: 91-3.

Tin'kova, S., Proshkin, A., Veretnova, T. & Vostrikov V. 2007. *Metallurgical heat engineering* (Krasnoyarsk: Siberian Federal University).

Zatsepin, E. 2009. The Influence of Electric Arc Radiation on the Lining of the Walls and Roof of Steel-Melting Furnace. *News of the higher educational institutions of Chernozemic Region.* 2 (16): 76-81.

Scientific and Practical Studies of Raw Material Issues – Litvinenko (Ed)

Modeling of operation modes of electrical supply systems with non-linear load

V.N. Kostin, A.V. Krivenko & V.A. Serikov
St. Petersburg Mining University, St. Petersburg, Russia

ABSTRACT: A power supply system with voltage up to 1000 V, comprising a power supply source – a transformer, linear load, a capacitor bank and non-linear load – a three-phase bridge rectifier as a generator of higher current harmonics, have been considered. To study non-sinusoidal operation modes of the power supply system, its three-phase computer model has been developed. To assess the computer model reliability, the method of physical simulation was used. Qualitative agreement of the results of computer and physical models has been obtained by all the basic parameters: the spectrum of higher harmonics, their values and patterns of spreading over the power supply system elements. Some quantitative differences in the simulation results are apparently associated with the basic assumption accepted, i.e. representation, via the computer model, of the power supply system parameters, calculated for the fundamental harmonic.

1 INTRODUCTION

Due to the mass introduction of semiconductor converters both in industry, and among consumers in urban utility electrical networks, one of the main tasks of improving the quality of electric power is to reduce the negative impact of higher harmonics on the operation of power supply systems (PSS) equipment (Abramovich & Sychev 2016, Shklyarskiy & Skamyin 2016, Temerbaev et al. 2013, Tulsky et al. 2013). To solve this task, it is necessary to have reliable information on the nature of higher harmonics generation, their spectrum, values and patterns of spreading over the PSS elements.

Currently, one of the most reliable tools to analyze and study different modes of the PSS operation is the computer simulation, providing the researcher with quick and cost efficient possibility to assess the PSS properties and behavior under any operation mode, while conducting the study using an actual object in the vast majority of cases is deemed to be hardly possible.

Computer simulation of the PSS non-sinusoidal modes is discussed in many scientific papers, comprising the analysis of harmonic currents spread using a simplified single-phase model of the PSS, wherein each higher harmonic is deemed to be a source of current with infinite capacity, i.e., a constant value (Shklyarskiy & Skamyin 2014, Skamyin & Dobush 2018, Zhezhelenko 2000). A number of other scientific papers (Kovalenko et al. 2017, Osipov et al. 2018) discusses a three-phase model, however, in these works, each harmonic is simulated by three phase-shifted current sources of the corresponding frequency, as well.

As is known, the main requirement for computer models is their conformity with actual processes that they simulate. However, actual objects of the PSS are so diversified that considering all the affecting factors in the computer model is impossible and even inexpedient. Any model reflects only an aspect of objective reality and therefore, it is simpler than the studied object. To build a computer model, the object of study must be formalized in a way to ensure that the object's inessential characteristics are not taken into account, while its main properties remain unchanged.

The authors are unaware of scientific papers that would deal with non-sinusoidal modes in the three-phase PSS with a valve converter – a direct source of these harmonics, as well as they are unaware of works, wherein studies would be performed using either actual PSS or its physical model.

The aim of the research is to study the generation, variation in value and spread of higher current harmonics in a three-phase system and to compare the results of the computer and physical simulation of non-sinusoidal operation modes of the PSS with non-linear load to assess the computer model conformity with actual physical processes.

The study of the PSS modes via a computer model was performed using the *Multisim* industrial standard software with an interactive schematic environment to instantly visualize and analyze electrical and electronic circuit behavior and an intuitive interface. The laboratory physical three-phase model of the PSS with non-linear load was considered as an actual object.

2 COMPUTER MODEL OF POWER SUPPLY SYSTEMS

A power supply system with voltage up to 1000 V, comprising a power supply source – a transformer, linear load, a capacitor bank and non-linear load (Figure 1, *a*). The PSS non-linear load was simulated by a three-phase bridge rectifier, assembled on virtual diodes D_1 ... D_6 and carrying some load, represented by the active resistance R_d (Figure 1, *b*). To smooth the rectified ripple current, the inductance L_d was connected in series with the resistance R_d. A capacitor bank (CB) was simulated by three virtual capacitances C.

The main difficulty during simulation is associated with the PSS elements that have mutual inductive coupling between phases; these are, in particular, transformers. Actually, the PSS element inductive resistance will depend on the frequency, i.e., on the harmonic number n. Thus, the transformer inductive resistance for the n-th harmonic is expressed by the following equations, given in Zhezhelenko (2000):

$$X_{Tn} = nX_T = n\frac{U_k[\%] \cdot U_{t\,nom}^2}{100S_{t\,nom}}10^3; X_{Tn} = X_T(0.75 + \frac{0.4}{\sqrt{n}})n; X_{Tn} = X_T n^{0.906} \qquad (1)$$

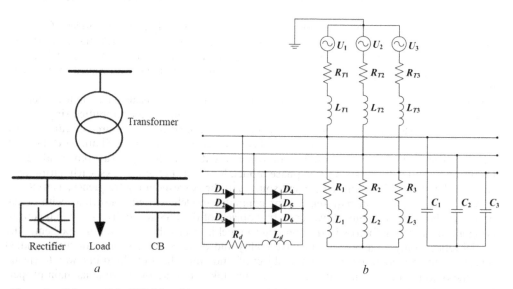

Figure 1. Scheme of the PSS (*a*) and its computer model (*b*).

where X_T is the transformer inductive resistance for the fundamental harmonic; U_k is short-circuit voltage (%); $U_{t\,nom}$ and $S_{t\,nom}$ are rated voltage and capacity.

From (1) it is seen that for the n-th harmonic the transformer inductive resistance increases by n or less than n times.

Higher harmonics, depending on their number, can have positive, negative, and zero sequences (Bessonov 1996). In the presence of a magnetic coupling between phases, the reactance is determined taking into account the mutual inductive coupling, depending on that which sequence of currents flows through the phases. For an element, whose magnetically coupled circuits are fixed relative to each other, the resistances of the positive X_1 and negative X_2 sequences are equal to each other. Accordingly, for transformers, power lines and reactors $X_1 = X_2$.

With the increase in frequency, active resistance of the PSS elements will increase due to the occurrence of the surface effect and proximity effect. As shown in Demoulias et al. (2004), Nan et al. (2003), Tofoli et. al (2006), with the increase in frequency up to 1000 Hz, active resistance of conductors with a cross section of 35 ... 400 mm 2 increases virtually linearly by 1.3 ... 3 times, respectively. In addition, hysteresis losses, proportional to frequency, and eddy current losses, proportional to the squared frequency, will also affect the transformer equivalent active resistance value (Nan et al. 2003, Tofoli et al. 2006).

Since in the studied PSS, a rectifier, as a non-linear element, will generate the entire spectrum of higher current harmonics, it seems impossible to take into account the values of resistances depending on the frequency of each harmonic. In connection therewith, the basic assumption accepted is the following: all active inductive elements (transformers, lines, loads) were accounted for by inductance and active resistance, calculated for the fundamental harmonic.

Thus, the transformer was simulated as a source of three-phase voltage U_1, U_2, U_3 (Figure 1, b), applied after inductance and active resistance:

$$R_T = \frac{\Delta P_k \cdot U_{t\,nom}^2}{S_{t\,nom}^2} 10^3; L_T = \frac{X_T}{314} \tag{2}$$

where ΔP_k are capacity losses (kW) under the short-circuit mode.

The linear load $S = P + jQ$ was accounted for as generalized and deemed to be active resistance and inductance, connected in series

$$R = \frac{U_{t\,nom}^2}{S} \cos \varphi; L = \frac{U_{t\,nom}^2}{314S} \sin \varphi \tag{3}$$

where φ – phase angle between current and voltage.

3   PHYSICAL MODEL OF POWER SUPPLY SYSTEMS

To compare the results of computer and physical simulation, a laboratory physical model of the PSS was created, according to the scheme in Figure 1, a. A dry-type transformer with the following parameters: rated capacity $S_{t\,nom} = 19$ kVA, rated voltage $U_{t\,nom} = 0.4/0.23$ kV, voltage and short-circuit losses $U_k = 10\%$, $\Delta P_k = 0.5$ kW was used as the PSS power source.

A three-phase bridge rectifier VS-26MT60 with diodes, designed for repetitive reverse voltage of 600 V and maximum average current of 25 A, was used as non-linear load. Wire rheostats, providing rectified current $I_d \approx 20$ A, were used as bridge load. Smoothing the rectified ripple current was implemented via a choke with inductance $L_d \approx 100$mH.

265

Wire rheostats and chokes, designed for current up to 20 A, were used as linear load. The assembled CB had a capacitance adjustment range of 10 … 320 μF.

During the study of the physical model modes, in addition to conventional laboratory instruments (ammeters and voltmeters), the Aktakom ASK-6205 digital storage oscilloscope and the Fluke 43B voltage quality meter were used.

For the purpose of visual clarity and generalization of the simulation results, all the controlled capacities were presented in relative units relative to the transformer rated capacity:

- non-linear load power $P^*_d = P_d / S_{t\,nom}$;
- linear load power $S^* = S / S_{t\,nom}$;
- the CB capacity $Q_{CB}{}^* = Q_{CB} / S_{t\,nom}$.

The amplitudes of currents of higher harmonics I_n were also presented in relative units $I_n{}^* = I_n/I_1$, where I_1 is the fundamental harmonic amplitude.

Studies, using the physical model, were carried out at loads $P^*_d \approx 0.25$ and $S^* \approx 0.2$.

It is known that for a three-phase bridge rectifier characteristic harmonics make up the series $n = 6k \pm 1$, where k = 1, 2, 3, The results of the transformer resistances calculations with the cable communication between the transformer and the rectifier accounted for, revealed that resonance phenomena at frequencies of the 5th, 7th, 11th, 13th, 17th and 19th harmonics should be expected at $Q_{CB}{}^* \approx 0.27$; 0.13; 0.052; 0.037; 0.022 and 0.017, respectively.

Since, in addition to non-linear loads, the supply mains (Skamyin & Baburin 2017) can be the source of higher harmonics, voltage quality at the dry-type transformer outputs, at non-linear load cut-off, was measured using the Fluke 43B instrument. Measuring results confirmed that the total coefficient of harmonic voltage components amounted to $K_U = 0.9\%$, whereas according to the state standard GOST 32144-2013, the values of this coefficient should not exceed 8%. Thus, a three-phase bridge rectifier was the source of higher current harmonics in the considered PSS.

4 RESULTS OF MODELING

4.1 Impact of the rectifier on voltage quality

In the computer model, the coefficient K_U was automatically determined using the *Multisim* software environment after the expansion into a Fourier series of the periodic phase voltage curve at the transformer output. In the physical model, as noted above, the coefficient K_U was measured with the Fluke 43B instrument. The simulation results are given in Figure 2. The permissible value

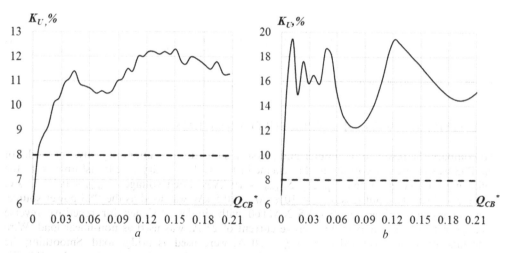

Figure 2. Dependences of the coefficient K_U on the CB capacity: a – physical model, b – computer model.

of the coefficient K_U is shown with the dashed line. As it can be seen, the CB connection reduces voltage quality. In its absence, the coefficient K_U has a minimum value and is within the acceptable limit ($K_U < 8\%$).

Qualitative agreement of the results of physical and computer simulation can be observed. Within the entire variation range Q_{CB}^* the coefficient K_U exceeds the permissible value. With an increase in capacity Q_{CB}^*, this coefficient has a general tendency to increase with local maxima at resonances and local minima in the modes between resonances. The computer model gives higher values of the coefficient K_U and more pronounced maxima of this coefficient at resonances at the 7th ($Q_{CB}^* \approx 0.13$) and higher harmonics.

4.2 Rectifier harmonics

Three-phase bridge rectifier is a generator of higher current harmonics. For engineering calculations, for example, according to a single-phase model, the amplitude of the n-th characteristic harmonic is set by the current source with the value $I_n = I_1/n$, where I_1 is the fundamental harmonic amplitude (Shklyahskiy & Skamyin 2011, Skamyin & Dobush 2018, Zhezhelenko 2000). In relative units, the amplitude of the current source in the single-phase model is $I_n^* = 1/n$.

To assess the impact of resonance phenomena on the values of higher harmonics in computer and physical models, the dependences between the amplitudes of these harmonics and the capacity value Q_{CB}^* were composed. The results are given in Figure 3.

In the computer model, in the inter-resonance modes, the amplitudes of higher current harmonics are close to the values $I_n^* = 1/n$. In particular, in Table 1, the amplitudes of higher harmonics are given for $Q_{CB}^* = 0.1$ (the mode between resonances at the 7th and 11th harmonics).

In the resonance modes, "dips" of harmonic currents values can be observed. Moreover, at resonance at one of the harmonics, all the other harmonics fall in "dips". In particular, Table 1 presents the amplitudes of higher harmonics for $Q_{CB}^* = 0.13$ (resonance at the 7th harmonic).

In the physical model, in the inter-resonance modes, the amplitudes of higher harmonics are less than values $I_n^* = 1/n$, wherein the higher the harmonic number, the more pronounced this difference (see Table 1). The physical model confirmed the presence of current harmonics "dips" in the resonant modes. In particular, Table 1 presents the amplitudes of higher harmonics for $Q_{CB}^* = 0.05$ (resonance at the 11th harmonic).

Thus, in resonant modes as well as in modes close to resonant ones, not considering current harmonics "dips" when performing calculations thereof for models with current sources with the amplitude $I_n^* = 1/n$, can give the results different from actual ones.

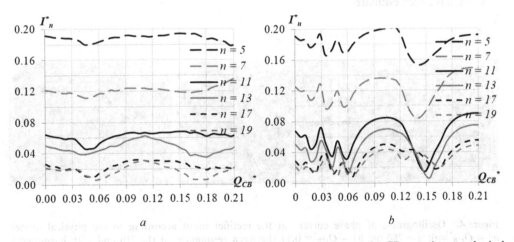

Figure 3.   Dependences of the amplitudes of the rectifier harmonics on the CB capacity a – physical model, b – computer model.

Table 1. The amplitudes of higher current harmonics in computer and physical models.

n	$I_n^* = 1/n$	computer model		physical model	
		I_n^* for $Q_{CB}^* = 0.1$	I_n^* for $Q_{CB}^* = 0.13$	I_n^* for $Q_{CB}^* = 0.05$	I_n^* for $Q_{CB}^* = 0.1$
5	0.2	0.2	0.15	0.18	0.19
7	0.14	0.136	0.084	0.11	0.12
11	0.091	0.084	0.014	0.045	0.065
13	0.077	0.07	0.01	0.04	0.06
17	0.059	0.05	0.02	0.01	0.03
19	0.053	0.043	0.021	0.005	0.027

4.3 Current switching angle between the diodes

The decrease in values of higher current harmonics, generated by the rectifier under resonant modes, is probably associated with variation in phase current waveform at the rectifier input when capacity Q_{CB}^* changes (Kostin & Serikov 2019). Figure 4 illustrates the oscillograms of the initial stage of the positive half-wave of phase current, made using the physical model in the absence of Q_{CB}^*, as well as at values of this capacity, corresponding to the inter-resonant and resonant modes. Figure 5 illustrates the same oscillograms, but made using the computer model.

From the oscillograms in Figures 4 and 5, it can be seen that current switching interval γ by the rectifier valves varies within fairly wide limits, thereby quantitatively varying the phase current harmonic composition. For example, according to the physical model, current switching interval varied from 400 to 900 μs, according to the computer model – from 100 to 1400 μs.

The occurrence of the high-frequency (units of kilo-hertz) oscillating component in the switching interval (oscillograms *b* and *c*) is due to the inductances of cables from 0.23 kV buses to the rectifier and the CB (Figure 1, *a*). The attenuation of this oscillating component is due to the active resistance of the above-mentioned cables.

It is worth noting that in the PSS with voltage up to 1000 V the resistances of cable communication between the transformer, the rectifier and the CB significantly affect current waveform at the rectifier input in the switching interval of its valves.

4.4 Capacitor bank currents

In the PSS, the CBs are the most sensitive to current overloads (Skamyin & Belsky 2017). Therefore, in this paper, total current, flowing through the CB, as well as its harmonic composition have been estimated.

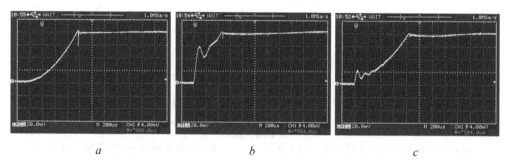

a   *b*   *c*

Figure 4. Oscillograms of phase current at the rectifier input according to the physical model: *a)* – $Q_{CB}^* = 0$; γ ≈ 780 μs; *b)* – $Q_{CB}^* = 0.11$ (between resonances at the 7th and 11th harmonics); γ ≈ 470 μs; *c)* – $Q_{CB}^* = 0.04$ (resonance at the 13th harmonic); γ ≈ 860 μs.

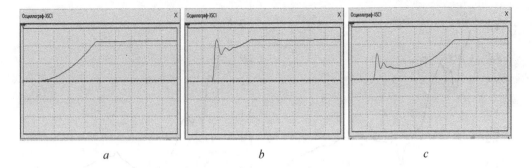

a   b   c

Figure 5. Oscillograms of phase current at the rectifier input according to the computer model:
a) $Q_{CB}^* = 0$; $\gamma \approx 780$ μs; b) $Q_{CB}^* = 0.11$; $\gamma \approx 440$ μs (between resonances at the 7th and 11th harmonics);
c) $Q_{CB}^* = 0.04$; $\gamma \approx 1000$ μs (resonance at $n = 13$).

As a result of computer and physical simulation, it was revealed that in operation modes with non-linear load and modes, close to resonant ones, total current, flowing through the CB, can reach twice the value relative to rated current (overload factor $K_{OVL} \approx 2$), and the amplitudes of higher current harmonics can be comparable to the amplitude of current of the fundamental harmonic.

Figure 6 illustrates the dependences of the relative values of higher current harmonics, flowing through the CB, on its capacity. For computer and physical models, these dependences qualitatively repeat each other. For each resonant capacity, specified above, $Q_{CB}^* \approx 0.13$; 0.052; 0.037; 0.022 and 0.017 the amplitudes of the 7th and higher harmonics reach a maximum. In a quantitative sense, the physical model provides lower values of higher current harmonics, than the computer model.

Figure 7 illustrates the dependences of the CB current overload factor K_{OVL} on its capacity. The horizontal dashed line corresponds to the CB permissible current overload ($K_{OVL} = 1.3$). It is apparent that there are inadmissible CB current overloads in the area of low power Q_{CB}^*. After capacity of $Q_{CB}^* > 0.08$ for both models, the CB overloads differ slightly and are within the overload factor limit.

During the pursuance of the research, it was revealed that in both computer and physical models linear load had virtually no impact on the simulation results.

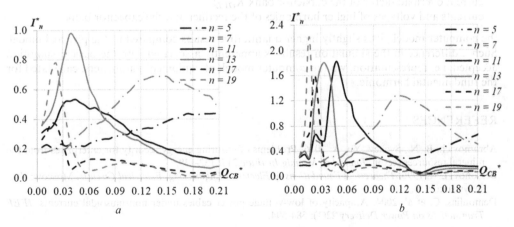

a   b

Figure 6. Dependences of amplitudes of the CB current harmonics on its capacity: a – physical model, b – computer model.

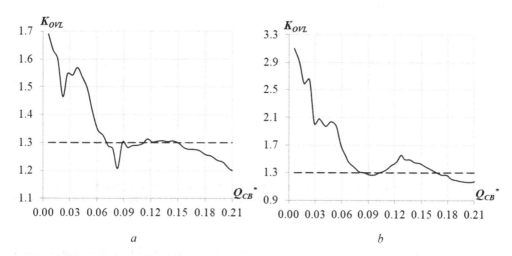

Figure 7. Dependence of the overload factor K_{OVL} on the CB capacity: *a* – physical model, *b* – computer model.

5 CONCLUSION

Non-sinusoidal modes of operation of a three-phase computer model of a power supply system with voltage up to 1000 V with non-linear load — a three-phase bridge rectifier, linear load and a capacitor battery have been considered. To confirm the computer model conformity with actual physical processes, similar non-sinusoidal modes have been considered using the physical laboratory model.

Qualitative agreement of the results of computer and physical simulation has been revealed. Physical simulation confirmed higher current harmonics "dips" in resonant modes as well as in modes close to resonant ones and variation in waveform and duration of current switching interval by the valves.

For both computer and physical models within the entire range of the capacitor bank capacity adjustment, voltage quality in the power supply system failed to meet the requirements of GOST, and the capacitor banks virtually always were overloaded with higher current harmonics.

By the basic parameters:

- total coefficient of harmonic voltage components K_U;
- current overload factor of the capacitor bank K_{OVL};
- currents and voltages of higher harmonics of the rectifier and the capacitor bank,

the computer model gives slightly higher quantitative results compared to the physical model. Such a difference in the simulation results is apparently associated with the basic assumption accepted, i.e. representation, via the computer model, of equipment parameters, calculated for the fundamental harmonic.

REFERENCES

Abramovich, B. N., Sychev, Yu. A. 2016. Problems of ensuring energy security for enterprises from the mineral resources sector. *Journal of Mininig Institute* 217: 132-139.

Bessonov, L.A. 1996. *Theoretical Foundations of Electrical Engineering. Electrical Circuits.* Moscow: Vysshaya shkola.

Demoulias, C. et al. 2004. Ampacity of low-voltage power cables under nonsinusoidal currents. *IEEE Transactions on Power Delivery* 22(1): 584-594.

Kostin, V.N. and Serikov, V.A. 2019. Computer simulation of operating power supply modes with non-linear load. *St. Petersburg polytechnic university journal of engineering science and technology* 25(01): 19–29.

Kovalenko, D.V., Kisselyov, B.Yu., Plotnikov, D.I., Shakenov, Y.Y. and Kulinich I.O. 2017. The calculation method of passive filters to compensate higher harmonics current in power supply systems of industrial enterprises. *International research journal* 1-4 (55):82-86.

Nan, Xi. and Sullivan, C. R. 2003. An improved calculation of proximity-effect loss in high-frequency windings of round conductors. *IEEE Annual Power Electronics Specialists Conference* 34(2): 853–860.

Osipov, D.S., Dolgikh, N.N., Gorovoy, S.A. and Poplavskaya V.E. 2018. The analysis of additional losses from higher harmonics in 380 V networks using packet wavelet transform algorithms. *The Journal Omsk Scientific Bulletin* 162(6): 76-81.

Shklyarskiy, Y. E. and Skamyin, A. N. 2011. Reduction methods of high harmonics influence on the electric equipment operation. *Journal of Mininig Institute* 189: 121-124.

Shklyarskiy, Y. E. and Skamyin, A. N. 2014. High Harmonic Minimization in Electric Circuits of Industrial Enterprises. *World Applied Sciences Journal* 30(12): 1767-1771.

Shklyarskiy, Y. E. and Skamyin, A. N. 2016. Compensation of the reactive power in the presence of higher voltage harmonics at coke plants. Coke and Chemistry 59(4): 163–168.

Skamyin, A.N. and Baburin, S.V. 2017. Algorithm for Selecting Cross-Capacitive Compensation Device Parameters with High Harmonics Accounted For. *International Journal of Applied Engineering Research* 12(6): 1049-1053.

Skamyin, A. N. and Belsky, A. A. 2017. Reactive power compensation considering high harmonics generation from internal and external nonlinear load. *IOP Conference Series: Earth and Environmental Science* 87(3),032-043.

Skamyin, A. N. and Dobush, V. S. 2018. Analysis of nonlinear load influence on operation of compensating devices. *IOP Conference Series: Earth and Environmental Science* 194(5), 052023.

Temerbaev, S. A., Boyarskaya, N. P., Dovgun V. P. and Kolmakov, V. O. 2013. Analysis of power quality in distribution grids 0,4 kV. *Journal of Siberian Federal University. Engineering and Technologies* 6: 107-120.

Tofoli, F. L. Sanhueza, S. M. R. and Oliveira, de A. 2006. On the study of losses in cables and transformers in nonsinusoidal conditions. *IEEE Transactions on Power Delivery* 21(2): 971-978.

Tulsky, V.N., Kartashev, I.I., Nasyrov, R.R. and Simutkin, M.G. 2013. The influence of current higher harmonics on the operation modes of cables in the 380 V distribution network. *Industrial power engineering* 5: 39-44.

Zhezhelenko, I.V. 2000. *Higher harmonics in power supply systems of enterprises*. Moscow: Energoatomizdat.

Scientific and Practical Studies of Raw Material Issues – Litvinenko (Ed)
© 2020 Taylor & Francis Group, London, ISBN 978-0-367-86153-7

The analysis of modern regulation devices of power streams on the basis of FACTS - devices

E. Shafhatov & E. Zhdankin
Saint-Petersburg Mining University, Saint-Petersburg, Russian Federation

ABSTRACT: Russia's Unified Electric Power System (UEPS) is a complex managerial and engineering facility managed hierarchically, which provides balance between generation, network distribution and consumption in a territorial section for ensuring energy security of regions. This enables the exchange of streams of power and energy in normal and emergency operation to increase energy efficiency.

1 INTRODUCTION

Russia's unified electric power system (UEPS) is a complex managerial and engineering facility managed hierarchically, which provides balance between generation, network distribution and consumption in a territorial section for ensuring energy security of regions. This enables the exchange of streams of power and energy in normal and emergency operation to increase energy efficiency.

Power links of high capacity comprising Unified National Power Grid, which is a part of Unified Energy Systems of Russia, ensure rational use of different energy resources (coal and hydropower resources, nuclear power plant and pumped storage power plant), unevenly distributed across Russia, as well as power supply of large consumption centers. Automatic control systems used in Unified Energy Systems of Russia, systems of relay protection and automatic equipment, ways and methods of regulation of excitation of synchronous generators and compensators and others contribute to reliable and steady work of UPS under normal and emergency conditions.

The electrical power system of Russia requires the renewal of nearly 50% of outdated equipment as well as the application of modern technology involving diagnostics and information systems and control systems. These technologies are supposed to provide reduction of losses in the process of electricity production and transmission, and to optimize reserve capacities.

The electrical network as the structure providing strong links between generation and consumers plays a leading role in the process of modernization of power industry based on the new principles. The latest technologies applied in networks, providing adaptation of equipment characteristics to various operation modes, active interaction with generation and consumers allows to create an effectively functioning system incorporating modern information and diagnostic systems, automation systems of controlling all elements included in processes of electricity production, transmission, distribution and consumption.

According to the draft of the Energy Strategy of Russia until 2030, the Smart Grid and FACTS technologies intended for energy consumption optimization and redistribution of electric power are also key trends of the electric power industry development. Smart Grids are upgraded channels of power supply involving information and communication technologies. Such systems ensure reliable functioning of the equipment through the introduction of remote control over serviceability of separate components. In its turn the active electrotechnical network equipment (FACTS) can be considered a subsystem of Smart Grid (Balabanov, 2015).

FACTS allows to change the characteristics of transmission or transformation of electric energy to optimize the modes according to several criteria: capacity, level of technological losses, stability, redistribution of power streams, quality of electric energy, etc. This allows to operate power line capacity, to redistribute active power streams between parallel power lines, optimizing them in steady modes and to redirect them to power lines remaining after accidents, without being afraid of system stability violation. Thus, increase in reliability of power supply to consumers is provided (Akagi, 1990).

The authors of the article (Tuhvatullin et al., 2015) analyzed the main FACTS devices used in electrical networks.

The article (Jamila & Abdelmjid, 2016) shows the possibility of using STATCOM in alternative energy.

The use of devices based on FACTS technology will become the basis for creating smart networks.

The new FACTS devices of the second generation (FACTS-2) (Akagi, 1990) include devices that regulate the operating parameters based on fully controlled power electronics IGBT (Inverter bridge with transistors), IGCT (Integrated Gate Commutated Thyristor) and others (Pospelova, 2006).

Figure 1 shows the frequencies of operation of converter devices based on different switching devices: mechanical switches, reverse-blocking thyristors, gate turn-off GTO thyristors, IGBT transistors and IGCT thyristors (Skamyin et al, 2019). Distinctive features of the characteristics of power semiconductor switches are presented in Figure 2.

The variety of FACTS devices is constantly updated and improved. This is caused by the rapid development of semiconductor technology; technologies for the production of radio components and electromechanical devices; methods of regulating the parameters of voltage, current and power in power grids (Akagi, 1990).

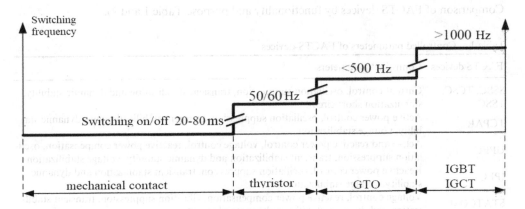

Figure 1. Switching frequencies of power switches.

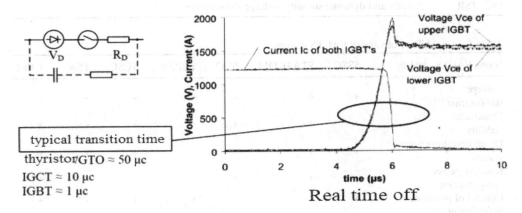

Figure 2. Switching times of power semiconductor switches.

2 DEVICES IMPLEMENTING TECHNOLOGY FACTS

Reactive Power Devices:

- Automatic capacitor banks and shunt capacitor banks – SCB;
- Thyristor switched capasitors – TSC;
- Control filter compensating device (automatic harmonic filter);
- Thyristor switched reactor – TSR;
- Thyristor controlled reactor – TCR;
- Circuit breaker switched reactors;
- Static VAR compensator–SVC;
- Static var generator – SVG;
- Static synchronous compensator – STATCOM;
- Static synchronous series compensator – SSSC;
- Interline power flow controller – IPFC.

Network parameter control devices:

- Thyristor controlled series capacitors – TCSC;
- Thyristor switched series capacitor – TSSC;
- Thyristor controlled phase angle regulator – TCPAR.

The device of longitudinal and transverse regulation:

- Unified power flow controller –UPFC

Comparison of FACTS devices by functionality and purpose(Table 1 and 2).

Table 1. Controlled parameters of FACTS-devices.

FACTS devices	Controlled parameters
SSSC, TCSC, TSSC	Current control, oscillation suppression, transient stabilization and dynamic stability, stabilization short circuit
TCPAR	Active power control, oscillation suppression, transient stabilization and dynamic stability, voltage stabilization
UPFC	Active and reactive power control, voltage control, reactive power compensation, oscillation suppression, transient stabilization and dynamic stability, voltage stabilization
IPFC	Reactive power control, oscillation suppression, transient stabilization and dynamic stability, voltage stabilization
STATCOM	Voltage control, reactive power compensation, vibration suppression, transient stabilization and dynamic stability, voltage stabilization
SVC, TCR, TSC, TSR	Voltage control, reactive power compensation, vibration suppression, transient stabilization and dynamic stability, voltage stabilization

Table 2. Capabilities of various FACTS devices.

Name FACTS- devices	SSSC	STATCOM	SVC	TCPAR	TCSC	TSSC	UPFC
Voltage stabilization	+	+	+	+	+	+	+
Dynamical stability	+	+	+	+	+	+	+
Oscillation suppression system	+	+	+	+	+	+	+
Reactive power compensation	+	+	+				+
Control of power flow	+			+	+	+	+
Softening of subsynchronous resonance	+			+	+		+

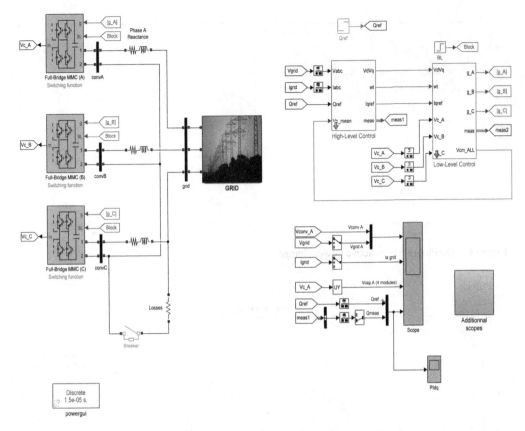

Figure 3.   STATCOM (detailed MMC model with 22 power modules per phase).

3   METHODOLOGY

Electrical compensation is used to control reactive power and voltage. In this example, a shunt compensation device, that is increasingly used in modern networks, is simulated: modular multi-level (MMC) STATCOM). MMC-STATCOM is built using the Full-Bridge MMC block to represent a power electronic converter of 22 modules per phase. To speed up the simulation while maintaining the accuracy of the simulation, the model of the switching function is selected (Figure 3).

4   OPERATING PRINCIPLE

STATCOM can absorb or generate reactive power. The transfer of reactive power through the phase reactance. The converter generates a voltage in phase with the mains voltage. When the voltage amplitude of the converter is lower than that of the mains voltage, STATCOM acts as reactive power absorbing inductance. When the voltage amplitude of the converter is higher than the bus voltage amplitude, STATCOM acts as a capacitor that generates reactive power.

In Figure 4, you can see that the simulation starts in a steady state and that STATCOM works in an inductive mode, following the given value of Qref (-5 Mvar). After 0.1 seconds, the given value changes from -5 to +10 Mvar. The STATCOM control system responds very quickly to the changes in the output voltage of the inverter in order to generate 10 Mvar of reactive power (capacity mode) (Figure 5).

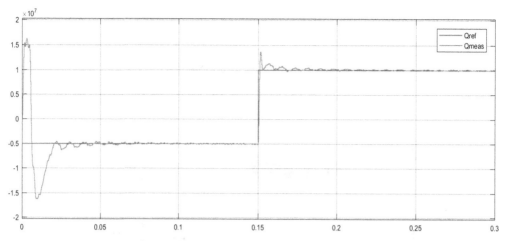

Figure 4. Oscillogram of reactive power change.

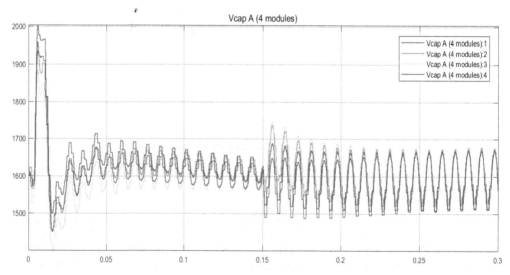

Figure 5. Oscillograms of the voltage change of the multilevel converter module.

5 CONCLUSION

The main property of STATCOM is the ability to generate a current of any phase relative to the voltage in the network, thereby ensuring the regulation of the output voltage and its phase. This regulation is possible due to the correction in reactive power consumed or supplied to the network, that the oscillograms show.

The analysis of existing flexible AC transmission systems showed that the passive compensation systems do not contain regulatory elements. Therefore, the degree of compensation in these systems is constant. This principle allows to reduce the resistance of power lines to compensate for reactive power and increase the power transmitted by wire.

In comparison with passive systems, active systems and devices have adjustable extent of jet power compensation in virtually any limits. It allows to increase stability of power supply system, to limit an overstrain during violations of working hours of network and to reduce fluctuations of tension during dynamically changeable loading. Technical developments in power electronics increase the efficiency and competitiveness of FACTS in terms of network application.

REFERENCES

Akagi H. 1990. Analysis and design of an active power filter using quad-series voltage source PWM converters/H. Akagi, Y. Tsukamoto, A. Nubae//IEEE Trans. Ind. Applicat. Vol 26: 93-98.

Balabanov, M. S. 2015. FACTS device. A choice at design of electric equipment of the enterprises: monograph/M. S. Balabanov, R. N. Hamitov. – Omsk: Publishing house of OMGTU:184.

Chakrabarti, A. 2010. Power System Analysis: Operation And Control 3Rd Ed/A. Chakrabarti, S. Halder//PHI Learning Pvt. Ltd.:1270.

Eda Uz. 2017. Design and implementation of thyristor switched shunt capacitors. [Electronic resource]// Information website. URL: http://etd.lib.metu.edu.tr/upload/12611616/index.pdf.

Gabriela, G. 2005 FACTS Flexible Alternating Current Transmission Systems./G. Gabriela//EEH – Power Systems Laboratory. – ETH Z"urich.

Jamilaa E., Abdelmjidb S. 2016 Comparative Study of the Performance of Static Synchronous Compensator, Series Compensator and Compensator/Battery Integrated to a Fixed Wind Turbine/E. Jamila and S. Abdelmjid//IJE TRANSACTIONS A: Basics Vol. 29, No. 4, (April 2016): 581-589.

Narain, G.H. 2017. Understanding FACTS: Concepts and Technology of Flexible AC Transmission Systems [Electronic resource]/G.H. Narain, G. Laszlo//Information website. URL: http://media .johnwiley.com.au.

Pospelova,T.G. 2006 Potential areas of use of FACTS and AFM in the Belarusian energy system/T.G. Pospelova//Energy and Management. - 2006. - № 4 (31): 37-43.

Sen, K.K. 2009. Introduction to FACTS Controllers: Theory, Modeling, and Applications/K.K. Sen, M. L. Sen. – England: The Institue of Electrical and Electronics Engineers: 523.

Singh, S.N. 2008. Electric power generation: transmission and distribution/S.N. Singh//PHI Learning Pvt.Ltd:452.

Skamyin, A.N., Dobush, V.S. & Rastvorova, I.V. 2019 Method for determining the source of power quality deterioration. Proceedings of the 2019 IEEE Conference of Russian Young Researchers in Electrical and Electronic Engineering, ElConRus 2019: 1077-1079.

Stalemate. 2631973 Russian Federation, MPK N03N 7/18. Way of control of the phase control device/ Panfilov D. M., Astashev M. G., Rozhkov A. N., etc., applicant and patent holder "Power Institute of G. M. Krzhizhanovsky" Joint-stock company statement. 24. 11. 2016; it is published. 29. 09. 2017, Byul. No. 28: 9.

Tuhvatullin M.M, Ivekeev V.S., Lozhkin I.A., Urmanova F.F. 2015 Analysis of the modern FACTS devices used for increasing the functioning efficiency of the Russian electric power systems/ Tuhvatullin M.M, Ivekeev V.S., Lozhkin I.A., Urmanova F.F//Electrotechnical systems and complexes 2015: 41-46.

Scientific and Practical Studies of Raw Material Issues – Litvinenko (Ed)
© 2020 Taylor & Francis Group, London, ISBN 978-0-367-86153-7

Author Index

Agafonov, Y. 165
Alekseenko, V.A. 143

Babenko, D. 103
Bazhin, V. 114, 251
Binder, A. 205
Blishchenko, A. 58
Boduen, A. 120
Bongaerts, J.C. 171
Bykova, M.V. 135

Cherdancev, G. 18

Drebenstedt, C. 34, 63, 129, 165, 191, 219
Dzyurich, D. 46

Eremeeva, A. 108
Erokhov, S. 232

Fadeev, D. 239
Fedorov, G. 165

Garipov, B. 244
Glazev, M. 114
Golubev, D.D. 75
Grafe, B. 34

Heide, G. 3
Hutwalker, A. 148

Ivanov, S. 239
Ivanov, V. 46
Ivanova, D. 182

Kholmskiy, A.V. 71
Kitcenko, A.A. 200
Kobylyanski, A. 120
Kokorin, A.V. 91
Kondakova, V.N. 11
Kondrasheva, N. 108
Korchak, P. 53
Kostin, V.N. 263
Kovyazin, V.F. 200
Krivenko, A.V. 263

Langefeld, O. 148, 205

Martynova, E. 251
Mukhina, A. 157
Mukminova, D. 82

Nazarova, M. 97
Nelkenbaum, K. 108
Nguen, T.T. 53
Nikiforova, V. 3
Nikolaeva, E. 3

Paschke, M. 213
Pashkevich, M.A. 135
Petrov, G. 120

Plett, T. 148
Pomortsev, O.A. 25
Pomortseva, A.A. 25
Pospekhov, G.B. 11
Puzanov, A.V. 143

Ritter, R. 63
Romanchikov, A.Y. 200

Schipachev, A.M. 91
Serikov, V.A. 263
Shafhatov, E. 272
Shepel, T. 34
Shvydkaya, N.V. 143
Sidorov, D.V. 71
Sultanbekov, R. 97
Suslov, A. 251

Talovina, I. 3

Ustinov, D. 244

Vilner, M. 53
Volohov, E. 82
Vu, Trong 219

Zhdankin, E. 272
Zhukova, V. 120